The Fine Structure of Algal Cells

The Fine Structure of Algal Cells

JOHN D. DODGE

Department of Botany,
Birkbeck College,
University of London

1973

ACADEMIC PRESS

London · New York

A Subsidiary of Harcourt Brace Jovanovich, Publishers

ACADEMIC PRESS INC. (LONDON) LTD.
24/28 Oval Road
London NW1

United States Edition published by
ACADEMIC PRESS INC.
111 Fifth Avenue
New York, New York 10003

Second Printing 1975

Library of Congress Catalog Card Number: 73-15686
ISBN: 0 12 219150 1

PRINTED IN GREAT BRITAIN BY
ABERDEEN UNIVERSITY PRESS, FARMERS HALL, ABERDEEN

Foreword

Part of the fascination that the algae have for researchers and students lies in their extraordinary diversity, both of morphology and cell chemistry. On the chemical side the various combinations of pigments, wall polysaccharides and storage materials have been discussed at great length and most of the phylogenetic schemes for the algae have been based mainly upon this information. On the structural side the gross morphological variations between the classes and cytology at the light microscopical level was concisely summarized by Fritsch (1935) long before the electron microscope had been used to examine algae. The present work is an attempt to bring together much of the information which has been gathered by electron microscopists over the past 20 years or so. This is a task which is becoming more and more difficult for, whereas there were perhaps two papers in this field published in 1952, there were 20 (+) in 1962 and 80 (+) in 1972 (with some 1972 journals not yet received). Altogether, just over 750 papers have been referred to and many other, mainly taxonomic papers, have not been included. This fine structural data is surely as important as the biochemical data for any consideration of the classification of algae and for attempting to analyse their phylogenetic relationships. However, it must be admitted that in some areas of algal ultrastructure very much more work is needed before any meaningful conclusions can be made.

It is impossible to write about algal fine structure without paying tribute to the hard work and inspiration which has been provided by Professor Irene Manton F.R.S., since the very beginning of the subject. Her first paper in this field in 1950 was about the male gamete of *Fucus* and since then, together with various colleagues, she has produced over 65 informative and beautifully illustrated papers on algal ultrastructure.

Fine structure is essentially a pictorial subject, a fact that many editors of scientific journals do not seem to understand, much to the anguish of the authors. In this book most of the main points have been illustrated but, as far as possible, the pictures have been assembled from micrographs taken at Birkbeck College. It is inevitable that they show a slight but I hope not too obvious bias towards the dinoflagellates. Perhaps at this point mention should be made of the main preparation methods which we have used. Apart from the direct preparations, which were mainly shadowed with gold/palladium alloy, the material was normally fixed in buffered glutaraldehyde, postfixed in osmic acid and embedded in Araldite or Spurr resin. All micrographs were taken on a Zeiss EM9A electron microscope. For much of the preparation and many of the pictures I am indebted to R. M. Crawford, B. S. C. Leadbeater, G. B. Lawes and B. T. Bibby. For the considerable amount of photographic work involved I am very grateful to P. Randall and D. Rogers. The research work, of which many of these micrographs formed a part, was generously supported by the Science Research Council, the Central Research Fund of London University, by Birkbeck College and in particular by C. T. Ingold, Emeritus Professor of Botany.

Many of the illustrations used in this book have already been published and I am grateful to the editors and publishers of the following journals for allowing me to use them again: *Archiv. für Mikrobiologie, Botanical Journal of the Linnean Society, Botanical Review, British Phycological Journal, Journal of Cell Science, Journal of Phycology, New Phytologist, Nova Hedwigia, Protistologica and Protoplasma.*

I am very grateful to T. Bråten, R. M. Crawford, B. S. C. Leadbeater, M. Neushul, E. Swift, J. A. West and P. A. Walne for very kindly providing micrographs to illustrate specific points. Thanks are due to the various authors (named in the legends) who have allowed me to reproduce their figures and to the publishers of the following journals, in which the figures originally appeared: *Advances in Botanical Research. Archiv. für Mikrobiologie, Journal of Cell Biology, Journal of Cell Science, Journal of Phycology, Journal of Ultrastructure Research, Nature, Phycologia, Planta, Plant Physiology, Protistologica and Protoplasma.*

I would like to thank my colleagues at Birkbeck and elsewhere who, by their discussion and description of their discoveries, have helped me to form the views expressed in this book. I sincerely apologize to any authors who feel that I have misrepresented their findings. Lastly, my grateful thanks are due to Mrs Jill Smith for patiently typing the manuscript.

Department of Botany.
Birkbeck College

JOHN D. DODGE

March 1973

Contents

INTRODUCTION

In the past twenty or so years the electron microscope has added a completely new dimension to our knowledge of the algae. To some extent this has complemented the traditional disciplines of taxonomy, morphology and ecology but, as the intricacies of algal ultrastructure have unfolded, 'fine-structure' has come to have a place of its own and has provided us with many vital keys to the understanding of the inter-relationships and phylogeny of the algae. Whereas the first two decades of algal fine structure work have been essentially devoted to the descriptive approach, the trend is now towards use of electron microscopy to try to understand the functions of cells and organelles under both normal and experimental conditions. Clearly this is a trend that will continue, especially as the number of genera whose normal ultrastructure has not yet been described rapidly diminishes.

History

It is interesting to note the correlation between the bursts of fresh information on algal fine structure and the development of new techniques. The first such burst commenced in 1945 when Brown published what appear to have been the first algal electron micrographs. He used the recently developed shadow-casting technique to examine the flagella of various algae. This technique was perhaps most successfully applied in the early 1950s by Manton and her colleagues who investigated the structure of flagella in numerous algae and lower plants. During this period there was some descriptive work on algal cell walls and particularly on diatom frustules, which required no special preparation other than thorough cleaning.

The next burst of information came in the late 1950s when, with the development of methacrylate embedding and fairly reliable ultramicrotomes, some details of the internal structure of algal cells began to be revealed. Unforunately polymerization of the embedding medium often caused much distortion of cell structure and in retrospect the work done during this period was only of limited value. However, a great improvement took place in the early sixties, first with the development of epoxy resins for embedding which resulted in improved stability of the tissue, and secondly with the introduction of aldehyde fixation which really stabilized most of the cell components. Now, with automatic ultra-microtomes and diamond knives generally available, the great period for fine structure began. Initially, this was almost entirely confined to unicellular organisms which could be fixed and embedded most readily. The revealing of the internal fine-structure of algal cells had now become limited less by technique and more by the availability of material and the inclination and time of the investigators.

The development of freeze-etching or freeze-fracturing during the mid sixties provided a technique which made it possible for the surfaces of cells, walls, membranes and organelles to be examined. Thus, we have now an increasing number of papers in which attempts are made to correlate the structure of membranes, etc., with their particular function in the cell.

The latest major development, scanning electron microscopy, has as yet made little general impact on our knowledge of algal ultrastructure. Certainly, it has already become a vital tool for diatom taxonomists and has been used for the study of the surfaces of desmids and dinoflagellates. As specimen preparation techniques are developed and as the resolution of the microscopes improves SEM will no doubt become of more general use.

Of the other fine structure techniques, negative staining, which is so valuable for virus work, has been little used on algae although it has proved of value in studies of flagellar hairs. Electron microscopical autoradiography and other cyto-analytical techniques have not yet been much used with algae, but no doubt will be employed more in the future.

Scope of the book

The present survey excludes the blue-green algae (Cyanophyta) for these procaryotic organisms are quite distinct from the true algae and they are, in any case, the subject of a recent review (Lang, 1969) and a number of books in preparation. At the other extreme the Charophyceae (stone worts) have not been considered in detail although they are occasionally referred to. The Craspedophyceae (Choanoflagellates) have also not been discussed in this book for their complete lack of chloroplasts seems to suggest that their affinities are more with the Protozoa than the algae.

The endeavour has been to provide a hybrid between a review and a comprehensive descriptive work. In other words to provide enough description of organelles, etc., to make it possible for the student to visualize and compare their structure and at the same time to provide enough references so that the research worker can enter the literature to find out more precise details from the original sources. The references given are not fully comprehensive but are intended to highlight the more important facts. It is hoped that the present work will complement the few specialized reviews of fine structure already published* and will perhaps make some of these known to a wider audience.

(* Euglenophyceae—Leedale, 1967; Eyespots—Dodge, 1969a; Chloroplasts—Gibbs, 1970; Pyrenoids—Griffiths, 1970; Nuclear division—Pickett-Heaps, 1969; Leedale, 1970; Pyrrophyta (= Dinophyceae)—Dodge, 1971.)

Classification of the algae

Almost every new algal text book that is published presents a revised system for the classification of these organisms. This has made for a situation, which must be highly confusing to the student and which makes teaching extremely difficult. In a work such as this it is essential to have some 'labels' at a higher level than that of the genus yet at the same time it is probably better to use a neutral classification scheme. Thus the largest units employed here are classes, most of which go back to Fritsch (1935, 1945), or earlier, and which, apart from the classes more recently designated (Prasinophyceae, Haptophyceae, Eustigmatophyceae) are the only stable feature which has persisted through the numerous classification schemes of the past few years. Phyletic terms such as 'Chlorophyta' will not normally be used in this book. The system, as used here, makes no implied allusions to the relationships between the classes although this subject is discussed in the text where appropriate and is summarized in Chapter 14.

The 13 classes covered in this book are as follows:

Chlorophyceae
Prasinophyceae
Chloromonadophyceae
Euglenophyceae
Eustigmatophyceae
Xanthophyceae
Chrysophyceae
Haptophyceae
Bacillariophyceae
Phaeophyceae
Dinophyceae
Cryptophyceae
Rhodophyceae

1. *A General Account of the Structure of Algal Cells*

The object of this chapter is to give a brief description of the main structural features of the various classes and to note the organelles present in typical cells. For the smaller classes reference is made to all known papers describing general cell structure but for the larger classes only representative papers are noted. Later chapters should be consulted for detailed descriptions of, and references to, the organelles mentioned. The classes are here arranged in alphabetical order.

I. Bacillariophyceae (Diatoms)

This is a very distinctive and natural group of organisms, for the possession of a silica wall or frustule distinguishes the members of this class from all other algae. The frustule has been the subject of a great deal of work, including a wide ranging survey using transmission electron microscopy (Helmcke and Krieger, 1953–1966) and several more recent studies using scanning microscopy (e.g. Ross and Simms, 1970, 1971, 1972).

In general, diatoms are unicellular organisms, although they may become joined together to form filaments or colonies. Each cell contains two or more chloroplasts; a single nucleus; several mitochondria and Golgi bodies; endoplasmic reticulum and ribosomes (Fig. 1.1). Most of the detailed studies of fine structure have been concerned with pennate diatoms (i.e. those which do not have radial symmetry) e.g. Drum (1963), Drum and Pankratz (1964), Stoermer and Pankratz (1964), Stoermer *et al.* (1964, 1965), Lauretis *et al.* (1968), and Taylor (1972). As yet very little is known about centric diatoms except for details of the nuclear divisions (Manton, Kowallik and von Stosch, 1969a,b, 1970a,b), wall structure and formation after cell division (Round, 1971a; Crawford, 1971, 1973) and the structure of spermatozoids. These motile cells possess a single hairy flagellum with an atypical internal structure (Manton and von Stosch, 1966; Heath and Darley, 1972). Attempts have been made to study the structures involved in the gliding movement of pennate diatoms (Drum and Hopkins, 1966; Drum and Pankratz, 1965b) which is thought to be brought about by an expanding mucilaginous propellant. Colchicine-induced polyploids of *Navicula* have been studied by Coombs *et al.* (1968b).

FIG. 1.1. A longitudinal section of the centric diatom *Melosira* showing the silica frustule (wall), the peripheral arrangement of chloroplasts, the flattened nucleus at the lower end of the cell, and the large central vacuole. (× 5,000) Micrograph by courtesy of R. M. Crawford (after Crawford, 1973).

II. Chloromonadophyceae (Raphidophyceae)

This is a small and rather obscure class the members of which are unicellular flagellates. The normal organelles are present and unusual features include a sheath of Golgi bodies around much of the nucleus and a fibrous root which appears to connect the flagellar bases and the nucleus. Of the two flagella, the anterior one bears stiff hairs and the posterior is smooth. Accounts of fine structure in the genera *Vacuolaria* and *Gonyostomum* have been provided by Schnepf and Koch (1966), Koch and Schnepf (1967), Mignot (1967), Heywood (1968, 1972), and Heywood and Godward (1972).

III. Chlorophyceae (Green Algae)

The old class Chlorophyceae contains such a diverse variety of organisms that, not only is it impossible to summarize their attributes in a few paragraphs, but there is considerable justification for accepting the recent proposals of Round (1971b) which divide the group (termed Chlorophyta by Round) into four classes. In the present book the term Chlorophyceae is used in the old sense (as in Fritsch, 1935), however, it may be helpful to employ the new scheme for the summaries of cell structure which follow. To date there have been well over 130 papers dealing with the fine structure of green algae and the references given here represent only a small selection.

A. ZYGNEMAPHYCEAE

This class contains filamentous algae and the unicellular desmids, the characteristic feature being sexual reproduction by conjugation. Generally the cells contain one or two chloroplasts and a single nucleus. No flagella are found at any stage. There is as yet no comprehensive description of the fine structure of any one organism although much can be ascertained from the detailed studies of conjugation in *Spirogyra* and *Closterium* (Fowke and Pickett-Heaps, 1969a,b, 1971; Pickett-Heaps and Fowke, 1970c, 1971). Kiermayer (1968, 1970) has studied cell division and the development of new half-cells in *Micrasterias* and Pickett-Heaps and Fowke (1969) cell division in *Closterium*. Various aspects of wall structure, etc., have been reported by Dawes (1965), Mix (1966), and Kies (1970a,b). Scanning microscopy has recently been used to examine the morphology of vegetative cells and zygotes of desmids (Lyon, 1969).

B. OEDOGONIOPHYCEAE

This group consists entirely of filamentous algae which generally have a reticulate chloroplast and a single nucleus. There is a unique form of cell wall extension during cell division which has been studied by Hill and Machlis (1968), and Pickett-Heaps and Fowke (1970a). Numerous flagella are present in the zoospores and gametes and their structure, together with that of the associated roots, has been studied in some detail (Hoffman, 1970; Hoffman and Manton, 1962, 1963). The process of zoospore formation and development into a sporeling has recently been described (Retallack and Butler, 1970b; Pickett-Heaps, 1971a; 1972b,c,d).

C. BRYOPSIDOPHYCEAE

This class contains the algae which were formerly in the orders Siphonales and Cladophorales. The unifying features are coenocytic organization (many nuclei per 'cell') and asexual and sexual reproduction by motile zoospores or gametes. Electron microscopical studies have, as yet, been mainly concerned with the structure of the motile stages, the cell walls (see below) and the plastids (Hori and Ueda, 1967). The large unicellular alga *Acetabularia*, which is of considerable interest to geneticists and molecular biologists, has been much studied by cytologists (Crawley, 1963, 1964, 1966; Werz, 1964; Puiseux-Dao, 1966, 1970; Mackie and Preston, 1968; Woodcock and Bogorad, 1970; Woodcock, 1971; Burr and West, 1971b; various authors in: Brachet and Bonotto, 1970) *Bryopsis* has also been studied in some detail (Burr and West, 1970, 1971a,b) as has *Caulerpa* (Dawes and Rhamstine, 1967; Dawes and Barilotti, 1969). The large ovoid coenocyte *Valonia* provided much of the early information on the structure of algal cell walls (Cronshaw and Preston, 1958; Roelofsen, 1959, 1966; Steward and Muhlethaler, 1953; Wilson, 1951).

D. CHLOROPHYCEAE (*sensu* ROUND)

This group contains a wide variety of morphological types, including the unicellular and colonial flagellates (Volvocales), coccoid algae (Chlorococcales), simple and branched filaments (Ulotrichales and Chaetophorales) and larger parenchymatous seaweeds (Ulvales). Typically, each cell contains a single chloroplast and nucleus (Fig. 1.2). Motile stages are found as vegetative cells, zoospores and gametes, and these generally possess two flagella. Many ultrastructural studies have been carried out on members of this class, of which the following are some of the more important:

Volvocales: survey (Lang, 1963a); *Chlamydomonas* (Lembi and Lang, 1965; Ringo, 1967b); *Volvox* (Joyon, 1963a; Kochert and Olson, 1970a,b); *Chlamydobotrys* (Merrett, 1969; Wiessner and Amelunxen, 1969a,b) *Dunaliella* (Trezzi *et al.*, 1964).

Chlorococcales: *Chlorella* (Murakami *et al.*, 1963), *Chlorococcum* (Deason, 1965); *Pediastrum* (Hawkins and Leedale, 1971; Gawlik and Millington, 1969); *Ankyra* (Swale and Belcher, 1971); *Golenkinia* (Mφestrup, 1972).

Ulotrichales: *Ulothrix* (Floyd *et al.*, 1972a), *Klebsormidium* (Floyd *et al.*, 1972b).

Chaetophorales: *Stigeoclonium* (Manton, 1964b; Floyd *et al.*, 1972a) *Fritschella* (McBride, 1970).

Prasiolales: *Prasiola* (Friedmann and Manton, 1960).

Ulvales: *Ulva* (Bråten, 1971); *Enteromorpha* (Evans and Christie, 1970).

Recently Schötz (1972) and Schötz *et al.* (1972) have constructed three-dimensional models of *Chlamydomonas* in an attempt to get a better understanding of the inter-relationships of the organelles.

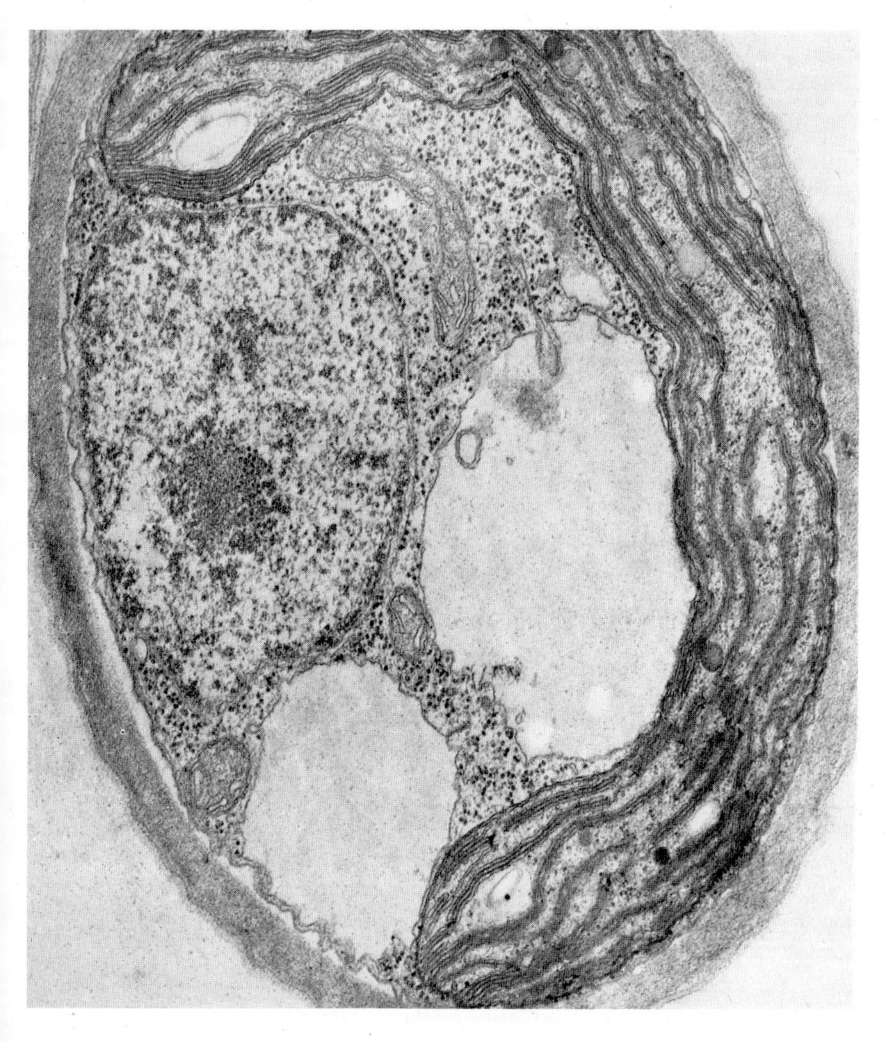

FIG. 1.2. A cross-section of *Ankistrodesmus* (Chlorophyceae) showing the nucleus, chloroplast, vacuole, cell wall, etc. (×29,400)

The strange intracellular asexual reproduction of some members of the Chlorococcales has been studied in *Hydrodictyon* (Marchant and Pickett-Heaps, 1970, 1971, 1972a,b) and in *Pediastrum* (Moner and Chapman, 1960; Gawlik and Millington, 1969; Millington and Gawlik, 1970; Hawkins and Leedale, 1971). The parasitism of the Chloroccalean alga, *Scenedesmus* by Chytrids has been investigated by Schnepf *et al.* (1971a,b).

IV. Chrysophyceae (Golden Algae)

The members of this class are unicellular or colonial organisms and are mostly flagellated. A typical cell (Figs 1.3 and 1.4) is somewhat elongated and has one hairy and one smooth flagellum, both inserted at the anterior end. There are two lateral chloroplasts and a central nucleus. The cell membrane may be covered with silica scales, which are of considerable taxonomic value (Harris and Bradley, 1956, 1960; Fott, 1962; Bradley, 1964; Harris, 1966; Kristiansen, 1969a; Pennick and Clarke, 1972; Wujek and Hamilton, 1972), or be surrounded by a lorica (Fig. 1.4) which has a characteristic form in each genus (Karim and Round, 1967; Kristiansen 1969b, 1972a,b).

The first chrysophyte to be studied in detail by electron microscopy was *Synura* (Manton, 1955) and this has been followed by studies of *Chromulina* (Rouiller and Fauré-Fremiet, 1958), *Ochromonas* (Gibbs, 1962b) *Sphaleromantis* (Manton and Harris, 1966), *Pedinella* (Swale, 1969), *Chrysococcus* (Belcher, 1969a), *Mallomonas* (Belcher, 1969b), *Synura* (Schnepf and Deichgräber, 1969), *Olithsodiscus* (Leadbeater, 1969), *Dinobryon* (Joyon, 1963a; Wujek, 1969) *Ochromonas tuberculatus* (Hibberd, 1970), *Chrysamoeba* (Hibberd, 1971), *Ankylonoton* (Van der Veer, 1970), *Apedinella* (Throndsen, 1971) and the silico-flagellate *Dictyocha* (Van Valkenburg, 1971a,b).

V. Cryptophyceae

This is a very distinctive class, consisting almost entirely of flagellated unicells which have an anterior or lateral depression from which the two flagella arise (Fig. 1.5). There are usually two chloroplasts, containing 2-thylakoid lamellae, and a single central nucleus. Apart from the other normal organelles, the cells contain a unique form of ejectile organelle termed ejectosomes (see Chapter 9). The structure of the cell covering is relatively simple, although more complex than was originally thought (Gantt, 1971; Hibberd *et al.* 1971). The first member to be described in detail was the apochlorotic *Chilomonas* (Anderson, 1962), since when, members of the following genera have also been described: *Chroomonas*, *Cryptomonas*, *Cyathomonas* and *Hemiselmis* (Joyon, 1963a; Mignot, 1965; Mignot *et al.*, 1968; Schuster, 1968; Dodge, 1969b; Lucas, 1970a,b; Hibberd *et al.* 1971; Taylor and Lee, 1971).

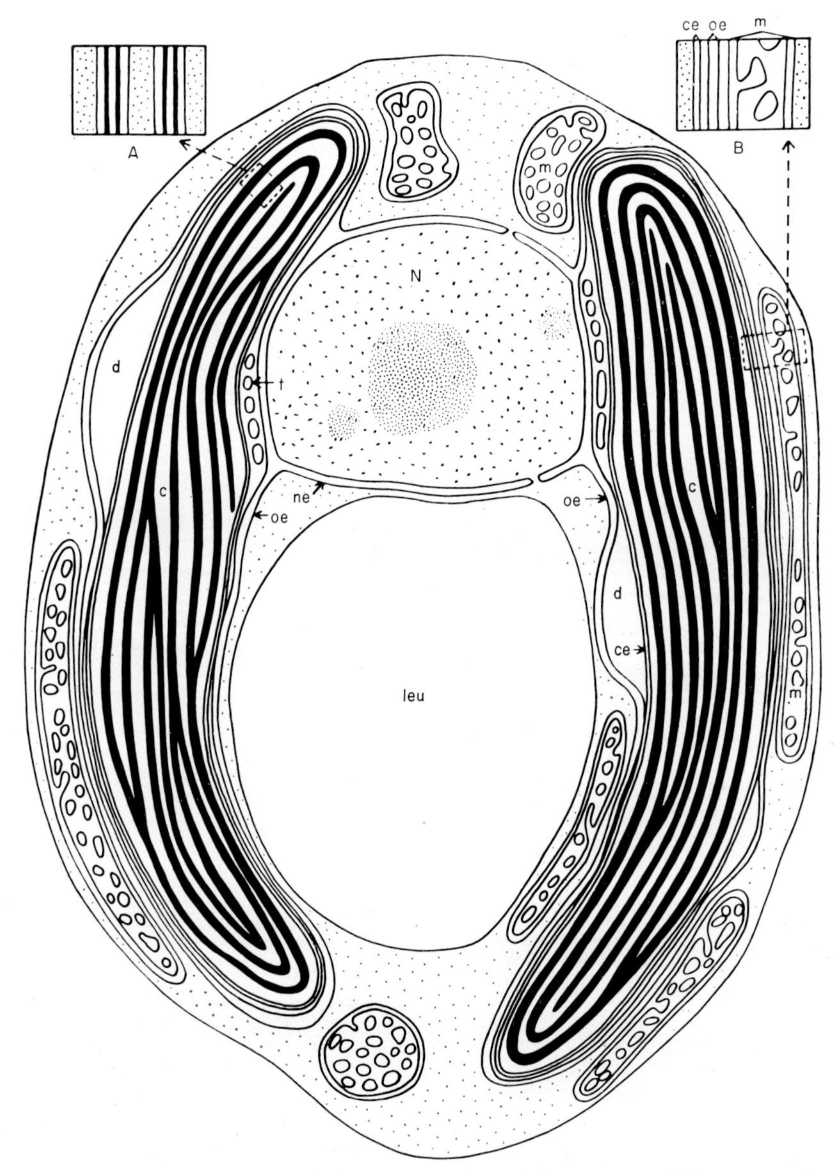

FIG. 1.3. A diagrammatic longitudinal section of *Ochromonas* (Chrysophyceae) showing the arrangement of the main cell components (apart from the flagella). Note especially the extension of the nuclear envelope (ne) which encloses (labelled oe) the chloroplasts. The cell contains a large leucosin vacuole (leu) and several mitochondria. Reproduced, with permission, from Gibbs (1962d).

VI. Dinophyceae (Dinoflagellates)

This class can readily be distinguished for its members, which are mostly unicellular flagellates, have a number of unique features. The most striking structure is the nucleus (Fig. 1.6), which has been termed a *mesocaryon* (Dodge,

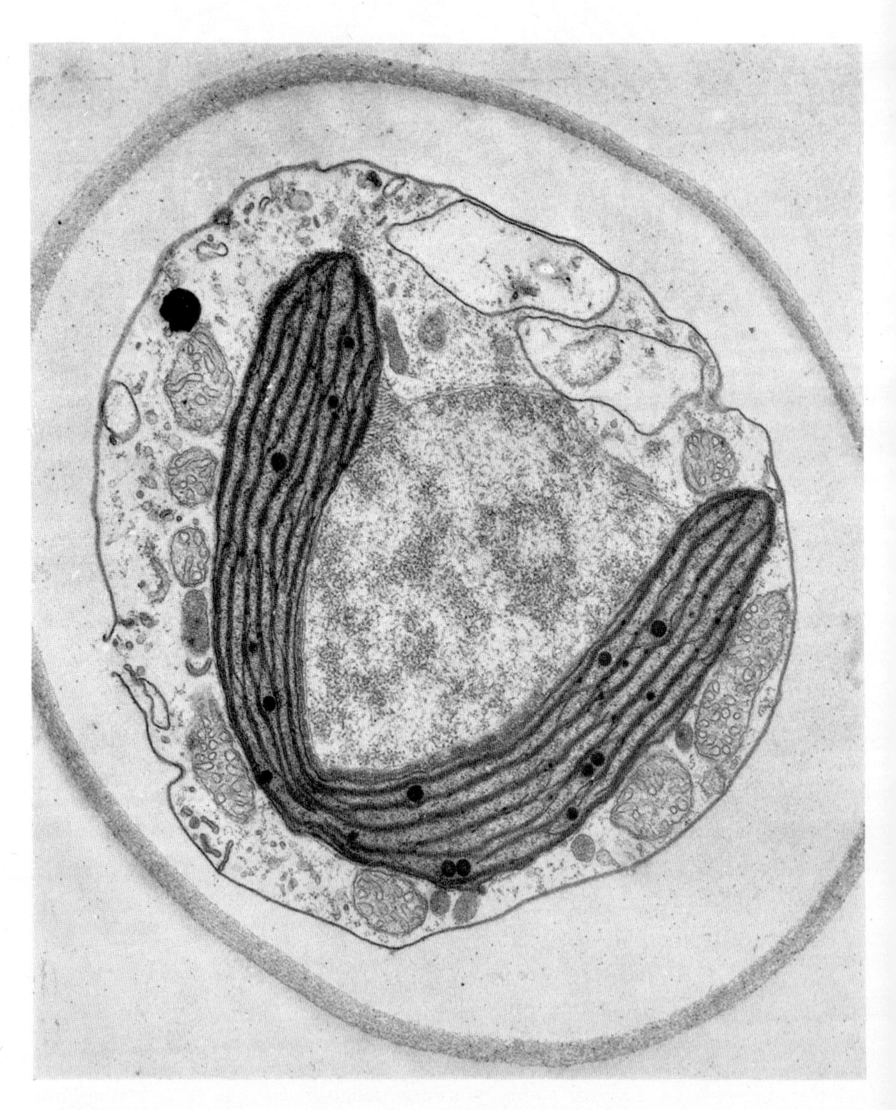

Fig. 1.4. A transverse section of *Dinobryon* (Chrysophyceae) showing the cell loosely enclosed by a fibrillar lorica (see Chapter 2) and containing nucleus, chloroplast, mitochondria, etc. (× 16,500)

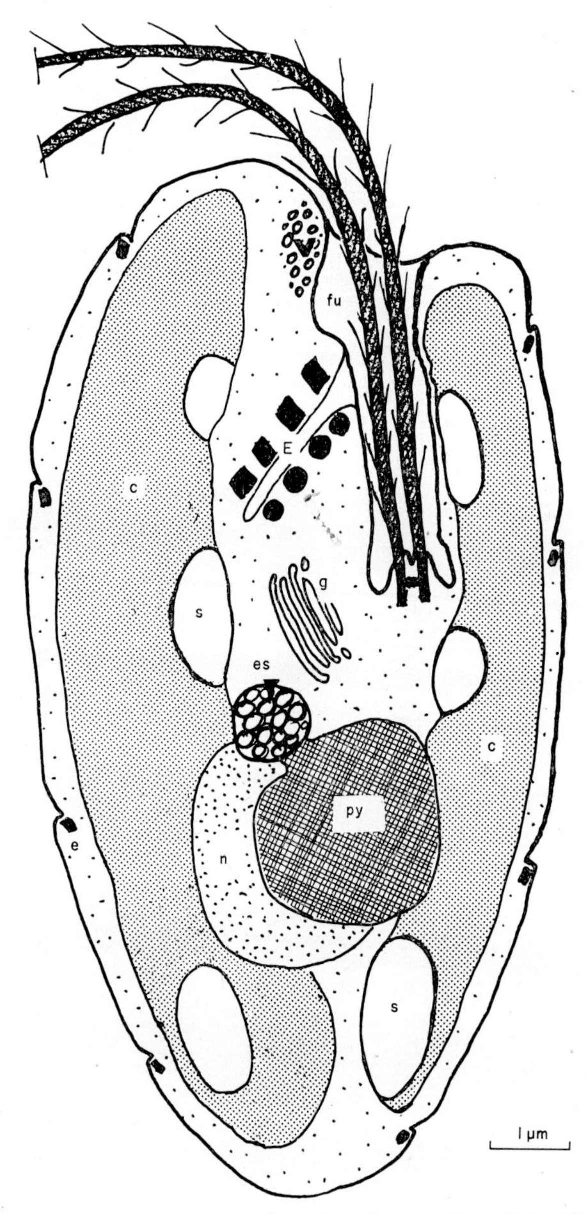

FIG. 1.5. Diagrammatic longitudinal section of a member of the Cryptophyceae showing the anterior furrow (fu) and the position of chloroplasts (c), pyrenoid (p), eyespot (es), starch (s), Golgi (g), large and small ejectosomes (E) and (e) contractile vacuole (V). (After Dodge, 1969b.)

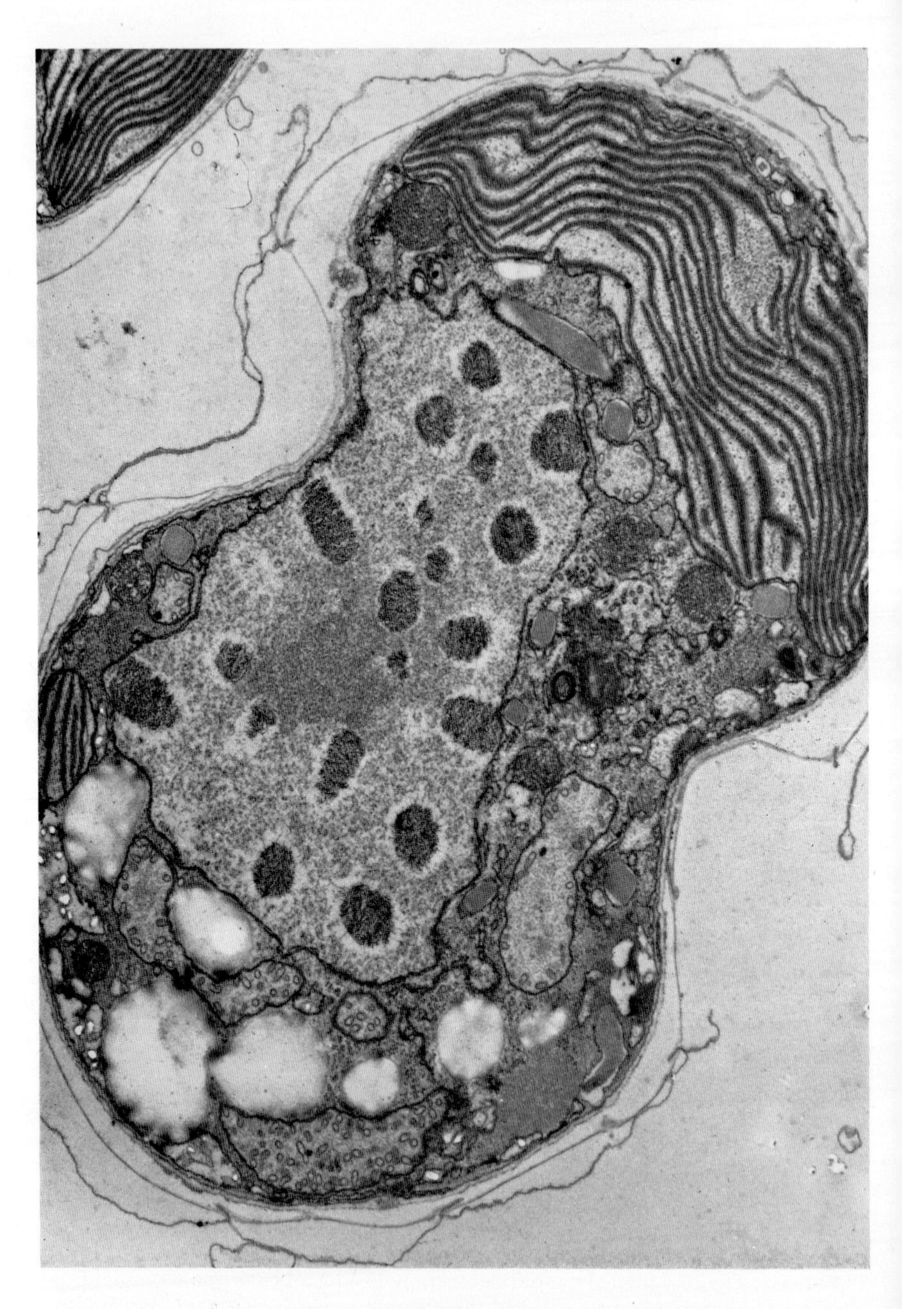

Fig. 1.6. A longitudinal section of *Glenodinium hallii* (Dinophyceae) showing the presence of the normal cell organelles together with the mesocaryotic nucleus characteristic of this class. Several starch grains are seen at the posterior end of the cell. (× 13,800)

1966), for the chromosomes lack histone and can be visualized at all stages of the nuclear cycle (see Chapter 7). The mitosis is also unusual (Leadbeater and Dodge, 1967b; Kubai and Ris, 1969; Ris and Kubai, 1972). There are two flagella, one of which is normal in construction and the other has a helical form (Leadbeater and Dodge, 1967a). The cell covering, which is generally termed the theca (or amphiesma), has a distinctive structure (Dodge, 1965; Dodge and Crawford, 1970a; Loeblich, 1970; Kalley and Bisalputra, 1970, 1971). The plates, which may form part of the theca, are now being studied by scanning microscopy (e.g. Taylor, F. J. R. 1971). Ejectile organelles termed trichocysts are found here (Bouck and Sweeney, 1966) (see Chapter 9) and there is a unique osmo-regulatory organelle called the pusule (Dodge, 1972).

The general fine structure in this class has been reviewed by Dodge (1971) and detailed accounts published of fine structure in the following genera: *Amphidinium* (Grell and Wohlfarth-Bottermann, 1957; Dodge and Crawford, 1968), *Aureodinium* (Dodge, 1967), *Ceratium* (Dodge and Crawford, 1970b), *Gonyaulax* (Schmitter, 1971; Gaudsmith and Dawes, 1972), *Gymnodinium* (Dodge and Crawford, 1969; Mignot, 1970; Schnepf and Deichgräber, 1972), *Nematodinium* (Mornin and Francis, 1967), *Oxyrrhis* (Dodge and Crawford, 1971b, 1972), *Peridinium* (Messer and Ben-Shaul, 1969) *Polykrikos* (Lecal, 1972), *Woloszynskia* (Leadbeater and Dodge, 1966; Crawford *et al.*, 1971), *Prorocentrum* (Dodge and Bibby, 1973).

Symbiotic dinoflagellates (Zooxanthellae) have been studied by Taylor (1968a,b, 1969a,b,c, 1971a,b) and Kevin *et al.* (1969) and encystment in a freshwater *Woloszynskia* species by Bibby and Dodge (1972).

VII. Euglenophyceae

All members of this class are unicells which generally possess flagella inserted in an anterior depression (the reservoir) (Fig. 1.7). There is an unusual nucleus with its own type of nuclear division (Leedale, 1968) and a unique form of cell covering termed the pellicle. The cells contain grains of paramylon carbohydrate food reserve, which is only found in this class. There is often a large eyespot (Walne, 1971; Kivic and Vesk, 1972) and this is situated adjacent to the reservoir and not associated with a chloroplast.

Members of this group were some of the earliest algae to be examined by electron microscopy and they provided the first information about the construction of algal flagella, chloroplasts, etc. (Houwink, 1951; Wolken and Palade, 1952, 1953; Wolken, 1956; Roth, 1959; de Haller, 1959; Frey-Wyssling and Mühlethaler, 1960). A detailed account of the fine structure of *Euglena spirogyra* has been provided by Leedale *et al.* (1965) and reviews of the structure in various genera by Mignot (1966), Leedale (1966, 1967) and Buetow (1968).

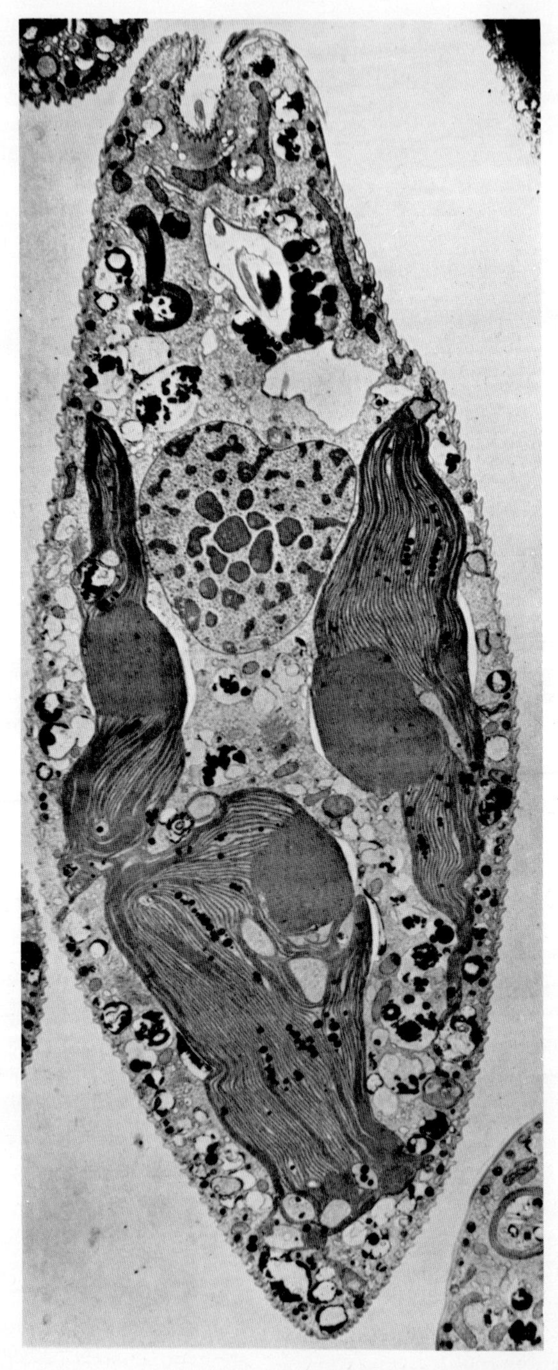

FIG. 1.7. Longitudinal section of *Euglena granulata* showing parts of the anterior reservoir, the nucleus with distinct chromosomes, chloroplasts and various granular inclusions. (×6,000) Micrograph by courtesy of P. L. Walne (from Walne and Arnott, 1967).

VIII. Eustigmatophyceae

This is a recently designated class (Hibberd and Leedale, 1970) containing organisms which were previously included in the Xanthophyceae. Distinguishing features found in the zoospore (Fig. 1.11A) include an anterior eyespot which is not membrane-bound and is independent of the chloroplast, a single emergent flagellum and absence of Golgi bodies. In vegetative cells there is a distinctive type of pyrenoid and the chloroplasts lack girdle lamellae. The fine structure of members of 12 genera has been described by Hibberd and Leedale (1972).

IX. Haptophyceae

The members of this class were included in the Chrysophyceae until the Haptophyceae was established by Christensen (1962). The main characteristic of the class is the presence, on motile stages, of an appendage termed the *haptonema,* in addition to the two flagella. Included here are organisms which bear organic scales and the coccolithophorids which produce calcite scales or coccoliths. These have been the subject of several taxonomic works employing direct observation by TEM or SEM, or carbon replicas (e.g. Halldal and Markali, 1955; Lecal, 1965, 1966; McIntyre and Bé, 1967; and on fossil coccoliths: Black and Barnes, 1961; Black, 1965).

The basic structure in this class is very simple (Fig. 1.8). Each cell normally contains two chloroplasts, one nucleus, a single Golgi body and a few mitochondria. The two anterior flagella are smooth. Scales are produced within Golgi vesicles and several studies have been made into the processes involved in this (see Chapter 2). Most of the considerable amount of fine structural information about this class has been provided by Manton and Parke and their colleagues. The most studied genus is *Chrysochromulina* where electron microscopy is essential for identification of the species (Green and Leadbeater, 1972; Leadbeater, 1972; Leadbeater and Manton, 1969; 1971, Manton, 1966a,b,c, 1972a,b; Manton and Leedale, 1961a; Manton and Parke, 1962; Parke, Lund and Manton, 1962; Parke and Manton, 1962; Parke *et al.*, 1955, 1956, 1958). The following other genera have also been described: *Coccolithus* (Manton and Leedale, 1969); *Cricosphaera* (Pienaar, 1969); *Crystallolithus* (Manton and Leedale, 1963b); *Hymenomonas* (Manton and Peterfi, 1969; Franke and Brown, 1971; Outka and Williams, 1971); *Isochrysis* (Billard and Gayrae, 1972); *Pavlova* (Green and Manton, 1972); *Phaeocystis* (Parke, Green and Manton, 1971); *Pleurochrysis* (Leadbeater, 1971); *Prymnesium* (Manton, 1966b; Manton and Leedale, 1963a) and *Syracosphaera* (Leadbeater, 1970). Some members of this class are known to have a benthic phase and this has been examined electron-microscopically in a few cases (Brown, 1969; Brown *et al.* 1970; Leadbeater, 1970, 1971a; Parke, 1971).

X. Phaeophyceae (Brown Algae)

The members of this class are filamentous or parenchymatous plants and here we find the largest and most highly differentiated algae—the giant kelps. There are often several distinct types of cell within a single plant and these range from the simple peripheral photosynthetic cell to the elongated conducting tubes of the medulla. These latter have certain similarities with sieve cells of higher

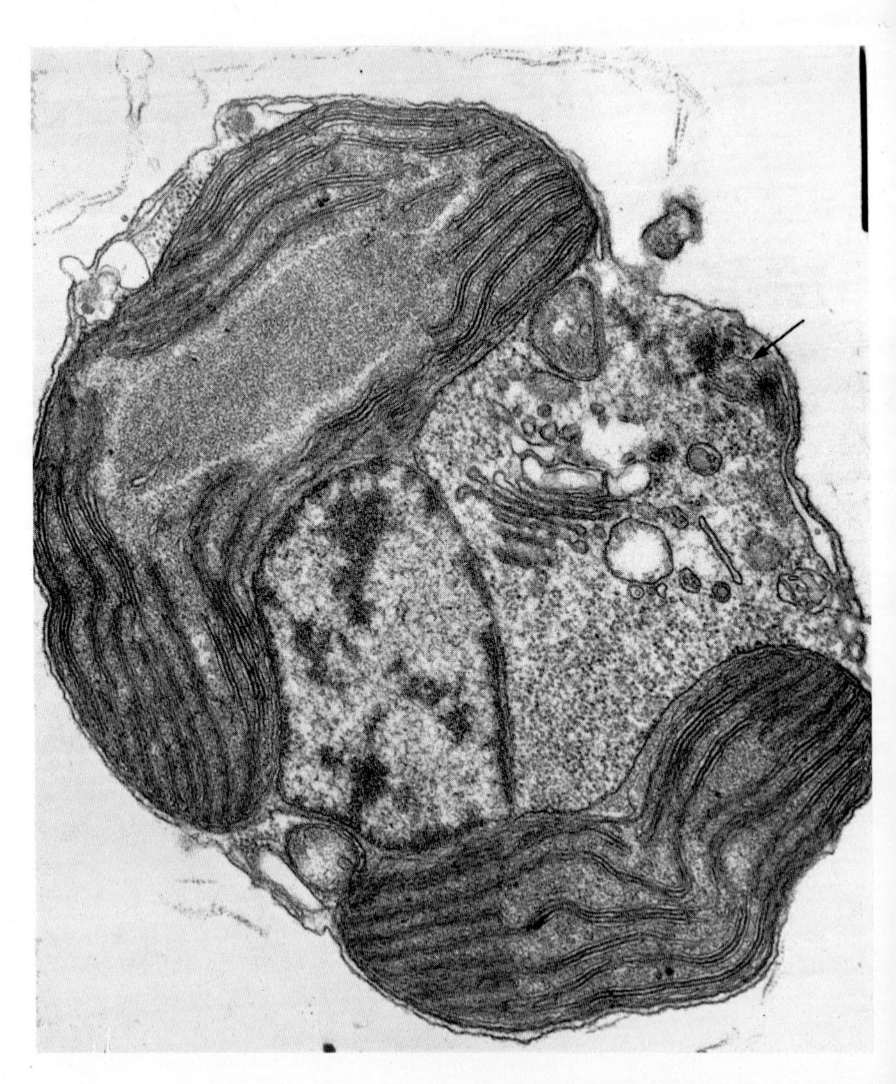

Fig. 1.8. Longitudinal section of *Chrysochromulina* (Haptophyceae). A flagellar base is visible at the anterior end of the cell (arrow). (×30,000) (After Dodge, 1970.)

plants and have therefore been studied in some detail (Parker and Philpott, 1961; Parker and Huber, 1965; Ziegler and Ruck, 1967). The smaller epidermal cells have been studied in *Fucus* (McCully, 1968) and there they are highly polarized with the nucleus and plastids in a basal position. There is a hypertrophied perinuclear Golgi system and a much convoluted plasmalemma but most of the cell is occupied by single-membrane bound vesicles which contain alginic acid, fucoidin and polyphenols.

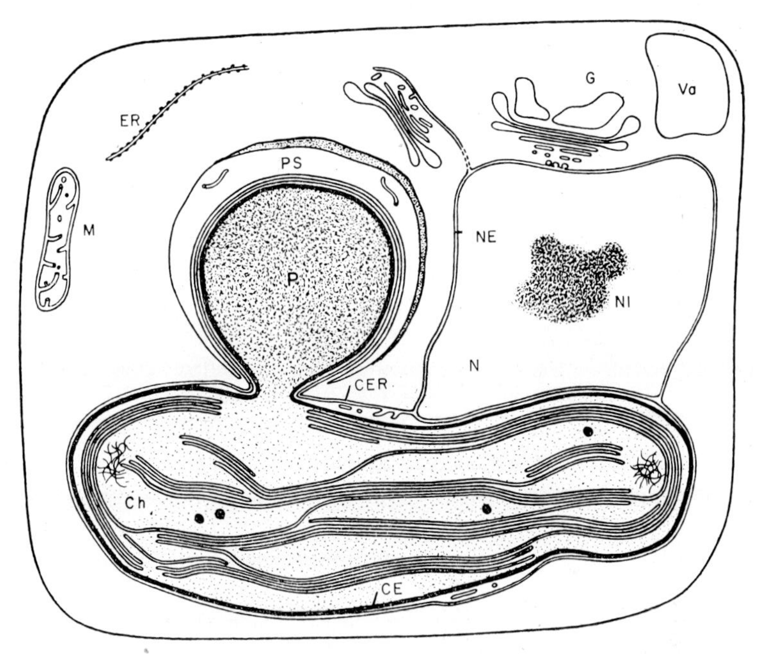

FIG. 1.9. A diagram of a hypothetical brown algal cell to illustrate some of the organelle associations. The outer chloroplast envelope (CER) is seen to be an extension of the nuclear envelope (NE). Around the pyrenoid (P) is a thick polysaccharide cap (P.S.). Reproduced, with permission, from Bouck (1965).

Normal cells in this class (Fig. 1.9) possess one or more chloroplasts which may have pyrenoids associated with them. An E.R. envelope, which is continuous with the nuclear membrane, ensheaths the chloroplasts and pyrenoids. Zoospores and motile gametes are heterokont, the anterior flagellum bearing stiff hairs and the posteriorly directed one being smooth. The early fine structure work on the Phaeophyceae was almost entirely concerned with the structure of these motile cells and particularly of their flagella (Manton and Clarke, 1950, 1951a,b, 1956; Manton, Clarke and Greenwood, 1953; Manton, 1957, 1959b, 1964a; Cheignon, 1964). The flagella have been examined more recently using high resolution techniques (Bouck, 1969). Several studies have been concerned with vegetative cells (Bouck, 1965; Cole, 1969, 1970; Cole and Lin, 1968; Bourne and Cole,

1968), with large apical cells (Neushul and Dahl, 1972a) and with the initial stages of gametophyte development (Bisalputra *et al.*, 1971).

XI. Prasinophyceae

The Prasinophyceae has similar biochemical characteristics to the Chlorophyceae but is distinguished from that class by morphological features, the most important being tiny scales which clothe the flagella. Here is a case where electron microscopy is essential for positive identification as a member of this class. In the present work a number of minute green uniflagellate organisms which Christensen (1962) separated into the class Loxophyceae are included in the Prasinophyceae (as in Parke and Dixon, 1968) for the taxonomic position of these organisms is by no means clear.

The basic organization in this class consists of motile unicells with a variable number of anterior flagella. A wide banded root connects the flagellar bases with the nucleus. There is a single cup-shaped chloroplast with a central pyrenoid which is usually penetrated by a number of cytoplasmic or nuclear tunnels. The cells may be covered with scales or enclosed in an amorphous wall or lorica. Many members of the class produce benthic or non-flagellated stages but, as yet, these have not received much attention from electron microscopists. As with the Haptophyceae most of our knowledge of the ultrastructure in this group has been obtained by Manton and Parke and their collaborators. The following genera have been studied: *Asteromonas* (Peterfi and Manton, 1968); *Brachiomonas* (Gouhier, 1970); *Heteromastix* (Manton *et al.*, 1965); *Mesostigma* (Manton and Ettl, 1965); *Micromonas* (Manton, 1959a; Manton and Parke, 1960); *Monomastix* (Manton, 1967a); *Nephroselmis* (Parke and Rayns, 1964); *Pedinomonas* (Ettl and Manton, 1964; Belcher, 1968b); *Platymonas* (Manton and Parke, 1965); *Prasinocladus* (Parke and Manton, 1965); *Pyramimonas* (Manton, Oates and Parke, 1963; Manton, 1966c, 1968b; Belcher, 1968a; Swale and Belcher, 1968; Maiwald, 1971).

XII. Rhodophyceae (Red Algae)

This class is both structurally and biochemically distinct from all other algal classes. Morphologically it is rather varied with unicells, filaments, parenchymatous plants and diverse forms of pseudoparenchymatous organization. There is some differentiation of cells, particularly in the rather complex reproductive organs. Basic cell structure is simple (Fig. 1.10), with one or many chloroplasts which have almost unique single-thylakoid lamellae, one nucleus, and a number of mitochondria and Golgi bodies. Grains of the polysaccharide reserve material, floridean starch, lie freely in the cytoplasm. No flagella are found in this group although a recent study (Simon–Bichard–Bréaud, 1971, 1972) has shown the presence of structures, which are thought to be a pair of flagella, in male gametocysts of *Bonnemaisonia*. Although these appear to be of about the same size as flagella, the photographs published do not allow one to see the essential tubules.

The unicellular members of this class, *Porphyridium* and *Rhodella*, have been studied by Brody and Vatter (1959), Speer *et al.* (1964), Gantt (1969), Gantt and Conti (1965), Gantt *et al.* (1968), Evans (1970), Neushul (1970), Guérin-Dumartrait (1970) and Wehrmeyer (1971). Cell structure of the more elaborate members of the class has been reported for *Lomentaria* (Bouck, 1962), *Compsopogon* (Nichols et al., 1966), *Smithora* (McBride and Cole, 1969), *Conchocelis* stage of *Porphyra* (Lee and Fultz, 1970), *Batrachospermum* (Brown and Weier,

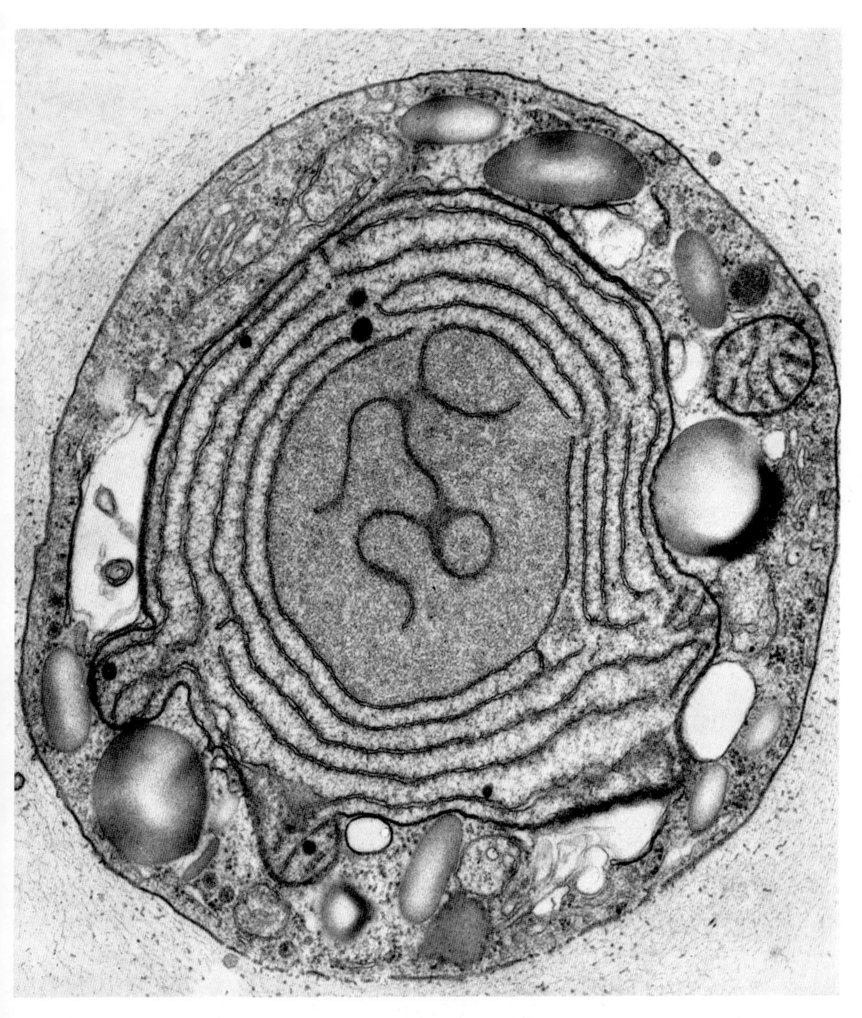

FIG. 1.10. A section through *Porphyridium cruentum* (Rhodophyceae) showing profiles of most of the organelles except the nucleus. Numerous starch grains lie in the cytoplasm. (×25,750)

1970). Some coralline algae have been studied by Bailey and Bisalputra (1970). The formation, release and germination of monospores has been studied in *Smithora* (McBride and Cole, 1971, 1972). Tumours in *Porphyra* have been examined by Kito (1972).

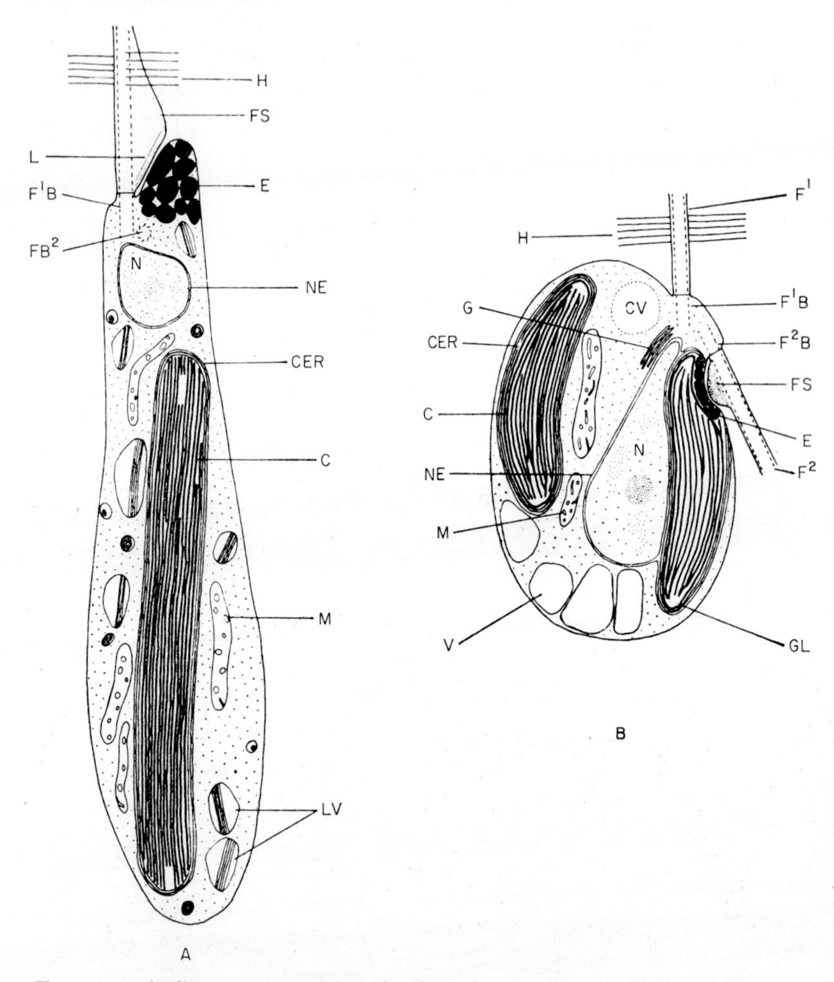

FIG. 1.11. A diagram comparing the fine structure of zoospores in the Eustigmatophyceae (A) with those of the Xanthophyceae (B). In particular note the differences in the flagella (F), eyespots (E) and the position of the nucleus (N). Reproduced, with permission, from Hibberd and Leedale (1972).

It is now generally accepted that the unicellular alga *Cyanidium caldarum* is a member of the Rhodophyceae but, owing to its formerly uncertain systematic position, this alga has attracted much attention from researchers (Rosen and Siegesmund, 1961; Mercer *et al.*, 1962; Staehelin, 1968; Seckbach, 1972;

Seckbach and Ikan, 1972). Amongst its more interesting features are chloroplasts which look very similar to unicellular blue-green algae (or the cyanelles of algae with blue-green symbionts).

XIII. Xanthophyceae (Yellow-green Algae)

The members of this class are mainly coccoid and filamentous organisms which are green but lack chlorophyll *b*. Chloroplast structure has definite similarities with that found in the Chrysophyceae and Phaeophyceae and, as in those classes, motile stages are heterokont (Fig. 1.11B). In most Xanthophyceans the cell structure is simple with one nucleus, one to several chloroplasts and the other usual organelles. A few members are coenocytic. In all but the coenocytic genera there is a characteristic 2-part construction of the wall of vegetative stages.

The most-studied genus is *Vaucheria* where zoospores (Greenwood *et al.*, 1957; Greenwood, 1959), spermatozoids (Møestrup, 1970b), cell walls (Parker *et al.*, 1963) the chloroplasts and pyrenoids (Marchant, 1972; Descomps, 1963a,b, 1972) have all been examined in some detail. Recent surveys by Falk (1967), Falk and Kleinig (1968), Massalski and Leedale (1969), Hibberd and Leedale (1971), Deason (1971a,b) have provided a clear picture of the fine structure of this class.

2. *The Cell Covering*

In the algae, besides naked uncovered cell membranes more typical of animal cells and cell walls similar to those of higher plant cells, we find a wide variety of other types of cell covering. Some, such as the pellicle and theca are characteristic of single classes whilst others are found in several classes (see Table I). In some organisms different types of cell covering are found at each stage of the life history.

I. Naked Membrane

Vegetative cells of members of the Chloromonadophyceae (Mignot, 1967; Heywood, 1968), some genera such as *Dunaliella* of the Chlorophyceae and, under certain conditions, *Ochromonas* of the Chrysophyceae are bounded on the outside by only their plasma-membrane (= plasmalemma or cell membrane). This membrane is normally 7 or 8 nm thick and quite often has a 'fuzz' of possibly proteinaceous material on the outside. The unicellular red alga *Rhodella maculata* (Evans, 1970) is bounded by a single membrane but there is a fairly thick sheath of mucilaginous material outside this. A recent study of another unicellular red alga, *Porphyridium* (Ramus, 1972) has shown that what appears as a wall in electron micrographs (Fig. 1.10) is in fact a capsule of soluble polysaccharide which gradually dissolves in the medium as it is pushed away from the cell by the deposition of new material. The fibres seen in micrographs are thought to result from dehydration during preparation of the material for electron microscopy.

Naked cells are found in many algal classes as zoospores or gametes, particularly spermatozoids. We find both of these in the Chlorophyceae, Xanthophyceae and Phaeophyceae, only zoospores in the Eustigmatophyceae and only spermatozoids in the Bacillariophyceae. In all cases these are normally very short-lived stages and, with zoospores, settlement is followed by immediate secretion of a cell wall. In the case of male gametes, these often have to penetrate an oogonial wall in order to carry out fertilization.

II. Modifications Within the Plasma-membrane

A. CRYPTOPHYCEAE—PERIPLAST

Until recently cells of the Cryptophyceae were thought to have only a single bounding membrane but now for two genera at least, *Cryptomonas* (Hibberd

TABLE I

A summary of the main types of cell covering found in the various algal groups

	Naked membrane		Modifications beneath membrane			Scales outside membrane			Silica frustule	Cell wall	
	Stage Only	Normal State	Periplast	Pellicle	Theca	Organic	Calcite	Silica		Incomplete (Lorica)	Entire Wall
Chlorophyceae	+	+									++
Prasinophyceae						++				+	
Euglenophyceae				++							
Chloromonadophyceae		++									
Eustigmatophyceae	+										++
Xanthophyceae	+										++
Chrysophyceae		+						++		+	
Haptophyceae						+	++				
Bacillariophyceae	+								++		
Phaeophyceae	+										++
Dinophyceae					++						+
Cryptophyceae			++								
Rhodophyceae		+									++

et al., 1971) and *Chroomonas* (Gantt, 1971) we now know that the covering or periplast is more complex. In these organisms (Fig. 2.1) there is indeed a continuous outer membrane and on the outside of this there is often a fuzz of granular or fibrillar material which Gantt regards as a separate layer. Under the plasma-membrane a layer of thin plates or membranes are found. These are said to be hexagonal in *Cryptomonas ovata* (Hibberd *et al.*, 1971) and more or

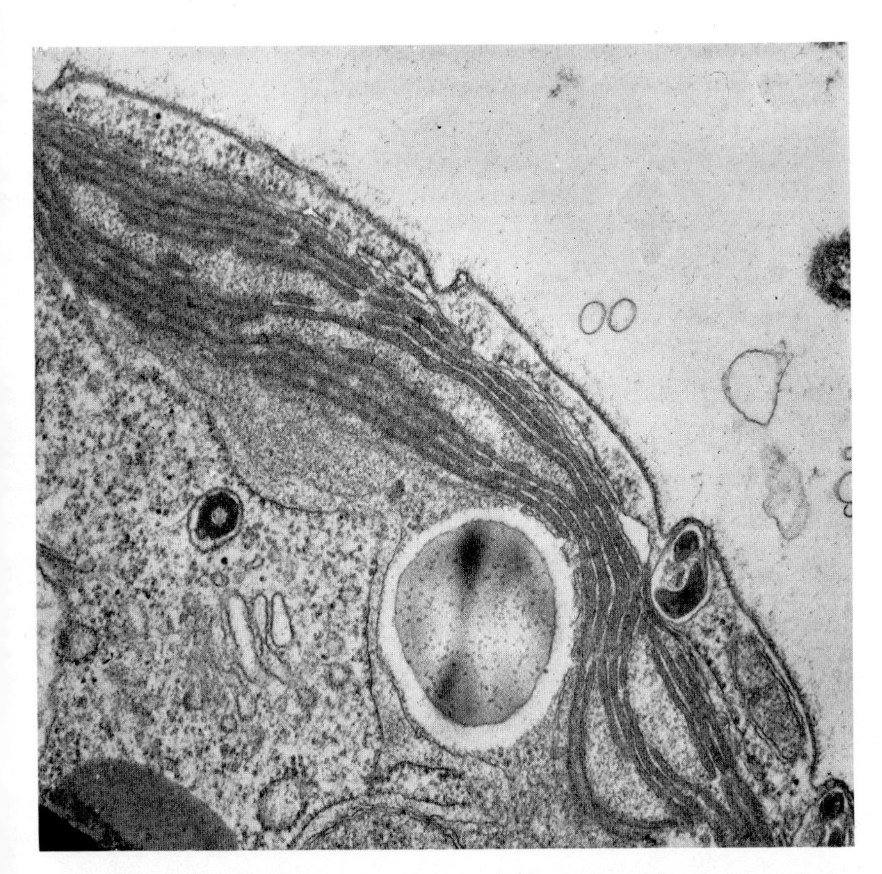

FIG. 2.1. The delicate ridged periplast of *Chroomonas* (Cryptophyceae). (× 36,000) (After Dodge, 1969b.)

less rectangular in *Chroomonas* sp. (Gantt, 1971). Here they are about 7·5 nm thick and are presumably proteinaceous in composition as they are removed by trypsin digestion. Occasional microtubules lie beneath the periplast.

B. EUGLENOPHYCEAE—PELLICLE

The most similar type of cell covering to that just described is the pellicle of the Euglenophyceae. In *Euglena* (Leedale, 1964, 1967; Sommer, 1965) and a

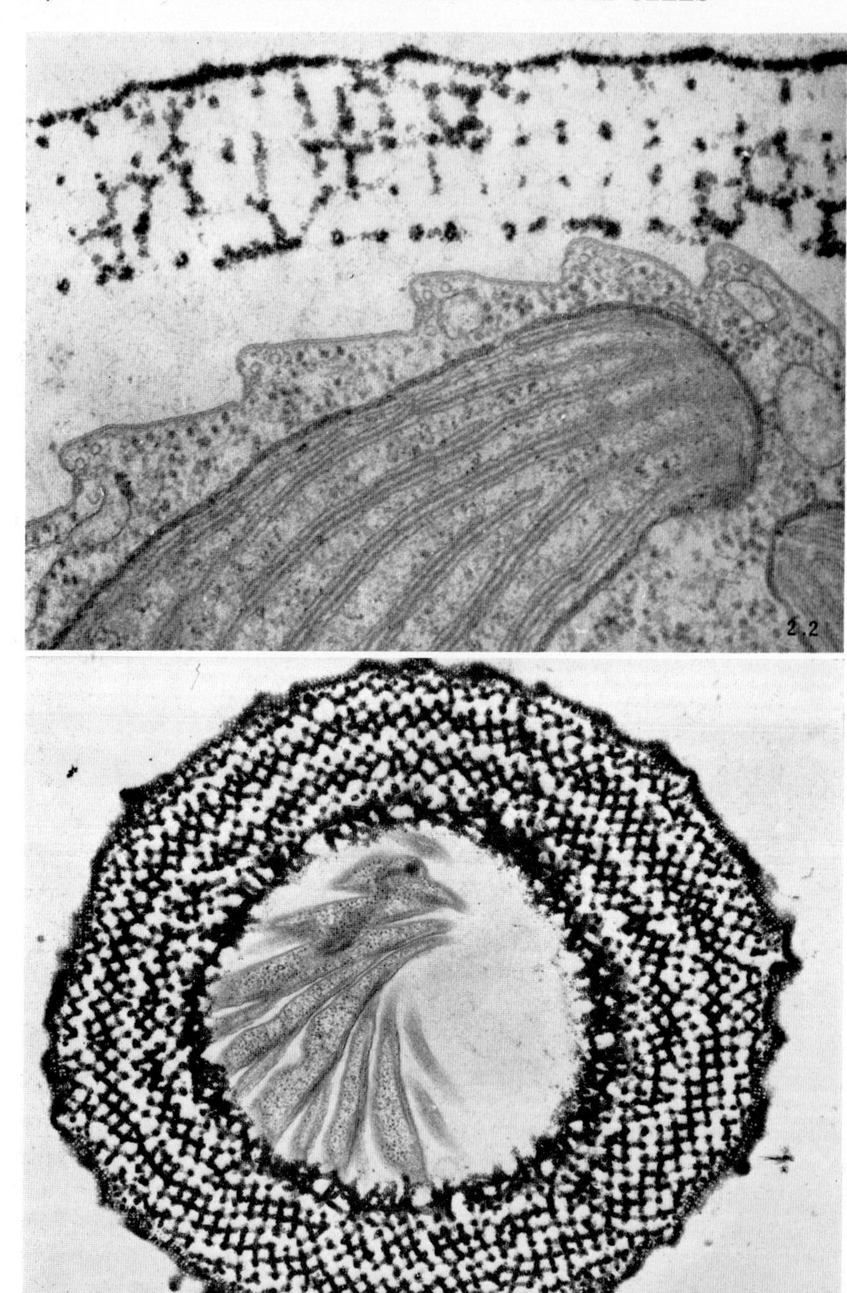

FIGS 2.2–2.3. The pellicle and iron-containing lorica of *Trachelomonas* (Euglenophyceae). (2.2) This is a transverse section and shows the characteristic ridges of the pellicle and the underlying microtubules. The lorica is seen to consist of an outer continuous layer and a wider loculate portion (× 59,000); (2.3) a glancing section from near the bottom of a cell showing several pellicular ridges (centre) and a tangential view of the lorica. (× 17,200)

number of other genera (Mignot, 1966) this consists of a continuous outer plasma-membrane which appears corrugated in transverse section. Beneath the ridges lie flat strips of proteinaceous material, about 70 nm thick, which run around the cell in a spiral course. These pellicular strips have a complicated form which allows them to articulate together and beneath them lie regularly arranged microtubules. Muciferous bodies are often found adjacent to the grooves of the pellicle with which they connect by narrow pores. Using freeze-etching, Schwelitz *et al.* (1970) found that the outer layer (the membrane?) of the pellicle is particulate, with 15 nm particles. The inner layer is striated with 5 nm bands at 3·5 nm intervals and a broad repeating pattern at 45 nm. This layer presumably corresponds to the pellicular strips composed of proteinaceous material. In some genera, such as *Trachelomonas* (Figs 2.2 and 2.3), the pellicular strips are missing but the remaining organization of the pellicle is as described. This organism has a rigid mineralized cup-shaped lorica outside the plasmalemma and some distance from it. Sommer and Blum (1964) have shown that after nuclear division and before cell division in *Astasia* new pellicular ridges arise as bands in the grooves of the old pellicle. Additional microtubules appear beneath the pellicle.

C. THE DINOPHYCEAE—THECA

In the Dinophyceae we find a complex cell covering, variously termed the theca or the amphiesma (Loeblich, 1970), which has a basic uniformity throughout most of the motile members of the class (Dodge and Crawford, 1970a). The standard construction consists of a single outer membrane which is continuous with the membrane around the flagella and therefore should be the effective plasma-membrane. Beneath this lies a single layer of flattened vesicles and under these there may be another single membrane. Finally there are subthecal microtubules which are sometimes evenly dispersed and sometimes arranged in groups. Within the class we find an interesting progression in which the thecal vesicles may be empty, may contain thin plates, thicker plates, or very thick plates. Correlated with increasing thickness of the plates their number decreases from several hundred to just two main ones (Fig. 2.4).

The most primitive arrangement is found in *Oxyrrhis* where the vesicles are irregularly arranged and separate from one another. In *Gymnodinium* and *Amphidinium* (Fig. 2.5) the vesicles are pressed together at their edges and thus have a polygonal outline. They may contain dark-staining amorphous material. In *Aureodinium* and *Katodinium* the vesicles contain thin plates and in *Woloszynskia* (Fig. 2.6) the plates are thick enough to be seen by light microscopy. We now find the introduction of angled sutures or joints at the edges of the plates (Fig. 2.7). In the genera *Heterocapsa* (Fig. 2.9), *Gonyaulax*, *Peridinium* (Figs 2.10 and 2.13) and *Ceratium* (Fig. 2.8) the number of plates is reduced to around twenty and these may be very thick and bear ridges or projections. Each plate has now a distinctive morphology depending on its position around the cell. The ultimate situation is found in the order Prorocentrales where the theca

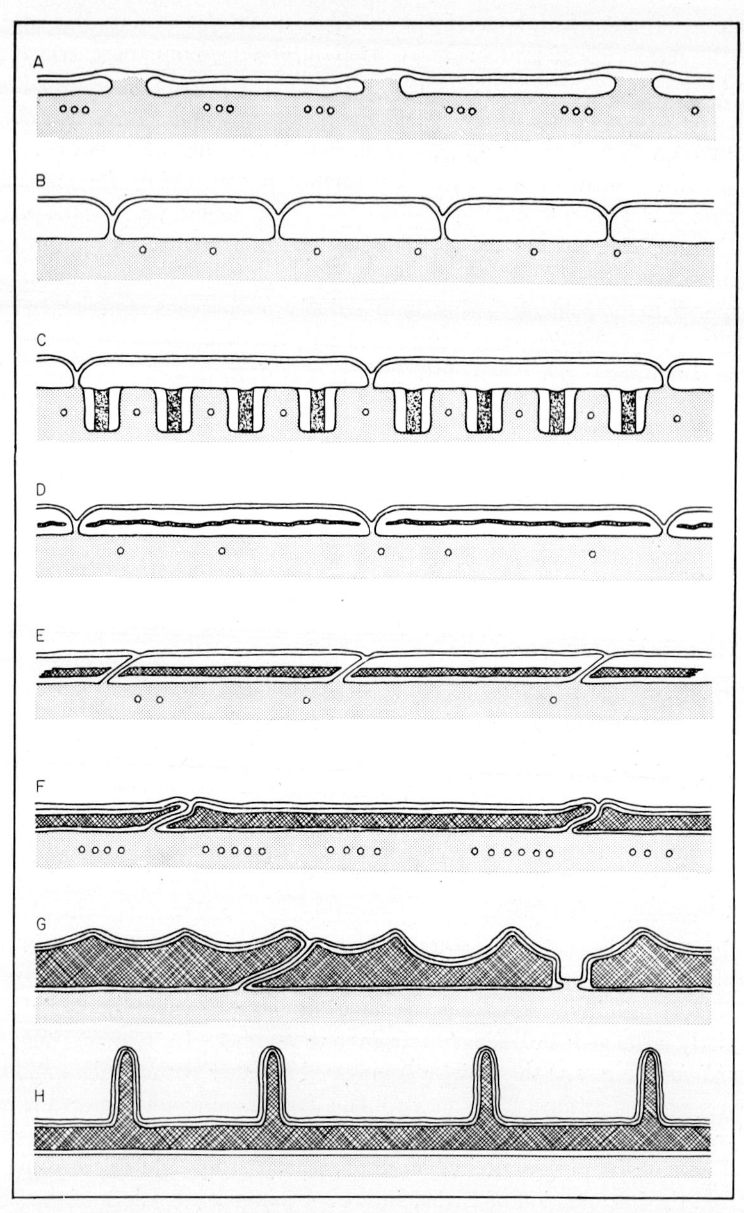

Fig. 2.4. Diagrams to illustrate the structure of various types of dinoflagellate theca, as seen in vertical section. A, as in *Oxyrrhis*; B, *Amphidinium*; C, some species of *Gymnodinium*; D, *Katodinium*; E, *Woloszynskia*; F, *Glenodinium* and *Heterocapsa*; G, *Ceratium* and some *Peridinium* species; H, *Prorocentrum* (= *Exuviaella*). (After Dodge and Crawford, 1970a.)

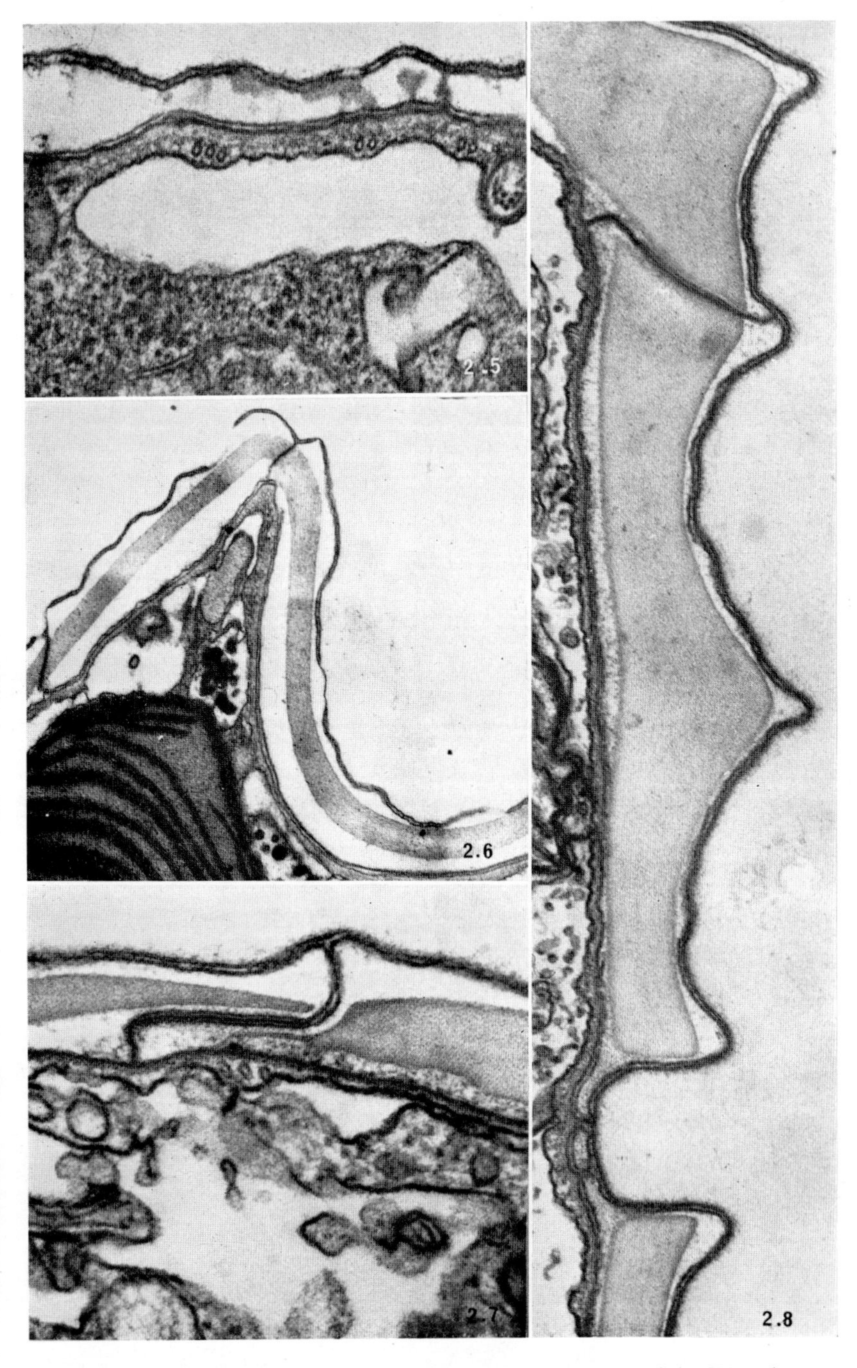

FIGS 2.5–2.8. Cross-sections of the theca in some members of the Dinophyceae. (2.5) *Gymnodinium* sp. showing empty thecal vesicles and underlying microtubules (× 59,000); (2.6) a girdle ridge of *Woloszynskia coronata* showing the plates of even thickness. (× 27,000); (2.7) *Ceratium*, a suture or junction between two plates (× 90,000); (2.8) *Ceratium*, a suture, trichocyst pore (at bottom) and the ridges or reticulations of the plates. (× 43,000) (2.5 after Dodge and Crawford, 1970a; 2.7 and 2.8 after Dodge and Crawford, 1970b.)

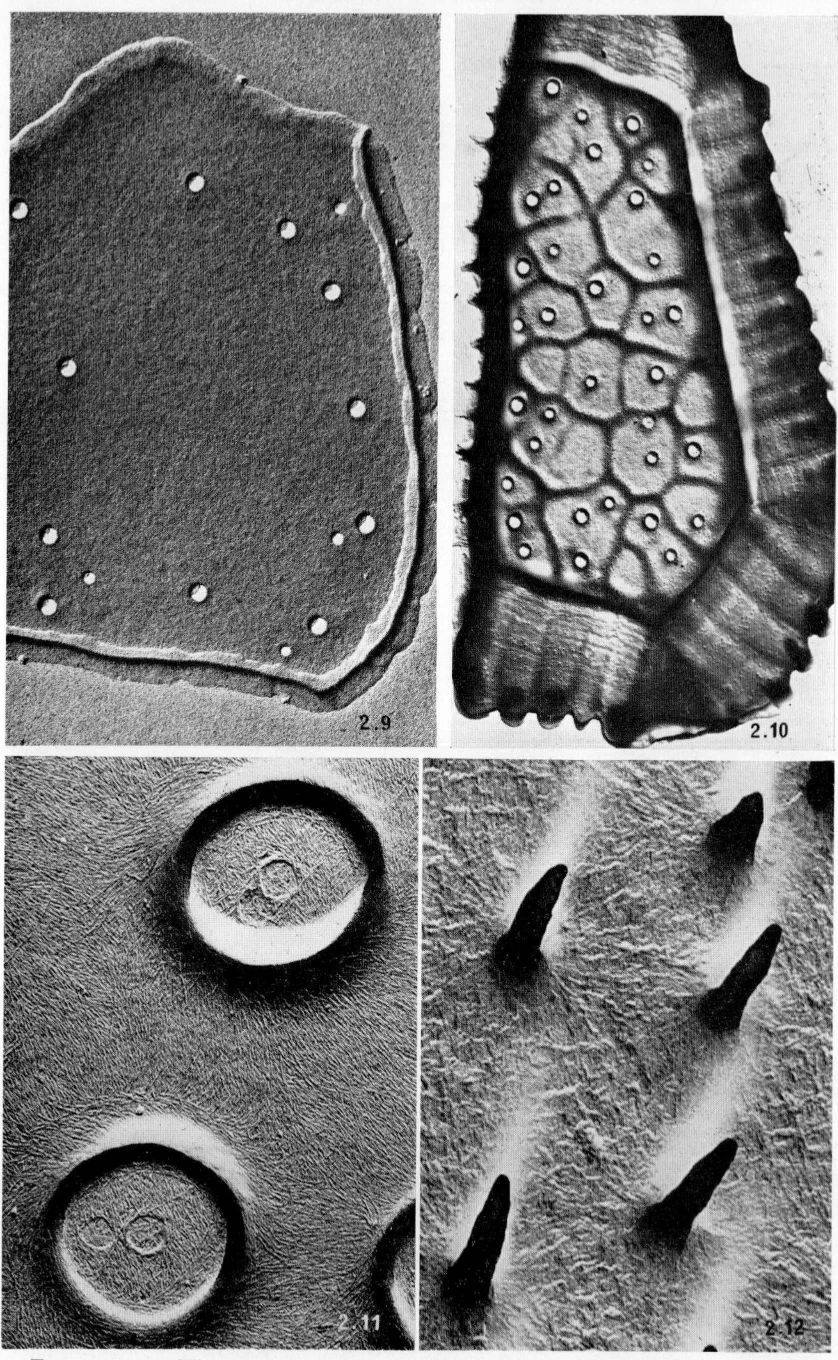

FIGS 2.9–2.12. Thecal plates from some dinoflagellates (shadowed preparations); (2.9) a smooth plate of *Heterocapsa*, showing only trichocyst pores, and a ridge around some of its sides (×14,000); (2.10) a reticulate and ridged plate from *Peridinium cinctum* (×5600); (2.11) part of a plate of *Gonyaulax poly dra* after cleaning and shadowing. Note the circular ridges around the trichocyst pores and the microfibrils of which the plate appears to be constructed (×21,000); (2.12) part of one plate of *Prorocentrum* (= *Exuviaella*) *mariae-lebouriae* showing the spines on the outer surface. (×49,000) (2.9, 2.10, 2.11, 2.12 after Dodge and Crawford, 1970a.)

contains two large saucer-shaped plates together with several minute plates which are situated around the flagellar pores (Dodge and Bibby, 1973).

In many cases the plates are covered with ridges or reticulations (Figs 2.10 and 2.13) and, in spite of being surrounded by membrane, they may have spines

FIG. 2.13. A scanning electron micrograph of the dorsal side of the dinoflagellate *Peridinium leonis* showing the arrangement of the thecal plates, the girdle, and the reticulations, etc. on the plates. (×1700)

or other processes (Fig. 2.12). Where the plates are thick microfibrils appear to provide the main component and these seem to show rather random orientation (Fig. 2.11) except in *Prorocentrum mariae-lebouriae* (Fig. 2.12) where they can be seen to radiate into the spines on the outer surface of the plates.

III. Scaly Covering Outside Cell Membrane

There are at least three distinct types of scales. Those constructed of silica are found only in the Chrysophyceae, those of calcite are formed by some members of the Haptophyceae, and organic scales may be found in the Prasinophyceae and Haptophyceae.

A. SCALES OF ORGANIC MATERIAL

Organic scales in the Prasinophyceae are usually of two types on any one organism. There are large body scales which are very varied in form and may be star-fish shaped as in *Heteromastix* (Manton *et al.*, 1965), square fretted scales as in *Pyramimonas* (Manton *et al.*, 1963), or like a filigree open-topped box as in *Mesostigma* (Manton and Ettl, 1965). The second type are the tiny scales, discoidal, square or pentagonal in shape, which clothe the flagella (see Chapter 3). In the Haptophyceae the scales may again take various forms. The genus *Chrysochromulina* exhibits a considerable variety of body scale form (Fig. 2.15) which now provides the main taxonomic characteristic for the species (Parke *et al.*, 1955; Manton and Parke, 1962, etc.). Here, there may be two types of scale on a single organism (Fig. 3.20) as for example in *Chrysochromulina camella* (Leadbeater and Manton, 1969, 1971) where there is an under-layer of flattened discoidal scales with a pattern of radial ribs and above this is found a layer of cup-shaped scales. Organic scales are also found as an under-layer on many organisms which produce calcareous coccoliths (see below) (Fig. 2.16).

The morphology of the organic scales in *Hymenomonas* has been described by Franke and Brown (1971). The scales are flat circular or oval discs made up of a network of radial and concentric fibrils. The margin consists of two distinct but closely associated fibrillar ribbons. The fibrils in general are all ribbon shaped and are $1–2{\cdot}2 \times 3–7{\cdot}5$ nm in section. In some of these organisms, as *Pleurochrysis* (Brown, 1969; Brown *et al.*, 1969, 1970; Leadbeater, 1971a) a thick wall consisting of numerous layers of closely appressed scales embedded in pectic material forms around the sedentary stages (Fig. 2.18). Here, as in *Apistonema* (Leadbeater, 1970) similar scales are also found on one of the motile stages. In this organism a recent study (Herth *et al.*, 1972) has shown that the main components of these scales is a cellulosic glucan to which is linked peptide material. It is therefore described as a cellulosic glycoprotein.

B. CALCITE SCALES

Calcareous scales or coccoliths are generally oval in shape but have very varied ornamentation. The most simple scales, such as those of *Crystalolithus* (Gaarder and Markali, 1956; Manton and Leedale, 1963b), consist of a number of calcite crystals similar in shape to those which can be obtained by chemical means.

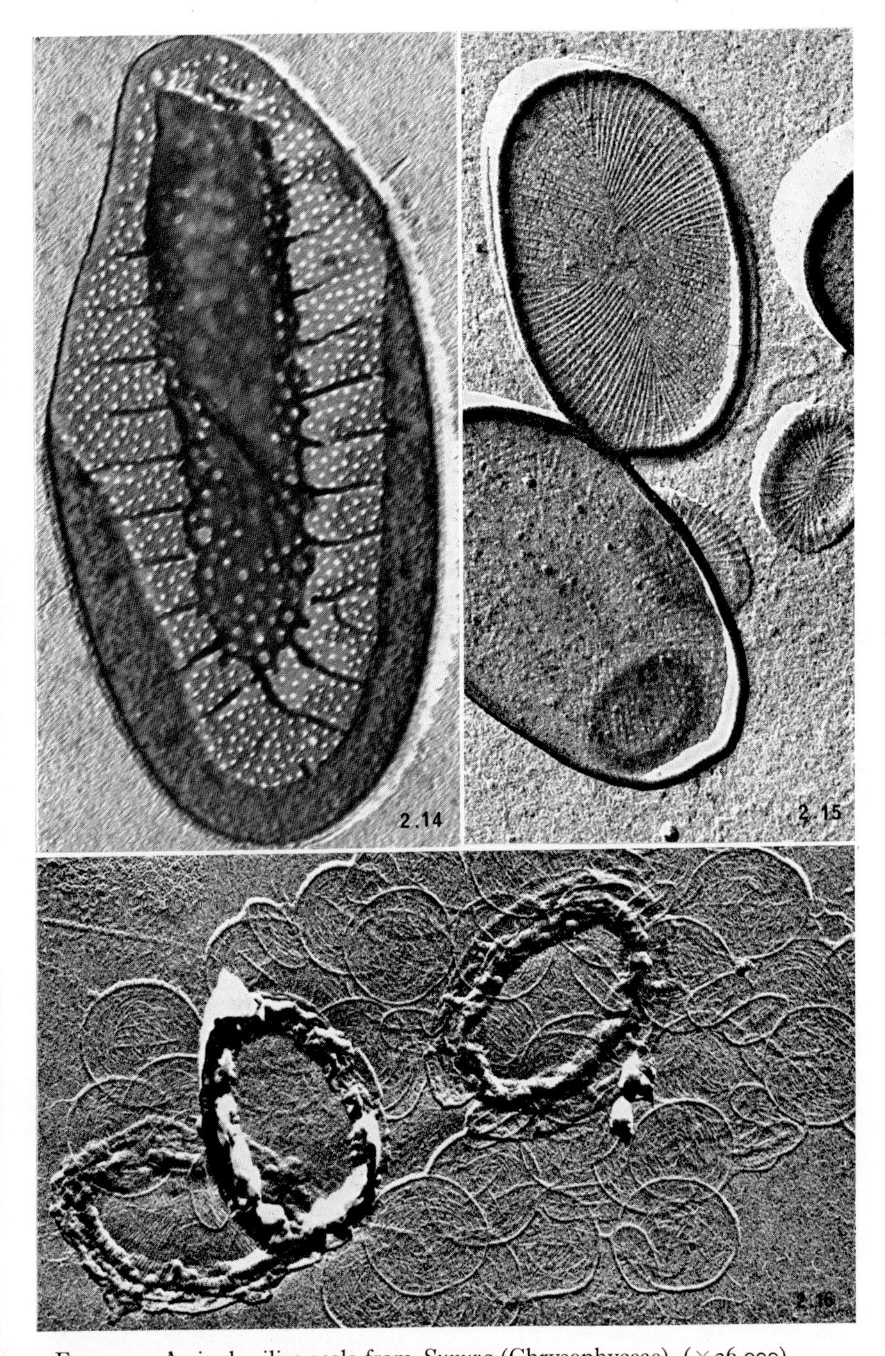

FIG. 2.14. A single silica scale from *Synura* (Chrysophyceae). (×26,000)

FIG. 2.15. Organic scales from a member of the Haptophyceae (*Chrysochromulina*?). (×39,600)

FIG. 2.16. Small organic scales and larger scales with calcified rims from *Coccolithus*. (×21,000)

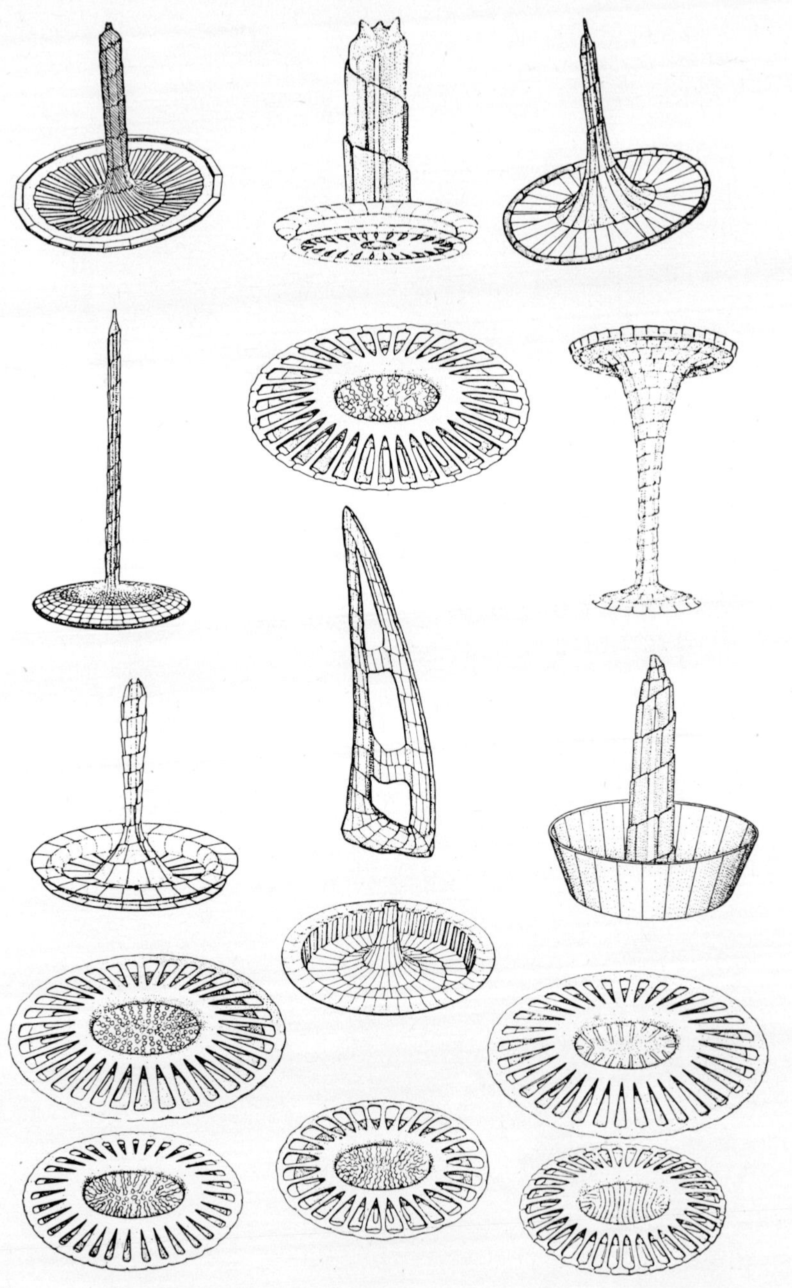

FIG. 2.17. Drawings to illustrate some of the great variety of form of the coccoliths. [Modified from Lecal (1965, 1966).]

In most coccoliths, however, the form of the crystals is much more elaborate and the individual crystals fit together to form the characteristic scale pattern for the species (Figs 2.16 and 2.17). Often, different stages of the life cycle have their own type of scale as was first shown by Parke and Adams (1960) for *Crystalolithus/Coccolithus* where scales described as placoliths alternate with those termed cricoliths. Most of these scales form first as a flat organic plate and later the crystals of calcite are deposited on this (see below). In the case of *Cricosphaera* (Manton and Leedale, 1969) the crystals only form around the rim and the organic component persists as a base. Cell types of *Coccolithus* have recently been described (Klaveness and Paasche, 1971) which are unable to form calcite scales and only organic scales are produced. Organisms which produce coccoliths are normally covered by an imbricated sheath consisting of perhaps 50–100 scales.

C. SILICA SCALES

These are formed exclusively in the Chrysophyceae. As with calcareous scales each cell is covered with a large number of scales but in some genera they may be of two types with, for example, one type located only at the ends of the cell. Such is the case in *Mallomonopsis* (Harris, 1966) and some species of *Mallomonas* (Harris and Bradley, 1960) where the scales are all oval and delicately patterned but those at the anterior and sometimes also the posterior ends of the cell have long spines attached to them. In *Synura* (Fig. 2.14) each species has its own distinct pattern of ridges and pores over the surface of the scale and there is usually a rim or flange around the edge.

D. SCALE FORMATION

In all cases which have been investigated the scales have been found to form within Golgi vesicles. In these organisms there is normally only one Golgi body which is relatively large and situated between the nucleus and the anterior end of the cell. Manton and Parke (1960) working with the small green flagellate *Micromonas* (= *Chromulina pusilla*), were the first to show that scales formed in vesicles within the cell rather than *in situ* on the outside. Later, Manton and Leedale (1961b), working with *Paraphysomonas* which has long silica spines, found that these spines also were formed within the cell. This organism had an extraordinarily large Golgi body. The involvement of the Golgi was finally resolved when Manton (1966b) found a regular sequence from immature to mature scales within the cisternae as they recede away from the Golgi in *Prymnesium*. This has since been confirmed in *Chrysochromulina* (Manton, 1967b,c) (Fig. 4.12) another member of the Haptophyceae. Here the fully formed scales were found to be liberated to the exterior of the cell at a point near the flagellar bases.

The process of scale formation has also been followed in the Prasinophyceae where, in *Pyramimonas* (Manton, 1966c) the scales for the flagella collect in a reservoir which opens to the exterior in the flagellar pit near the flagellar bases. The larger body scales appear to be liberated onto the body surface following the fusion of the vesicle which surrounds them with the plasmalemma.

The involvement of the Golgi apparatus in the formation of scales has been confirmed in studies on the haptophycean alga *Pleurochrysis* which in its vegetative phase has a wall composed of numerous layers of organic scales (Brown, 1969; Brown *et al.*, 1970; Leadbeater, 1971). Following cell division (Fig. 2.18) the single large Golgi body becomes located near to the newly forming transverse wall and inflated cisternae containing scales are orientated parallel to the cell surface. After the transverse wall is completed the cell protoplast rotates and completes a new wall around the cell. It was deduced (Brown, 1969) that between 41 and 82 Golgi generations were required to synthesize the cell wall of an actively growing cell.

The various stages of coccolith formation have been studied by Manton and Leedale (1969) and Pienaar (1969, 1971a) but the most complete account is that for *Hymenomonas* by Outka and Williams (1971). Here, the proximal two-thirds of the large Golgi body consists of typical smooth stacks of rather flattened cisternae. The distal third consists of more inflated cisternae which are spread further apart. Scales and parts of coccoliths are seen in the expanded ends of the cisternae and stages in coccolith assembly are found in vesicles which have become detached from the Golgi body. The process of assembly begins with the formation of the delicate oval base plates. These are composed of organic material and are patterned with spokes which radiate from a line passing through the long axis of the plate. Coccolithosomes, formed of 12 or so granular subunits each about 7 nm in size, collect in large numbers in ear-shaped vesicles which form around the developing base plates. Eventually they are released into the base vacuole where they become associated with the rim of the plate and gradually form an outline of the anvil-shaped subunits which eventually make up the rim. It is presumed that the coccolithosomes provide precursor material for the matrix of the rim, for they appear to be used up as the coccoliths mature. The rim of the coccolith now achieves its typical shape and calcification takes place with crystallization of calcium carbonate in the matrix of the rim. It is possible that the vesicular membrane plays some part in determining the eventual form of the coccolith for, at the beginning of rim formation, its profile is remarkably similar to that of the mature scale.

IV. The Diatom Frustule

The form of the diatom frustule has been extensively studied by normal transmission microscopy of acid-cleaned material, by the use of carbon replicas and, with most success, by scanning microscopy. The variations in shape,

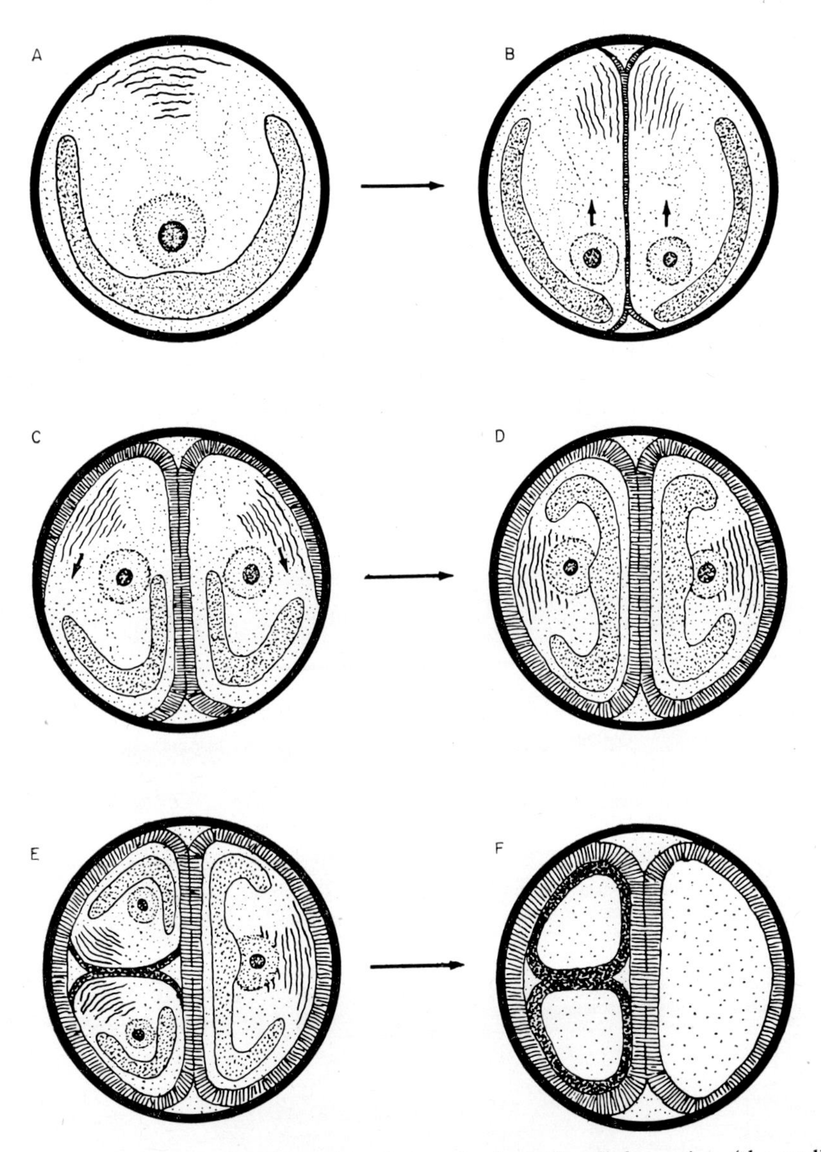

FIG. 2.18. A diagram illustrating the mechanism of wall formation (the wall consists mainly of organic scales) during cell division in *Pleurochrysis* (Haptophyceae). A, Vegetative cell with single Golgi (top) nucleus and chloroplast; B, the cell divides and a cleavage furrow forms. The Golgi bodies first synthesize the transverse wall; C–D, the protoplasts rotate and the Golgi produce wall material for the remainder of the cell wall: E–F, a second division, and wall forming operation, takes place. Reproduced, with permission, from Brown (1969).

ornamentation and perforation of the silica frustule are so many that it is impossible to summarize them here. Reference should be made to taxonomic works such as those of Helmcke and Krieger (1953–1966), Desikachary (1956), Ross and Simms (1970, 1971, 1972), Hasle and Heimdal (1970). A few examples of frustule structure are illustrated in Figs 2.19–2.22.

The formation of the silica wall has now been studied in several diatoms. The normal arrangement in these organisms is that each daughter cell takes one of

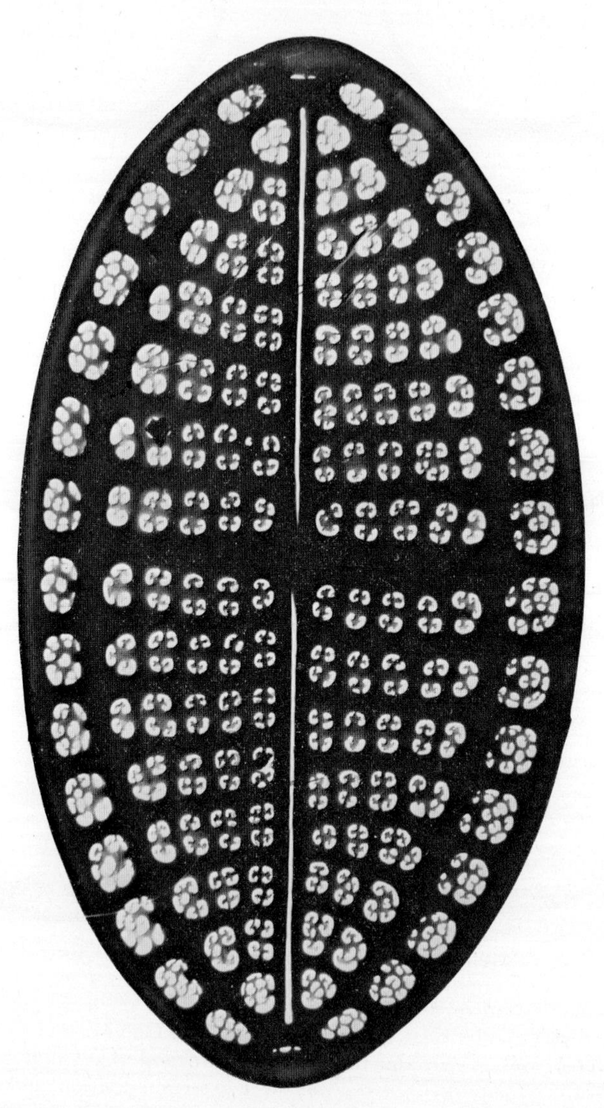

FIG. 2.19. One valve of the frustule of *Cocconeis* (Bacillariophyceae). Organic components of the cell have been cleaned away leaving only the silica. (×8000)

FIGS 2.20–2.22. Patterns of the silica in the walls of some diatoms. (2.20) Scanning micrograph of *Stephanodiscus*. (×2800) (After Dodge, 1970); (2.21) part of a valve from the pennate diatom *Gyrosigma* (×17,500); (2.22) a small portion of the cylindrical frustule of *Rhizosolenia*. (×17,500)

the original overlapping valves of the frustule and each has to make a new valve. Thus, following nuclear and cell division, the parent frustule enclose two daughter cells each of which is surrounded by a single plasma membrane. The work of Reimann (1964), Drum and Pankratz (1964), Stoermer *et al.* (1965), Lauritis *et al.* (1968), has shown that silicon deposition vesicles first collect beneath the plasmalemma at the centre of the site of a new frustule. As the vesicles fuse together the membranous component forms a continuous sac, the *silicalemma*, which eventually surrounds the area where the new frustule is forming (Fig. 2.23A). The silica, which is deposited rapidly, becomes arranged into the characteristic form of the frustule (Fig. 2.23B). When deposition has ceased the silica appears to be tightly bound to the silicalemma. At some stage, which is not clear from studies so far carried out, the portion of plasmalemma and any cytoplasm outside the new frustule must be abandoned and a new continuous cell membrane formed beneath the frustule, for in mature cells the functional cell membrane appears to lie within the wall. It is presumably formed from the fusion of the inner portion of the silicalemma with the portion of old plasmalemma which lies beneath the parent valve.

It has been suggested (Stoermer *et al.*, 1965) that the silica precursors are formed in the vicinity of Golgi bodies but, as yet, definite evidence is lacking. Reimann (1964) has stated that silica cannot be recognized until it arrives in the developing wall. Studies on the formation of new frustules have shown that the process is inhibited to some extent by colchicine (Coombs *et al.*, 1968a) in *Navicula* but in *Cyclotella* (Oey and Schnepf, 1970) cell division is inhibited but new valve formation is not. Agents which block DNA synthesis also result in inhibition of new valve formation (Oey and Schnepf, 1970) and the converse is also true in that silicon starvation results not only in inhibition of valve formation but also inhibition of mitosis and DNA synthesis (Darley and Volcani, 1969). In some diatoms the frustule is thought to contain a considerable amount of organic material (Reimann *et al.*, 1965) and in *Cylindrotheca* this is said to enclose the silica portions of the wall.

V. The Cell Wall

As in higher plants, the algal cell wall is always formed outside the plasmalemma. There are two basic types, with numerous variations.

A. INCOMPLETE WALLS (SUCH AS THE LORICA)

In several algal groups certain species with flagella have an incomplete cell wall in the form of a bottle or cup-shaped structure which is usually termed a lorica or tunica. The cell is usually fairly loosely fitted inside this type of wall. In the Prasinophyceae *Platymonas* (Manton and Parke, 1965) has a wall (often termed a theca in this genus) which is formed by the coalescence of many small

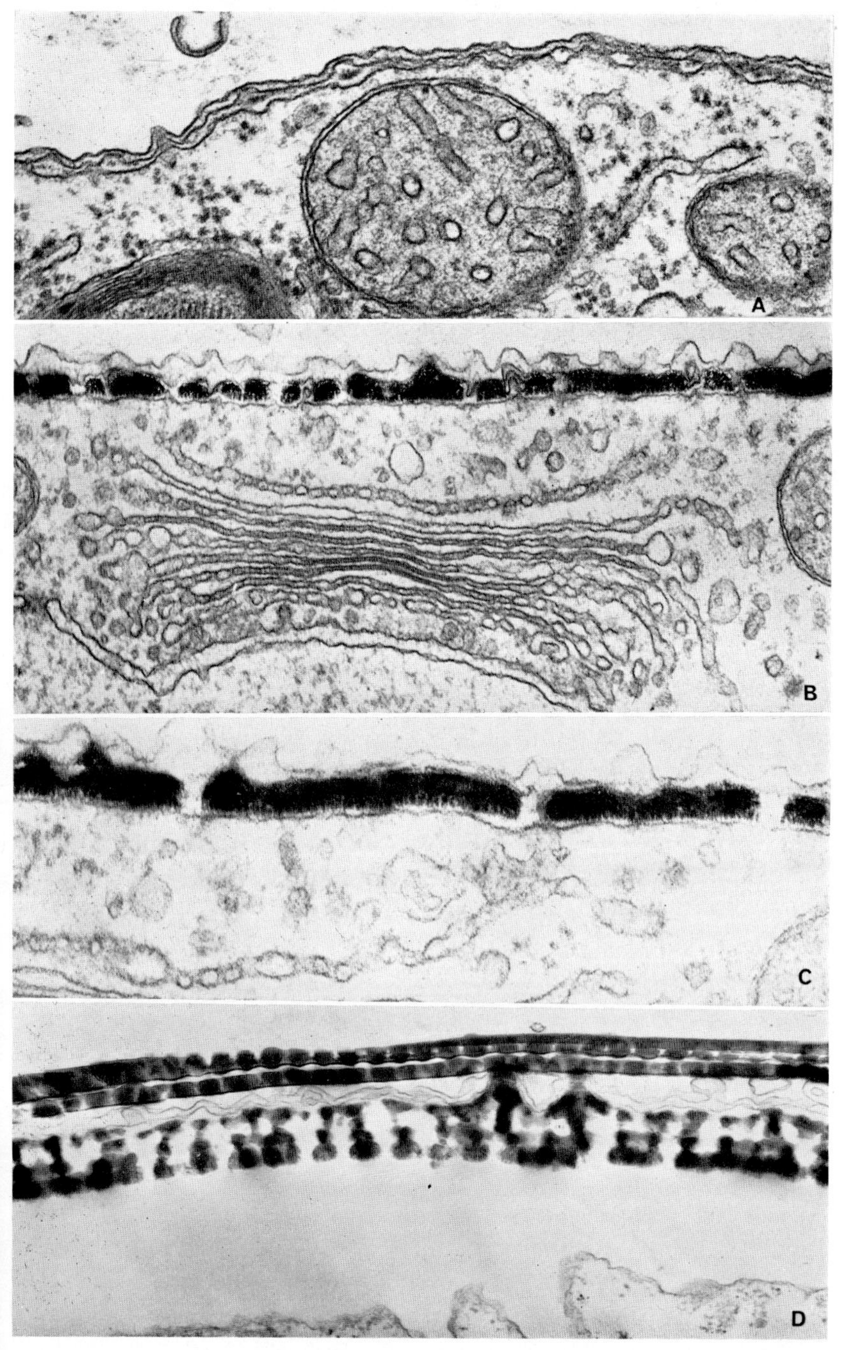

FIG. 2.23. Frustule formation in *Melosira*. A, Recently formed daughter cell bounded by three membranes, the outer being plasmalemma and the other two belong to the silicalemma (×46,000); B, the silicalemma now encloses a considerable amount of silica and the morphology of the wall is becoming evident. Note the 'active' Golgi body between the nucleus and the forming frustule (×46,000); C, a valve in which silica deposition is almost complete. The membrane beneath the valve will soon become the functional plasmalemma when the outer membranes are lost (×75,000); D, part of the side wall of the cell showing the overlapping girdle bands of the parent cell (top) enclosing a recently formed new frustule of the daughter cell. (×29,000) (Micrographs by courtesy of R. M. Crawford.)

stellate particles. These are produced in Golgi vesicles and are then released into the space surrounding the sister protoplasts of a recently divided cell but within the lorica of the parent cell. Coalescence begins at a predetermined place on the surface of each protoplast which is usually adjacent to the pyrenoid. The final lorica completely covers the cell except for the small gap through which the flagella protrude.

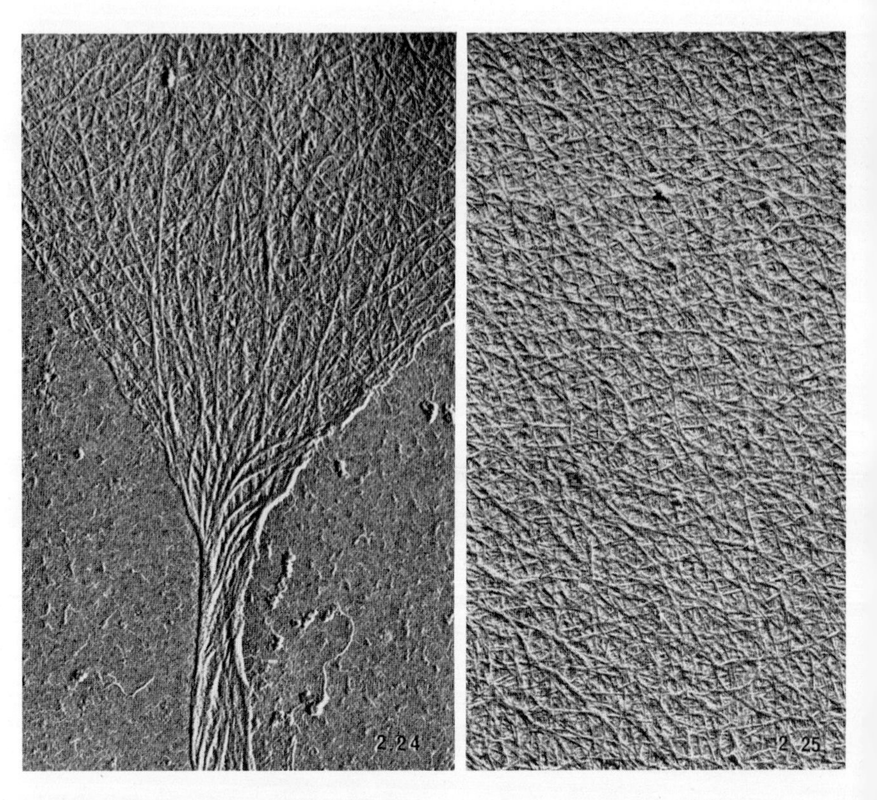

FIG. 2.24. Part of the delicate stalk and lorica of *Ochromonas* (Chrysophyceae) composed of cellulose fibrils. (×15,000)

FIG. 2.25. A small portion of the much more substantial lorica of *Dinobryon* in which the fibrils are arranged more or less circumferentially. (×28,000)

In the Chrysophyceae many organisms have loricas which are constructed of microfibrils, generally thought to be cellulosic in nature. The arrangement of the fibrils differ from organism to organism but in all cases they are probably formed *in situ* outside the cell. In *Ochromonas* (Fig. 2.24) the lorica is very delicate and consists of a long stalk and narrow cup. The structure consists of long cellulose fibrils which run in a spiral up the stalk and around the cup (Schnepf *et al.*, 1968). The fibrils have been found to consist of 1–2 nm thick elementary fibrils which are very loosely grouped into bundles (Kramer, 1970).

By washing the cells of active organisms out of their loricas it was possible to stop fibril synthesis suddenly and then study the form of growing ends. These Kramer found to be of several types: split ends where several subunits appear to be joining together; blunt and club-shaped ends; tapered ends which reduce to about 1 nm diameter. These variations are thought to be explicable if the polysaccharide chains grow separately and are later 'zipped' together.

A somewhat similar lorica is found in *Dinobryon* (Karim and Round, 1967) (Fig. 1.4). The fibrils are 7–15 nm in diameter and appear to be arranged in a series of shallow horizontal loops (Fig. 2.25), which may indicate that they are mainly synthesized around the advancing edge of the lorica as it increases in size. Much the same arrangement has been described from *Pseudokephyrion* (Belcher, 1968c) although in this organism mineral material is deposited over the microfibrils as the lorica matures. Rather more elaborate loricas have been reported from *Chrysolykos* (Kristiansen, 1969b) and from *Bicoeca* and its relatives (Kristiansen, 1972a). These latter organisms have had a varied taxonomic history but are probably closely related to the Chrysophyceae, if not actually genuine members of the class.

B. COMPLETE CELL WALLS

In the Chlorophyceae, Xanthophyceae, Phaeophyceae and Rhodophyceae some or all of the organisms consist of cells entirely surrounded by cell walls which are in many respects similar to those of higher plants. In general the wall consists of a microfibrillar framework and a large amount of amorphous material. Incrusting material such as silica or sporopollenin may also be present. Owing to the number of variations on the basic theme it will be best to examine each class separately, starting with the ones which at present appear to be most simple and uniform.

1. *Rhodophyceae*

Very few red algal walls have been studied in detail but, of those which have, *Ptilota* and *Griffithsia* (Myers *et al.*, 1956; Cronshaw *et al.*, 1958) have 20–25% of the wall composed of cellulose microfibrils which appear to be randomly arranged. In *Rhodymenia* cellulose make up only 7% of the wall. In *Porphyra*, which belongs to a distinct section of the Rhodophyceae, the Bangiales, the situation is rather different (Frei and Preston, 1964b). In the walls of this alga there is no cellulose, nevertheless microfibrils are present. In the main part of the cell walls these are composed of β1–3 linked xylan (as in certain green algae, see below). The outer layer, or cuticle, consists mainly of β1–4 linked mannan which does not appear as microfibrils but is granular. The xylan microfibrils are randomly arranged in most walls except in rhizoids where they tend to lie parallel to each other and to the long axis of the cell. Frei and Preston noticed short cross-connections between the xylan microfibrils and there is clear suggestion of these in sections of the wall of *Falkenbergia* (Fig. 2.26). This alga, which is not at all closely related to *Porphyra* also shows an amorphous outer layer the chemical nature of which is not known but which clearly could consist of mannan. A similar layer has been reported from *Laurencia* (Bisalputra *et al.*, 1967) where it is said to contain both pectic and proteinaceous material.

A structure found in the cross walls separating cells of many red algae is the pit connection (Fig. 2.27). Fine structure studies (Myers *et al.*, 1959; Bouck,

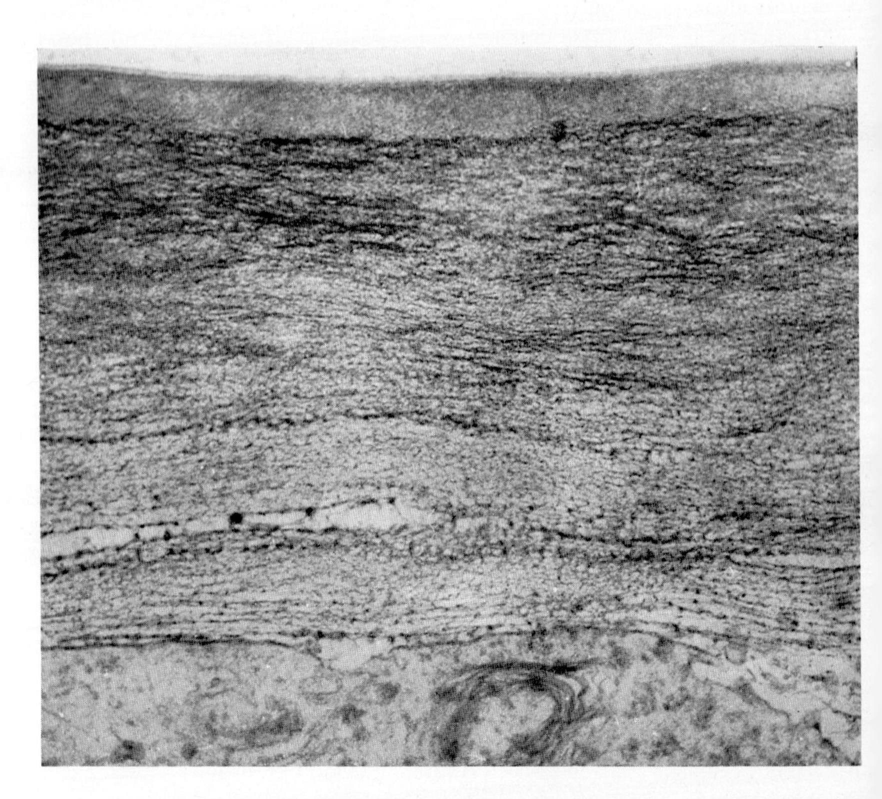

FIG. 2.26. Section through the wall of *Falkenbergia* (Rhodophyceae) showing the outer cuticular layer (top) and the thick microfibrillar layer in which there appear to be interconnections between the fibrils. (× 59,000)

1962; Bisalputra *et al.*, 1967; Ramus, 1969a,b; Brown and Weier, 1970; Bourne *et al.*, 1970; Sommerfeld and Leeper, 1970; Lee, 1971a) have shown that this structure is in fact an intercellular plug which fits neatly into a pore in the wall. It is wheel or lens-shaped, 0·3–2·5 μm in diameter, and mostly consists of granular material (5–10 nm diameter particles) which appears to be proteinaceous. The cell plasma-membrane passes around the sides of the plug and is therefore continuous from cell to cell. Over the ends of the plug is a 10–30 nm thick plug cap which often contains electron-dense material. During development of cross walls the septum grows in as an annular ingrowth from the existing side walls. This does not completely separate the two daughter cells but leaves a central aperture in which the plug forms (Ramus, 1969a). The structure of the pit connection has been compared with the dolipore of the Basidiomycetes (Lee, 1971a).

2. *Phaeophyceae*

In all brown algae that have been investigated the walls contain microfibrils which appear to consist of cellulose, but often these only make up a very small

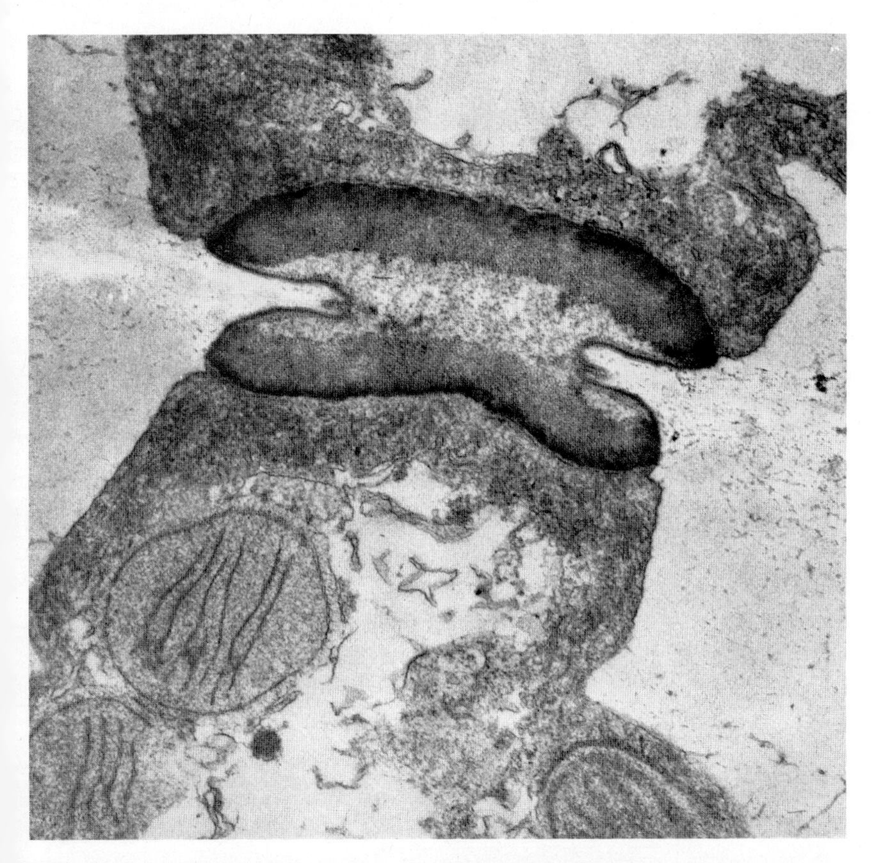

FIG. 2.27. An intercellular pit (or plug) in *Antithamnion* showing the thick cell wall (light) at either side and the granular nature of the plug material. (×24,500)

percentage of the wall (Cronshaw *et al.*, 1958). Thus, in *Pelvetia* cellulose is said to represent only 1·5% of the wall and in *Ascophyllum* about 7%. A more normal figure of about 20% is found in *Laminaria*. In general the microfibrils appear to be randomly arranged (Fig. 2.29) but in cells which have seen vast increase in length, such as the hyphal cells in the Laminariales, these show a predominantly longitudinal orientation of the fibrils (Fig. 2.30) (Dodge, unpublished). In the uniseriate filaments of *Ectocarpus* (Fig. 2.28) the predominant orientation of the

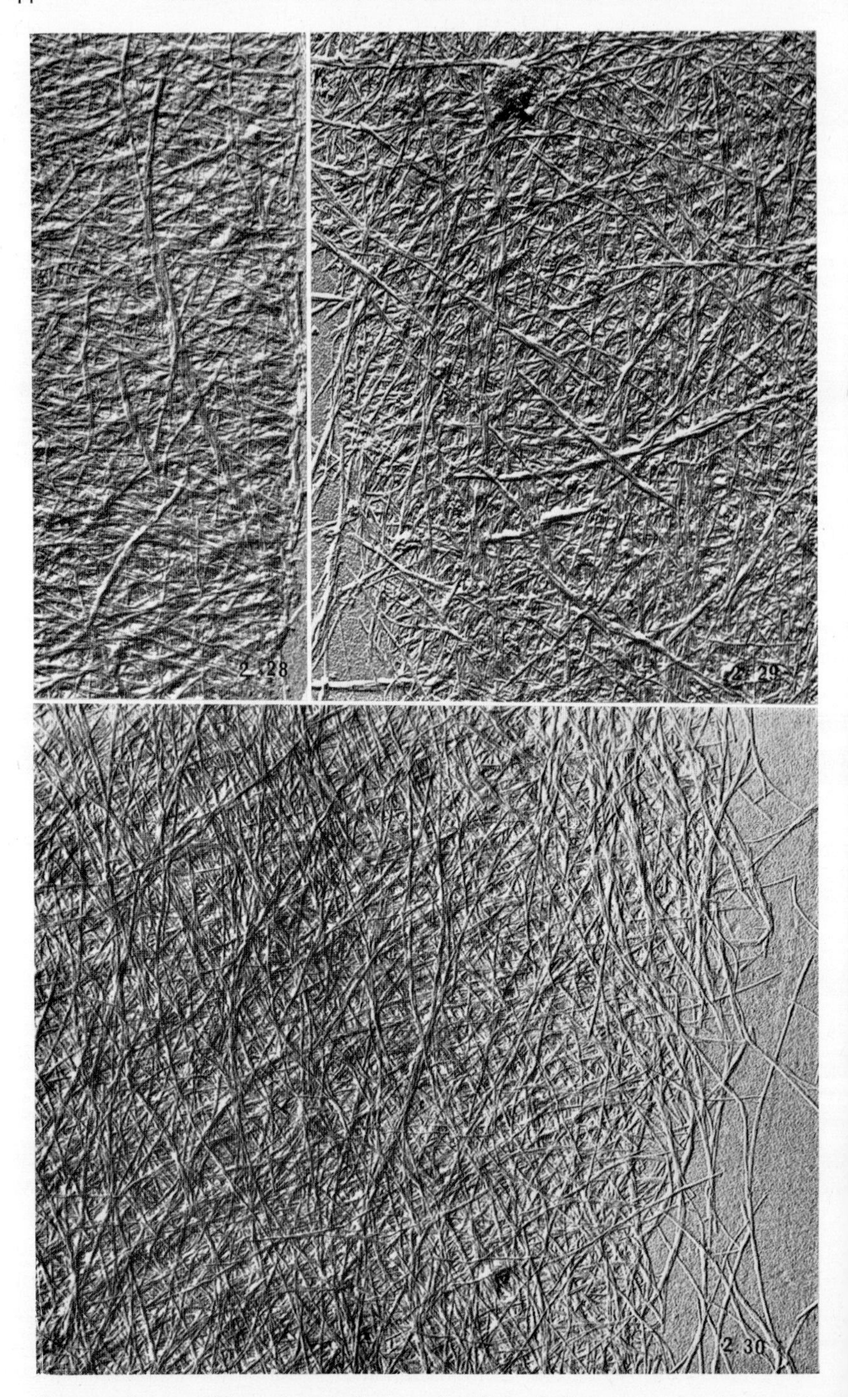

FIGS 2.28–2.30. Cleaned and shadowed cell walls from two members of the Phaeophyceae. (2.28) The side of a short cell from the filamentous alga *Ectocarpus* showing mainly transverse orientation of the fibrils. (× 18,000); (2.29) the side of a cubical cortical cell from *Laminaria* (× 21,000); (2.30) the edge of a long filamentous cell from the medulla of *Laminaria* showing a predominantly longitudinal orientation of the fibrils. (× 21,000)

microfibrils appears to be transverse when the cleaned walls are examined from the outside. In a detailed comparison by freeze-etching and thin sectioning techniques Bailey and Bisalputra (1969) found that in *Ectocarpus* the wall consists of an inner zone of compacted fibrils and an outer zone of interwoven bundles of fibrils. The outer surface of the cells is pilose as the individual fibrils fray out from the bundles. By way of contrast in *Elachista* the surface of the cell wall is smooth and the wall consists of a compacted reticulum of microfibrils.

Rather more specialized orientation of the fibrils is seen in the pit-fields of the Dictyotales (Dawes *et al.*, 1961) and the trumpet cells of *Laminaria* (Ziegler and Ruck, 1967). In these cases the microfibrils are so arranged that they skirt the circular or oval apertures of the pits. The result is rather similar to the primary pit field of the Angiosperm cell wall. In section (Parker and Philpott, 1961), the sieve areas of *Macrocystis* look like ordinary cell walls which are much perforated by plasmodesmata-like tubes.

In sporelings of the parenchymatous brown alga *Petalonia* Cole and Lin (1970) have reported the presence of plasmalemmasomes. These are membranous structures, formed in pockets between the cell wall and the plasma-membrane, and consisting of vesicles, tubules and flattened cisternae. Similar structures in fungi and higher plants, termed lomasomes (Moore and McAlear, 1961) or paramural bodies (Marchant and Robards, 1968), have been thought to be involved in wall formation.

3. *Xanthophyceae*

At present the walls of very few of the many filamentous members of the Xanthophyceae have been examined by electron microscopy. In the siphonous filaments of *Vaucheria* the walls contain randomly arranged cellulose microfibrils (Parker *et al.*, 1963) and the microfibrils also appear to be random in the more precisely shaped wall of *Tribonema* (Fig. 2.31). Sections of the wall of the vesiculate alga *Botrydium* (Falk, 1967) have shown several layers, in each of which the microfibrils all appear to be lying in one direction.

4. *The cell wall of a dinoflagellate cyst*

In the dinoflagellate *Pyrocystis* the main phase in the life cycle consists of cells which are enclosed within an entire wall. Opinions differ as to whether this should be regarded as a coccoid form or a cyst stage, however, in form and structure this wall is more similar to a normal cell wall than to the typical dinoflagellate theca. Swift and Remsen (1970) studied the wall by various X-ray and electron microscopical techniques. They found that in *P. lunula* there was an outer layer of rather amorphous osmiophilic material 15–20 nm thick (Fig. 2.33). Beneath this the main wall consists of some 24 layers of crossed cellulose microfibrils (Fig. 2.32) making up a thickness of about 170 nm. Lastly the inner wall consists of a layer about 30 nm thick made up of randomly arranged microfibrils. The plasma-membrane is inside the wall.

5. *Chlorophyceae*

In this class a great variety of wall organization has been discovered. Some examples were reviewed by Dawes (1966). This range of structure may give support to recent suggestions that the group should be divided into a number of classes (Round, 1971b).

(a) *Wall consisting mainly of cellulose microfibrils.* This type of construction is found in *Cladophora* and *Chaetomorpha* (Cladophorales) and *Valonia* and

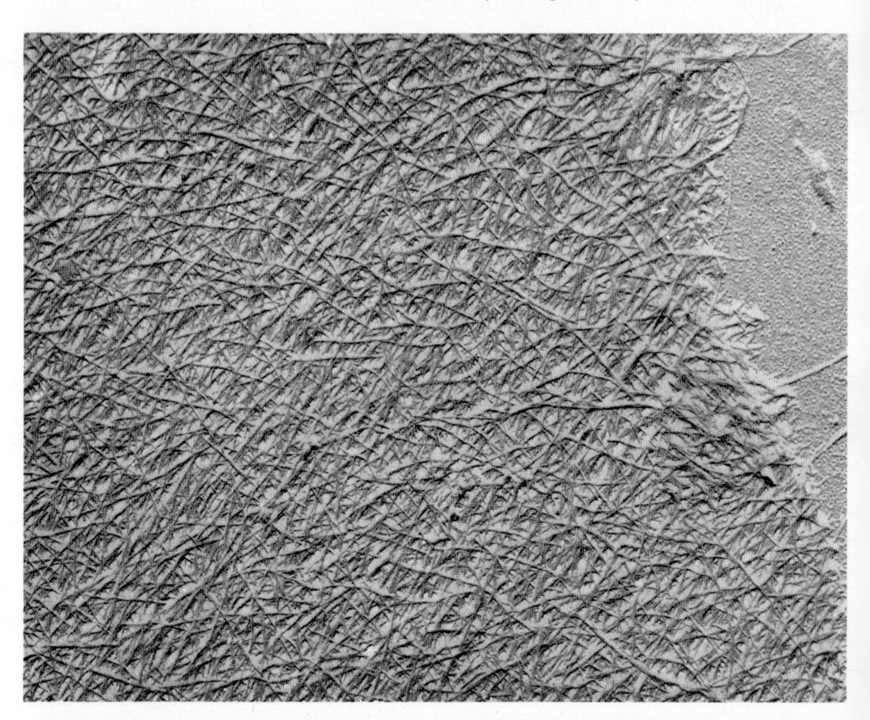

FIG. 2.31. Part of the cleaned wall of *Tribonema*, a filamentous member of the Xanthophyceae. (×21,000)

Apjohnia (Siphonocladales). Because of the small amounts of amorphous material the microfibrils can be readily observed and their arrangement has been extensively studied both by X-ray diffraction and electron microscopy (Preston and Kuyper, 1951; Wilson, 1951; Nicolai, 1957; Nicholai and Preston, 1959; Cronshaw and Preston, 1958; Cronshaw, Myers and Preston, 1958; Roelofsen, 1959, 1966; Steward and Mühlethaler, 1953; Hanic and Craigie, 1969; Dawes, 1969). In *Cladophora* and *Chaetomorpha* there is said to be an outer protein-rich cuticular region consisting of several alternating microfibrillar and amorphous layers. Beneath this the main part of the wall consists of numerous lamellae in which the cellulose microfibrils are arranged more or less parallel to each other.

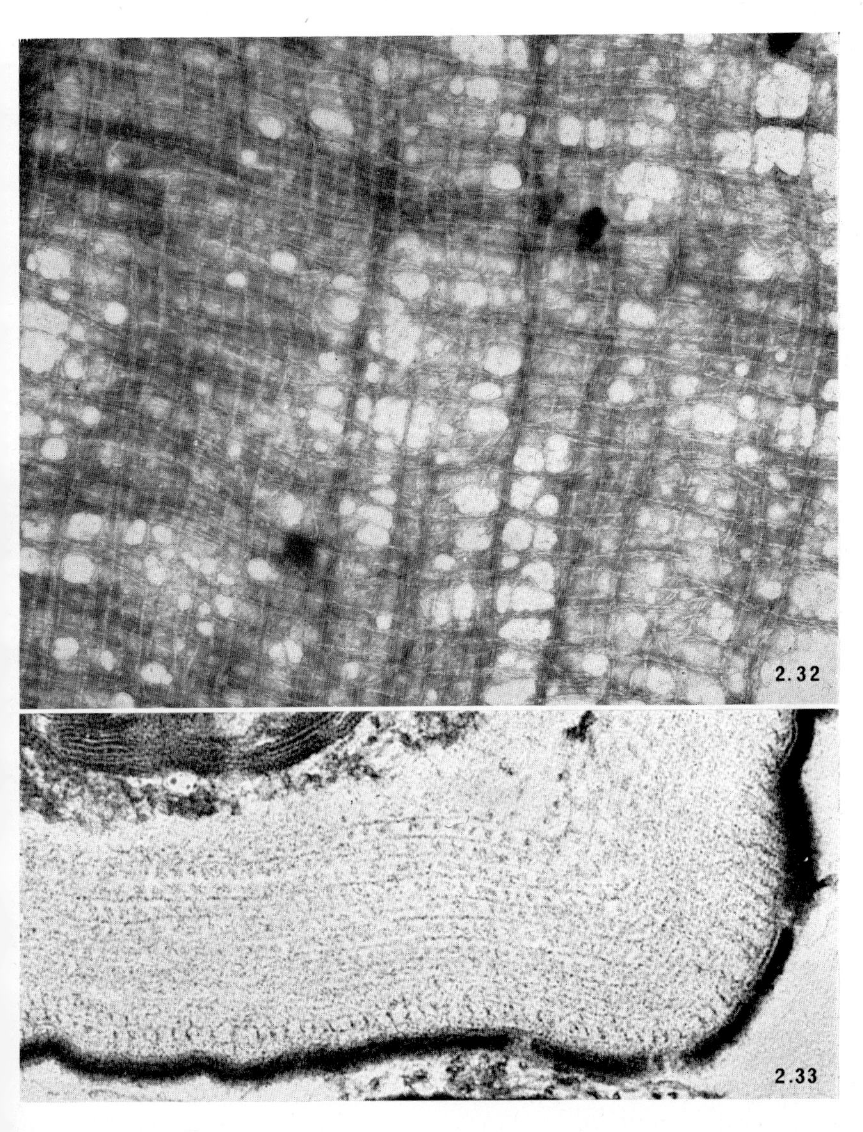

FIGS 2.32–2.33. The wall of the cyst stage of *Pyrocystis* (Dinophyceae). (2.32) A small fragment of wall separated by maceration and negatively stained with phosphotungstic acid. Note the horizontal and longitudinal orientations of the microfibrils. (×60,000); (2.33) the *Pyrocystis* wall in section showing the outer cuticular layer (dark), the thick layer of crossed-orientation fibrils and the inner random layer. (×7,000) (2.32–2.33 provided by E. Swift after Swift and Remsen, 1970.)

In adjacent lamellae the fibrils are orientated approximately at right angles. Amorphous material is present between the lamellae. When a new wall is formed by a recently settled zoospore the first layer consists of randomly arranged microfibrils. A 'pole' then develops and succeeding wall layers radiate from this or alternatively are formed concentrically around this.

Valonia consists of a single globular coencytic cell which may be up to 4 cm in diameter. Here the basic construction (Figs 2.34 and 2.35) is an outer lamella

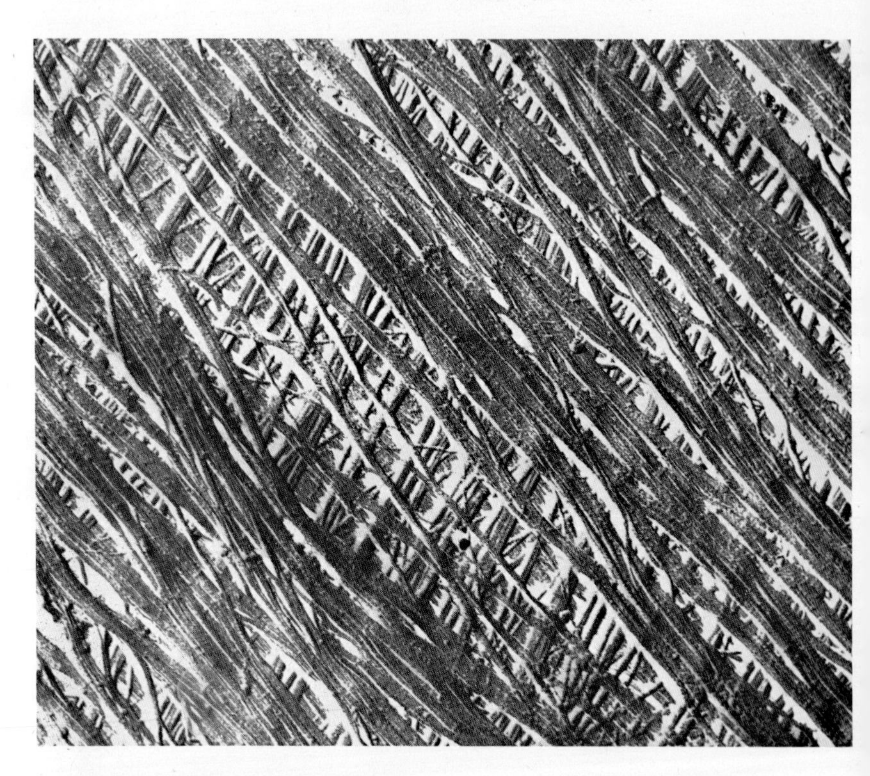

FIG. 2.34. A portion of the wall of *Valonia* (Chlorophyceae) showing two successive layers in which the fibrils are orientated at almost 90° to each other. (× 17,500)

of randomly arranged microfibrils followed by alternate layers at right angles to each other. In addition there are intermittent thinner layers which have a helical orientation (Cronshaw and Preston, 1958). These layers cross the main layers at an obtuse angle. This construction must make for an exceedingly strong wall.

(b) *Walls with less distinct organization of cellulose fibrils and with much amorphous material.* In *Ulva* and *Enteromorpha* (Cronshaw *et al.*, 1958) the wall has been found to consist of randomly arranged cellulose microfibrils embedded in amorphous matrix. In section, walls of this type have a rather uniform appearance.

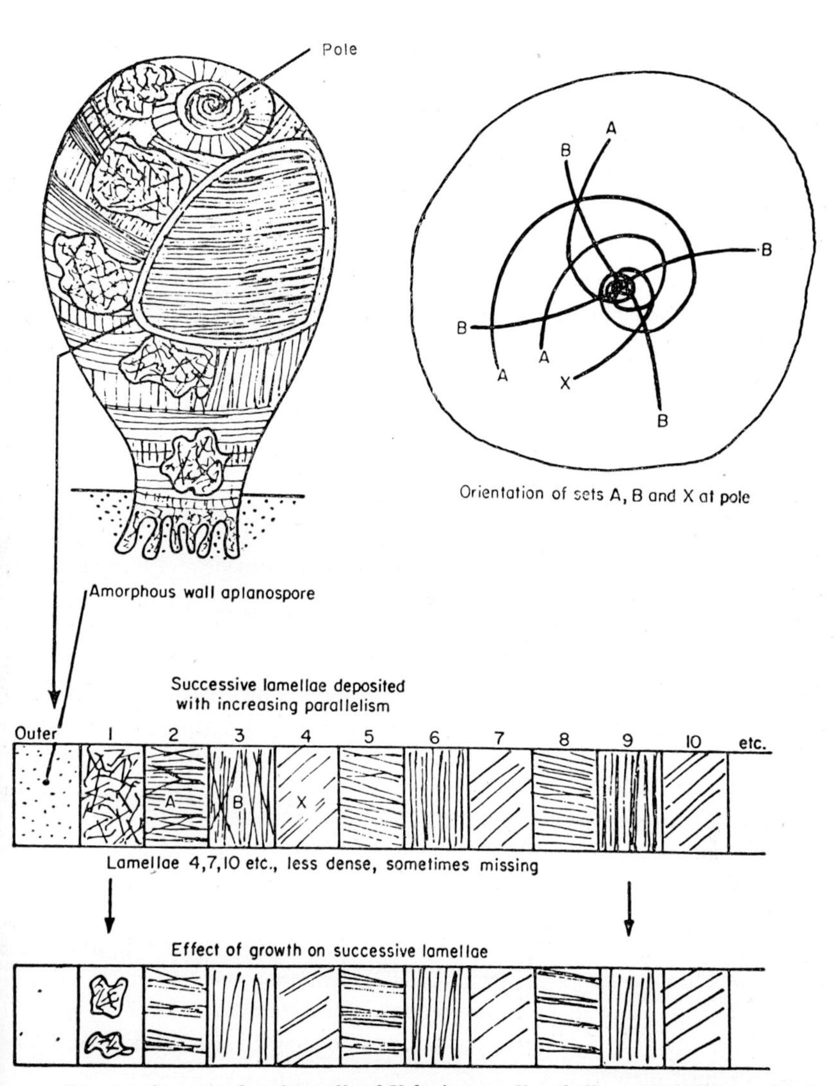

Structural type 4, found in cells of *Valonia spec. div.*, thallus cells of *Dictyosphaeria* and related genera.

FIG. 2.35. Diagrammatic illustration of the construction of the cell wall of the large coenocytic alga *Valonia*. *Top left* shows a cut-away drawing of the alga and *top right* the main orientations of fibrils as seen at the anterior end, or pole, of the plant. The lower diagrams show the main orientations of successive layers of the wall both before and after expansion of the alga. Reproduced, with permission, from Roelofsen (1966).

A similar type of wall appears to be present in *Oedogonium* and is most likely to be found in members of the Ulotrichales and Chaetophorales although none of these appear to have been studied in detail yet. In the Conjugales it has been shown that in *Spirogyra* (Dawes, 1966) and certain desmids (Mix, 1966) the main part of the wall consists of flat bands of microfibrils of varying width which wrap around the cell. The mean orientation of the fibrils is at random. In these organisms there appears to be an outer layer of wall which consists of a felt of individual randomly arranged fibrils (Mix, 1966, 1972). In the desmid *Penium spinulosum* this outer layer consists of electron-dense material which is perforated by numerous regularly arranged pores and it can also project outwards to form the spines (Gerrath, 1969). In *Cosmocladium* it forms connecting strands between adjacent cells (Gerrath, 1970). A similar layer has recently been reported from *Gonatozygon* (Mix, 1972). In *Micrasterias* Kiermayer and Stahelin (1972) have reported that the outer surface of the wall consists of a fibril-free layer which cleaves in much the same way as lipid bilayers. Using sections, Jordan (1970) has found three layers of mucilaginous material outside the cell wall proper of *Spirogyra*. The inner layer is 1–3 μm wide and appears coarsely fibrillar but, like *Porphyridium* (Ramus, 1972), these fibrils probably result from dehydration of the slime. The next layer is 1 to 0·3 μm wide and finely fibrillar, and the outer layer is only 10 nm thick and is very electron-dense.

In *Pediastrum* the inner layer of the wall contains cellulose fibrils which are probably arranged randomly. The narrow outer layer of the wall contains regularly arranged rods of electron dense material which have been shown to consist of silica (Millington and Gawlik, 1967; Gawlik and Millington, 1969).

The coccoid green algae have been the subjects for several investigations of wall structure. Both in *Chlorella* (Soeder, 1964, 1965; Staehelin, 1966; Sassen *et al.*, 1970), *Ankistrodesmus* (Mayer, 1969) and *Scenedesmus* (Burczyk *et al.*, 1971) the wall was shown to have three layers of which the thicker middle layer contained cellulose microfibrils. In a more recent survey of a number of these algae Atkinson *et al.* (1972) found that about half the species of *Chlorella*, *Scenedesmus* and *Prototheca* examined had a 14 nm thick trilaminar layer outside the cell wall proper (see Fig. 2.36). This layer is extremely resistant and is believed to consist of a polymerized carotenoid material like sporopollenin. Such material has not previously been discovered in algae although it is common in the spore walls of higher plants.

Earlier studies of *Scenedesmus* (Bisalputra and Weier, 1963, 1964a; Komarek, 1970), mainly involving shadowed preparations, had shown that the outer layer of the wall consisted of a delicate hexagonal meshwork together with complicated tubular 'props' and spines. The structure of bristles on the wall of some species of *Scenedesmus* has been studied by other workers (Trainor and Massalski, 1971: Marčenko, 1973) with no general agreement on the nature of these structures although a very regular transverse banding was clearly shown.

A recent investigation of the wall of *Chlamydomonas* (Horne *et al.*, 1971; Roberts *et al.*, 1972) has suggested that instead of being composed of a random mesh of cellulose microfibrils, as was thought to be the case, the wall has a complex layered and lattice-structure which is more like the walls of bacteria in being composed, at least in part, of proteinaceous material. Detailed studies reveal seven layers outside the plasmalemma. It seems that the central triplet section (known as W2–W6) is the section which reveals a lattice-structure in negatively stained preparations. The lattice is readily disrupted by dilute KOH,

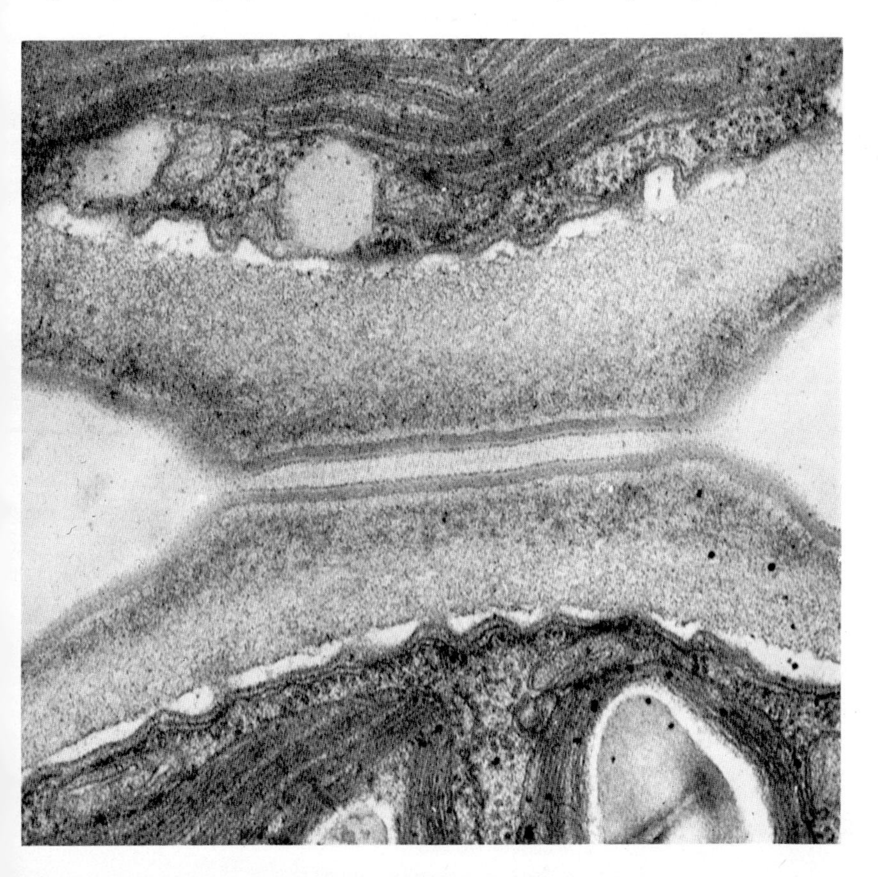

FIG. 2.36. A section through part of the walls of two cells of *Scenedesmus*, where they adjoin, showing the distinct outer layer thought to consist of sporopollenin and the wide fibrillar layer. (×59,000)

by proteases, and is soluble in detergent (SDA). Several glycoproteins are found in this wall, including hydroxyproline which has already been implicated in the cross-linking between cellulose microfibrils in the walls of filamentous algae (Thompson and Preston, 1968; Lamport, 1970).

The thin wall of *Furcilla* (Volvocales) (Belcher, 1968d) appears to consist of a delicate hexagonal meshwork, the composition of which has not been determined.

(c) *Walls not based on cellulose microfibrils.* In the 'group' of algae formerly known as the Siphonales there is much variety in the composition of the main structural component of the wall. Some organisms, such as *Valonia* (described above) have walls in which cellulose is the major component. The work of Preston and his collaborators has revealed that other algae in this 'group' make use of polymers of xylan and mannan in their walls. Xylan is found in *Bryopis*, *Caulerpa*, *Udotea*, *Halimeda*, *Penicillus* and *Dichotomosiphon* (Frei and Preston, 1964a) and in these organisms it forms microfibrils which often appear to be joined together by cross-linkages. The orientation of the fibrils varies from horizontal via random to almost perfectly longitudinal alignment. Amorphous material may also be present in these walls.

In *Codium*, *Acetabularia* and *Batophora* preliminary studies (Frei and Preston, 1961) suggested that the mannan, which forms the main component of the wall, was not in the form of true microfibrils, for the micrographs showed a rather granular wall. However, more recent studies which involved the use of a much more gentle cleaning technique have shown that this mannan is in the form of microfibrils of from 5 to 25 nm in width (Mackie and Preston, 1968). Incrusting material appeared to be firmly attached to these microfibrils. Parker and Leeper (1969) have shown that the mannan crystals, which make up the microfibrils to *Acetabularia* are 2·5 nm long and 1·0 nm in width and are arranged in longitudinal arrays. By way of contrast the microcrystals of cellulose and xylan from other algae were found to be 2–2·7 nm and 2·2–2·4 nm wide respectively.

C. CALCIFIED ALGAL WALLS

Studies have been made of the development of calcification in two types of coralline algae. In the green alga *Halimeda* Wilbur *et al.* (1969) have shown that crystals of aragonite (a form of $CaCO_3$) begin to form amongst a fuzz of fine fibrous material which clothes the outer surface of the cell walls. Gradually the spaces between the utricles (the divisions of the coenocytic alga) fill up with randomly orientated crystals which form a sheath around each utricle. There is no organic matrix between the crystals and no crystals are found either within the cells or in the cell walls of the alga. The calcium carbonate is deposited entirely in an extracellular position.

A rather different situation exists in some red coralline algae, *Corallina* and *Calliarthron*, where Bailey and Bisalputra (1970) found that crystalline material is mainly deposited within the matrix of the cell walls. They noted that in the apical regions of the algae Golgi systems are well developed and these produce numerous large vesicles. In regions of calcareous deposition vesicles containing amorphous or slightly granular contents were seen to be fusing with the wall and it is suggested that these were supplying wall components, some of which might later crystallize as calcite. Calcification seems to be initiated in the outer part of the wall and subsequently to extend inward. In addition to calcite crystals the walls contain randomly arranged cellulose microfibrils and pectic materials.

D. PLASMODESMATA IN ALGAL CELL WALLS

A few algae possess plasmodesmata which connect adjacent cells. In *Bulbochaete* Fraser and Gunning (1969) found the plasmodesmata to consist of cylindrical connections between the plasma-membranes of adjacent cells. At each end the cylinder is constricted to orifices of around 10 nm diameter and within the cylinder, where it is some 40–45 nm diameter, the inner face of the plasma-membrane is lined with particles which appear to show a helical arrangement. Unlike plasmodesmata of higher plants there is apparently no internal structure such as a desmotubule or a derivative of the cell endoplasmic reticulum.

In the filamentous green algae *Ulothrix* and *Stigeoclonium* Floyd *et al.* (1971) have described plasmodesmata which they regard as being similar to those of higher plants. These plasmodesmata are lined by plasma-membrane and have an electron-dense core. Groups of plasmodesmata forming 'pit-fields' have been described from meristoderm cells of two brown algae, *Fucus* and *Eggregia* (Bisalputra, 1966). These again appeared similar to plasmodesmata of Angiosperms and consisted of pores about 37 nm in diameter through which passed a tube of plasmalemma. No endoplasmic reticulum or other cell structures were seen in these pores.

E. CELL WALL FORMATION

As with higher plants, it would appear that Golgi bodies are involved in the production of wall material but, as yet, this has been little studied in the algae. In *Chlorella* (Bisalputra *et al.*, 1966) it was found that after nuclear division a pair of large Golgi bodies came to lie along the plane of the former spindle equator. The partition membrane was then observed to form between the Golgi bodies and it is suggested that it may form from the fusion of some of the numerous small vesicles, probably derived from the Golgi, which are found in this region. The partition membranes eventually fuse with the plasmalemma at the sides of the cell and by this time a thin electron-lucent layer is seen between the two membranes. This wall layer gradually gets thicker but, apart from their close proximity, it has not been possible to prove the direct involvement of the Golgi.

Several recent studies by Preston's laboratory have been directed towards trying to locate the mechanism, at the cell membrane, which might be responsible for the production and orientation of microfibrils. Much of this work has been carried out, using freeze-etching, on zoospores of *Cladophora* and *Chaetomorpha*. These are naked when motile but they begin to develop a cell wall immediately they settle onto a substrate in order to form a new filament. In *Cladophora* (Barnett and Preston, 1970) arrays of granules were found on the outer face of the plasmalemma. These were often in the form of a lattice but no definite evidence was found to link them with cellulose production. In another study of zoospores (Preston and Goodman, 1968) it was found that the naked cell had microtubules present beneath the plasmalemma. As soon as wall formation commenced the microtubules seemed to disappear, as though they were only required whilst the cell was naked. This seems in disagreement with work on higher plants where microtubules are present during secondary wall formation, when their orientation is similar to that of the wall microfibrils.

In a later study of the zoospores of *Cladophora* and *Chaetomorpha* Robinson and Preston (1971a) again found granules on the outer surface of the plasmalemma but now they appeared to be associated with fibrillar bodies which were thought to be microfibril initials. It is concluded that the granules are cellulose-synthesizing enzyme complexes and that they are equivalent to the units postulated in Preston's (1964) ordered-granule hypothesis. In zoospores which were in the process of forming a wall there was considerable Golgi activity. Numerous vesicles lay beneath the cell membrane and many were apparently being extruded through it by a form of reverse pinocytosis. The later stages of wall formation have been studied by Robinson *et al.* (1972). Nine hours after their liberation, swarmers have lost their flagella and developed a characteristic fibrous layer over the plasmalemma. These first fibrils are randomly orientated and are followed by a layer of more ordered microfibrils which have a granular texture. Linear arrays of granules up to 4 μm long were seen occasionally. After five days of settlement a thick wall of transversely orientated microfibrils has been formed and at this stage longitudinally orientated microtubules reappear after being absent during the earlier period of wall formation.

In addition to the work of zoospores described above, wall formation has been studied in the coccoid green alga *Oocystis*. Here, Robinson and Preston (1972) found 8·5 nm granules in rows and pairs on the outer face of the cell membrane. These appear to be orientated in similar directions to the major orientations of the microfibrils which, in this small spherical alga, were almost identical to those previously described from *Valonia* (Robinson and White, 1972).

In *Cyanidium* (? Rhodophyceae) Staehelin (1968) has used freeze-etching to study the changes in the structure of the plasmalemma during the life cycle. During cell division the plasma membrane is covered with randomly arranged 5·5 and 8 nm particles. Sometimes 4 nm fibrils can be seen leading from the 8 nm particles into the wall. Just after cell division arrays of hexagonally packed particles appear in depressions on the surface of the membrane and there are corresponding particle-studded humps on the cell wall. During the next stage of development the hexagonally patterned depressions become transformed into 30–35 nm wide folds and on the wall the humps change into ridges. In old cells the plasmalemma is characterized by long striated folds, but, just prior to cell division all folds and differentiations of the cell membrane break down so that the membrane is undifferentiated and 'embryonic' again. The significance of the changes in membrane structure in relation to cell wall synthesis has not yet been established but it would seem likely that there is a relationship.

There is a difference of opinion over the origin of amorphous cell wall material. Bisalputra (1965) working with *Scenedesmus* thought that the 'pectic' layer of the wall originated as a result of the activities of the nuclear envelope. In cells that were in the 4-nucleus stage of development dense vesicles were seen to bud off the nuclear envelope and these apparently migrated to the plasmalemma where the contents were discharged to the exterior by reverse pinocytosis. This material was thought to form the basis of the spines and the net-like structure which constitutes the outer layer of the wall (Bisalputra and Weier, 1963, 1964). More recent studies (Atkinson *et al.*, 1972) suggest that this outer layer is in fact composed of sporopollenin and it may be, therefore, that the nuclear envelope is the site of synthesis of this material. Working with *Spirogyra* Jordan (1970) found considerable Golgi activity in dividing cells. The large vesicles which budded off the Golgi bodies, and apparently fused with the plasma-membrane, were thought to contain the muco-polysaccharide material (slime) which constitutes the outer layers of the cell covering.

To sum up, there would seem little doubt that in the formation of algal cell walls the materials required are mainly collected into Golgi vesicles which then pass them through the plasma-membrane. On this membrane there is a pattern of granules which may consist of enzyme complexes responsible for synthesis of microfibrils, probably in a predetermined direction. How the orientation of fibril synthesis is switched at intervals has yet to be determined.

3. *Flagella and Associated Structures*

In many algal classes the main vegetative phase consists of unicells or colonies which are propelled by flagella. In several other classes only gametes and asexual zoospores are provided with flagella. The number of flagella per cell may vary within a class but in general the type of construction does not show much variation. Thus, because there are some differences from class to class (Table II), flagella type has long provided an important taxonomic criterion in the gross classification of the algae. Most of the variations in flagellar construction affect the structures external to the extension of the plasma-membrane which bounds the flagellum. There are a few distinct variations in internal structure and probably also some which affect the basal bodies or kinetosomes but, as yet, these have not been very thoroughly investigated in many organisms. Root systems are also proving to be both complex and varied and, for example, in four dinoflagellates from different genera the arrangement has been found to differ in each case (see below). Various aspects of flagella structure in the algae and Protozoa have been reviewed by Manton (1952, 1954, 1965), Pitelka and Schooley (1955), Pitelka and Child (1964) and Joyon and Mignot (1969). The first algal electron micrographs were probably those of the flagella of *Euglena*, *Ochromonas* and *Chilomonas* taken by Brown in 1945.

I. External Features

A. SMOOTH FLAGELLA

Simple flagella which are normally unadorned by hairs or processes and which therefore have a smooth outline (Fig. 3.1) are usually found in the Chlorophyceae, Haptophyceae and as the posterior or trailing flagellum in the Chrysophyceae, Xanthophyceae, Chloromonadophyceae and Phaeophyceae. In some cases the tip of the flagellum ends rather bluntly, as is usually the case with posterior flagella, but often it tapers to a point as in *Chlamydomonas* (Lewin and Meinhart, 1953). This latter type is called an acronematic flagellum in the terminology of Deflandre (1934). Very careful preparation (Ringo, 1967b) has shown that the supposedly smooth flagella of *Chlamydomonas*, may have very delicate hairlike processes attached to them in what is probably a bilateral array. Very fine hairs have also been observed on the posterior flagellum of *Ochromonas* (Bouck, 1971). There may, in fact, be no such thing as a smooth flagellum.

TABLE II

A summary of the main types of flagellum found in the various algal classes

	±Smooth	Hairy		Scaly + Short hairs	Paraflagellar Rod or strand	Transition zone		Notes
		Fine hairs	Stiff hairs			Stellate	Spiral	
Chlorophyceae	2(+)					+		
Prasinophyceae				2(+)		+		
Euglenophyceae		1(+)			+			
Chloromonadophyceae	1		1					
Eustigmatophyceae			1				+?	
Xanthophyceae	1		1				+?	
Chrysophyceae	1		1				+	
Haptophyceae	2					+		+Haptonema
Bacillariophyceae			1					
Phaeophyceae	1		1					
Dinophyceae	1	1			+			
Cryptophyceae			2					
Rhodophyceae								No flagella

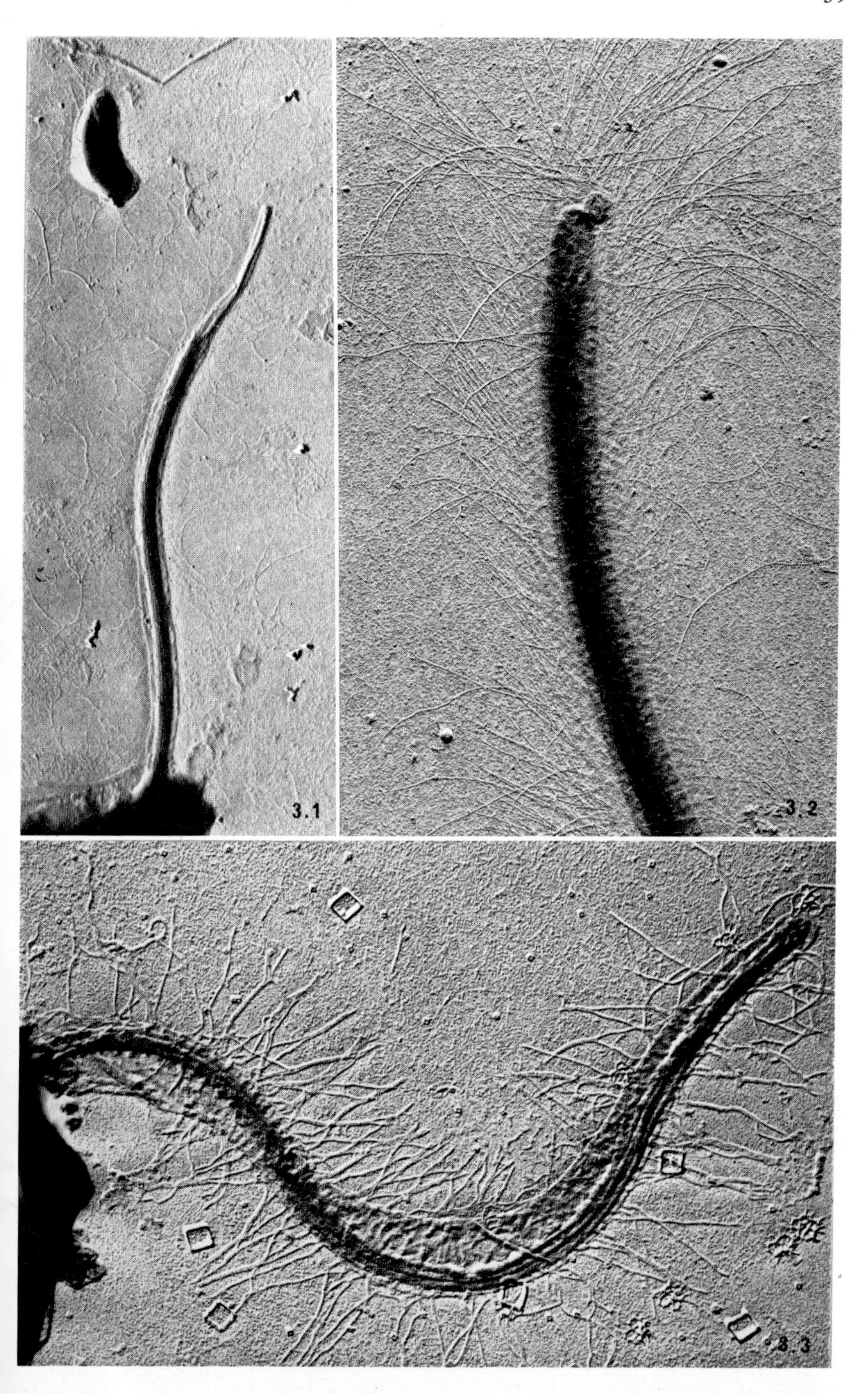

FIG. 3.1. A smooth acronematic flagellum from an unidentified alga (× 7200); (3.2) the flagellum of *Trachelomonas* (Euglenophyceae) showing the row of fine hairs along one side and the tuft of hairs at the tip (× 10,800 ; (3.3) the flagellum of *Pedinella* (Chrysophyceae) with stiff hairs and expanded flagellar sheath. (× 10,800)

B. FLAGELLA WITH SINGLE ROW OF HAIRS (Stichonematic)

In the Euglenophyceae the emergent flagellum (Fig. 3.2) normally bears a single row of very fine hairs 2–3 μm long (Leedale, 1967). In addition there may be a felt of short fibrous material coating the flagella and a tuft of slightly stiffer hairs at the tip. The transverse flagellum in the Dinophyceae (Fig. 3.12) bears a single row of fine hairs, *c.* 2 μm long, which arise from short stiff bases (Leadbeater and Dodge, 1967a). In *Pedinella* (Fig. 3.3) (Chrysophyceae) the one, rather unusual, emergent flagellum has a single row of rather stiff hairs arranged spirally around it (Swale, 1969). Very fine short hairs are found on the narrow tip region of the posterior flagellum of some dinoflagellates.

C. FLAGELLA WITH TWO ROWS OF HAIRS (Pantonematic)

In many classes of algae—Xanthophyceae, Chrysophyceae, Phaeophyceae, Eustigmatophyceae, Chloromonadophyceae and the Cryptophyceae—generally one flagellum bears rather stiff hairs of a uniform type (Figs 3.4 and 3.5). The presence of these hairs was first reported by Loeffler, using light microscopy, as long ago as 1889 but only recently have details of their structure and origin become available. These hairs have been variously termed *flimmer* (Fischer, 1894), *mastigonemes* (Deflandre, 1934) and *tinsel.* To avoid any of the misconception which later workers have applied to these terms they will here be termed '*hairs*'. Later in this chapter will be found an account of recent studies concerning their origin within the cell. Hairs of this type are generally arranged in two rows, one on either side of the flagellum. To date the only exception known is *Cryptomonas* (Cryptophyceae), where the longer flagellum has hairs as above, which are about 2·5 μm long, and the other flagellum has much shorter hairs along one side (Hibberd *et al.*, 1971). There is also a swelling near the base of the longer flagellum and this bears a tuft of hairs.

The stiff hairs appear to consist of at least three parts. There is a short, and probably flexible, basal portion about 0·25 μm long which is often directed forward at an angle of about 80° to the long axis of the axoneme (Heywood, 1972). Some preparations of disrupted flagella (Fig. 3.4) do suggest that the hairs may be attached to peripheral doublets of the axoneme as was suggested by Manton *et al.* (1953), but this has not been confirmed by study of sections which seem to indicate that the hairs are attached to the flagellar sheath. After the base comes the main stiff shaft of the hair which is tubular, about 15 nm in diameter, up to 2·5 μm long and may (in *Vacuolaria*) lie almost parallel to the axoneme (Heywood, 1972). Finally there are one or more very delicate terminal fibres up to 0·7 μm long (Fig. 3.5) (Leedale *et al.*, 1970; van de Veer, 1970).

The construction of the main shaft of the hair in members of the Phaeophyceae and Chrysophyceae has been investigated in detail by Bradley (1965, 1966), Bouck (1969, 1971) and Barton *et al.* (1970), using negative staining techniques. They found that the tubular structure was made up of 13 longitudinal rows of

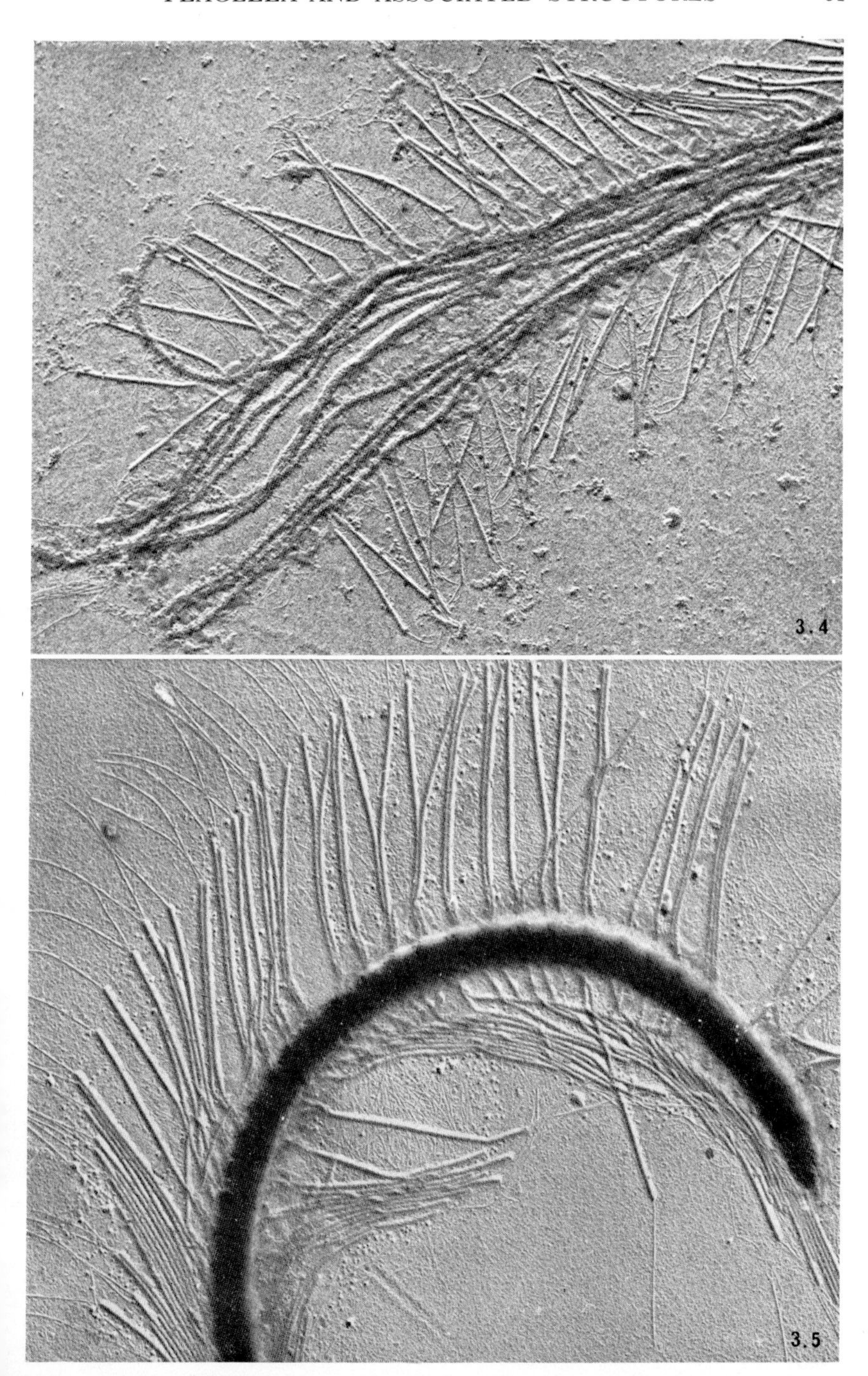

FIG. 3.4. The dismembered tip of the anterior flagellum of *Ochromonas* (Chrysophyceae) showing the 9 axonemal fibres (doublets) and the two rows of stiff hairs (× 17,500); (3.5) entire flagellum of *Synura* a member of the Chrysophyceae, in which the three main parts of the hairs, base, shaft and thin terminal fibres can be clearly seen. (× 21,000)

globular subunits, probably of proteinaceous nature. The terminal hairs appear to consist of a single row of globular units. In *Ochromonas* Bouck (1971) analysed the composition of isolated hairs and found large quantities of a single polypeptide which was thought to make up the globular units mentioned above. It was not the same polypeptide as is found in cellular microtubules. Mignot *et al.* (1972) used the Thiéry cytochemical technique to show that the hairs consist of glycoprotein. It is of interest here that Heywood (1972) found that when cells of *Vacuolaria* were treated with colchicine, spindle microtubules were destroyed but presumptive flagellar hairs were not affected. Although both these structures are proteinaceous and tubular there clearly must be differences in their construction. Bouck also found a small amount of polysaccharide which he thought might have come from the lateral filaments, 40–200 nm long, which were seen attached to the hairs (Fig. 3.6). Such structures appear to be present on flagella of many members of the Chrysophyceae. Studies on a variety of other organisms, including the spermatozoids of various brown algae (Loiseaux and West, 1970) have shown some variation in the thickness of the main part of the stiff flagellar hairs in this class and a range of 1·3–1·7 nm has been reported. Their lengths ranged from 0·7–1·45 μm.

D. FLAGELLA BEARING SPINES

In the spermatozoids of three brown algae the anterior flagellum bears one or more spine-like appendages. *Himanthalia* (Manton *et al.*, 1953) and *Xiphora* (Manton, 1956) both have a single spine and *Dictyota*, which only has a single flagellum, bears a row of short spines in addition to two rows of stiff hairs (Manton *et al.*, 1953; Manton, 1959b). In all cases the spines appear as dense material which is borne on the uppermost peripheral doublet of the axoneme and is surrounded by the extended flagellar membrane.

E. SCALY FLAGELLA

One algal class, the Prasinophyceae, is characterized in part by the possession of flagella which are covered with tiny scales (Christensen, 1962). There may also be short stiff (cadaucous) hairs (Figs 3.7 and 3.8) which in *Heteromastix* have been found to show transverse striations (Manton *et al.*, 1965). Flagellar scales were first reported from *Micromonas squamata* (Manton and Parke, 1960) and have since been found in a number of other organisms which, like *Micromonas*, were also at that time of uncertain taxonomic position. It was later found that these organisms had chlorophyll *b* pigment and starch as a food reserve. Thus the class Prasinophyceae was delimited to accommodate them. In general the scales are very small, of the order of 60 nm diameter, have a distinct form in each species and are arranged on the outside of the flagellar membrane in two relatively compact layers. Detailed studies (Manton, 1966c) have shown that scales are synthesized within golgi vesicles. The vesicles then migrate to a

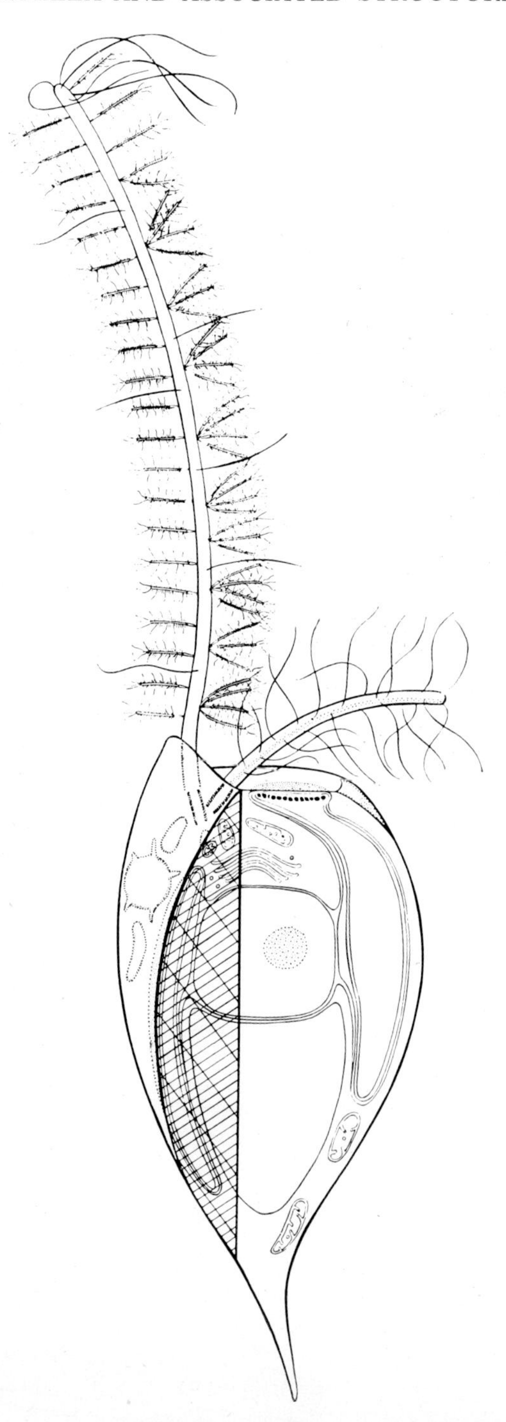

FIG. 3.6. Diagram of *Ochromonas* showing the arrangement and structure of the hairs on the two flagella. The thick hairs on the anterior flagellum are depicted covered with lateral filaments. Other details of cell structure are also shown. Reproduced, with permission, from Bouck (1971).

depression at the base of the flagella and from here they are in some way extruded and arranged on the flagella. Perhaps this can only take place when new flagella are being formed. There is no obvious function for these scales, in fact they would appear to make it more difficult for the flagella to beat.

FIG. 3.7. The four flagella of *Pyraminonas* (Prasinophyceae) covered with scales and bearing brittle (cadaucous) hairs (×10,000); (3.8) the tip region of a flagellum from an unidentified member of the Prasinophyceae showing the close-packed arrangement of the tiny scales and some of the stiff hairs. (×44,000)

Tiny scales, which are very readily detached, have been found on the flagella of several members of the Chrysophyceae including *Mallomonas* and many species of *Synura* (Bradley, 1966; Schnepf and Deichgräber, 1969; Hibberd, 1973). In *Sphaleromantis* (Manton and Harris, 1966) the scales on the two flagella are similar to those which cover the body. In completely distinct groups of algae, regularly arranged rows of scales have now been found on the flagella of spermatids of *Chara* and *Nitella* (Charophyceae) (Turner, 1968; Pickett-Heaps, 1968b; Møestrup, 1970a); both flagella of the phagotrophic dinoflagellate *Oxyrrhis* (Clarke and Pennick, 1972); one flagellum of *Pavlova* (Green and

Manton, 1970; van der Veer, 1972) which is probably a member of the Haptophyceae. In *Pavlova* both flagella are clothed with a very fine tomentum and the longer of the two also bears a regularly arranged layer of knob-like dense bodies *c.* 30×25 nm in size with a slight median constriction. These bodies are apparently produced in the Golgi bodies, but only during the night.

II. Internal Structure of Flagella

A. THE FREE PART

The main component of the free-moving part of most algal flagella is the 9+2 structure (Manton, 1952) or axoneme. This basic structure (Figs 3.9 and 3.10), which is common to all plant and animal flagella, consists of a ring of nine pairs of tubules (or fibres), often called doublets, and two separate central tubules. The doublets are 20 nm by up to 35 nm in cross section and the individual tubules are slightly oval. The 'A' tubule of each pair generally possesses two short projections or side arms which point towards the next doublet. Very high resolution studies on *Chlamydomonas* (Hopkins, 1970) have shown the presence of short hammerhead shaped structures which are also attached to the 'A' tubules but directed along the radius to the centre of the flagellum (Fig. 3.9). The heads were formerly thought (from transverse sections) to be continuous structures and were termed secondary fibres. There may be a delicate structure which surrounds the two central tubules. In the spermatozoid of the diatom *Lithodesmium* (Manton and von Stosch, 1966) there are no central tubules. A similar situation exists in some 'paralysed' mutants of *Chlamydomonas* (Warr *et al.*, 1966). In spermatozoids of the green alga *Golenkinia* there is but a single central tubule but the flagella appear to move in the normal manner (Mφestrup, 1972).

Also working with *Chlamydomonas* Ringo (1967a) has investigated the structure of the doublets. He finds that each tubule is composed of nearly spherical proteinaceous subunits some 4·5 nm in diameter which are arranged in longitudinal arrays (protofilaments). When seen in cross-section there are 13 units per tubule, the same number as constitute cell microtubules (Ledbetter and Porter, 1963). In the doublets adjacent tubules are thought to share two or three of the subunits in their common wall.

A more detailed study of flagellar components in *Chlamydomonas* has recently been carried out by Witman *et al.* (1972a,b) using various biophysical techniques as well as electron microscopy. They found that the delicate hairs were made up of a single row of ellipsoidal subunits, of glycoprotein, joined end to end. The matrix of the flagellum contained several proteins and the tubules were composed of two proteins with M.W. of 56,000 and 53,000. The axoneme was found to be remarkably stable but in experiments with detergent the tubules were solubilized in a regular sequence. The first to go was one of the central pair, then the second central tubule. This was followed by the outer wall of the B tubules (cf. Fig. 3.9) of the peripheral doublets. The remainder of the 'B' followed and then the outer wall of the 'A' tubule. The most resistant part of the axoneme was the three protofilaments which form the septum between the 'A' and 'B' tubules. These were found to consist of a protein termed 'tubulin 1' whereas the other parts of the doublets consisted of both tubulin 1 and tubulin 2. The synthesis and assembly of flagellar proteins during growth of new flagella has been studied by Rosenbaum *et al.* (1969) who found that assembly takes place at the tip of the growing flagellum and that when flagella are artificially amputated from cells the new flagella begin to grow out immediately, reaching almost their original length within an hour.

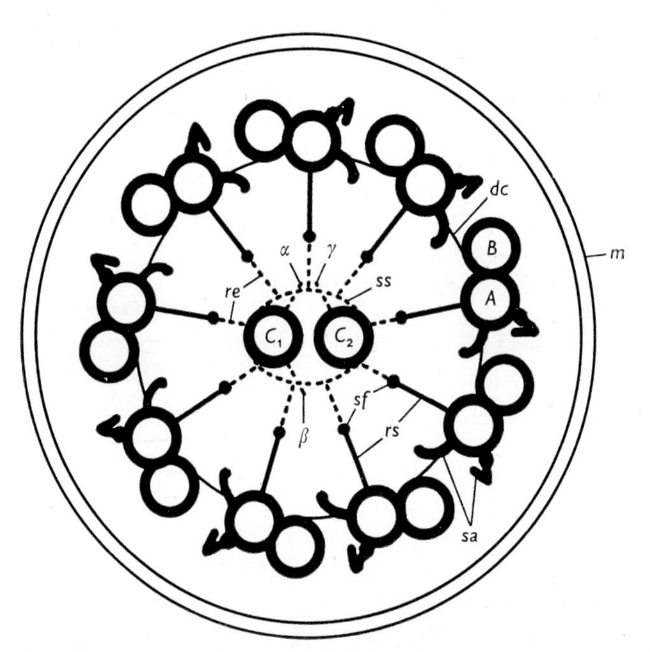

FIG. 3.9. A diagrammatic representation of the usual components of the flagellar axoneme, as seen in transverse section. Note the membrane (*m*), ring of doublets composed of A and B tubules, interdoublet connection (*dc*), side arms (*sa*), radial spokes (*rs*), possible continuation of radial spokes (*re*), secondary fibrils (*sf*), central tubules (C_1, C_2), central sheath (*ss*), and various projections (α, β, γ). Reproduced, with permission, from Hopkins (1970).

Ringo (1967b) has also studied the tip region of the flagellum. In *Chlamydomonas* the central two tubules are longest and these continue into the point at the tip of the flagellum. As the diameter of the flagellum decreases the doublets each gradually lose one of their tubules and then the remaining single tubules disappear leaving only the central pair.

In a few types of organism the free part of the flagellum contains other structures besides the axoneme. Thus, in the Euglenophyceae there is frequently

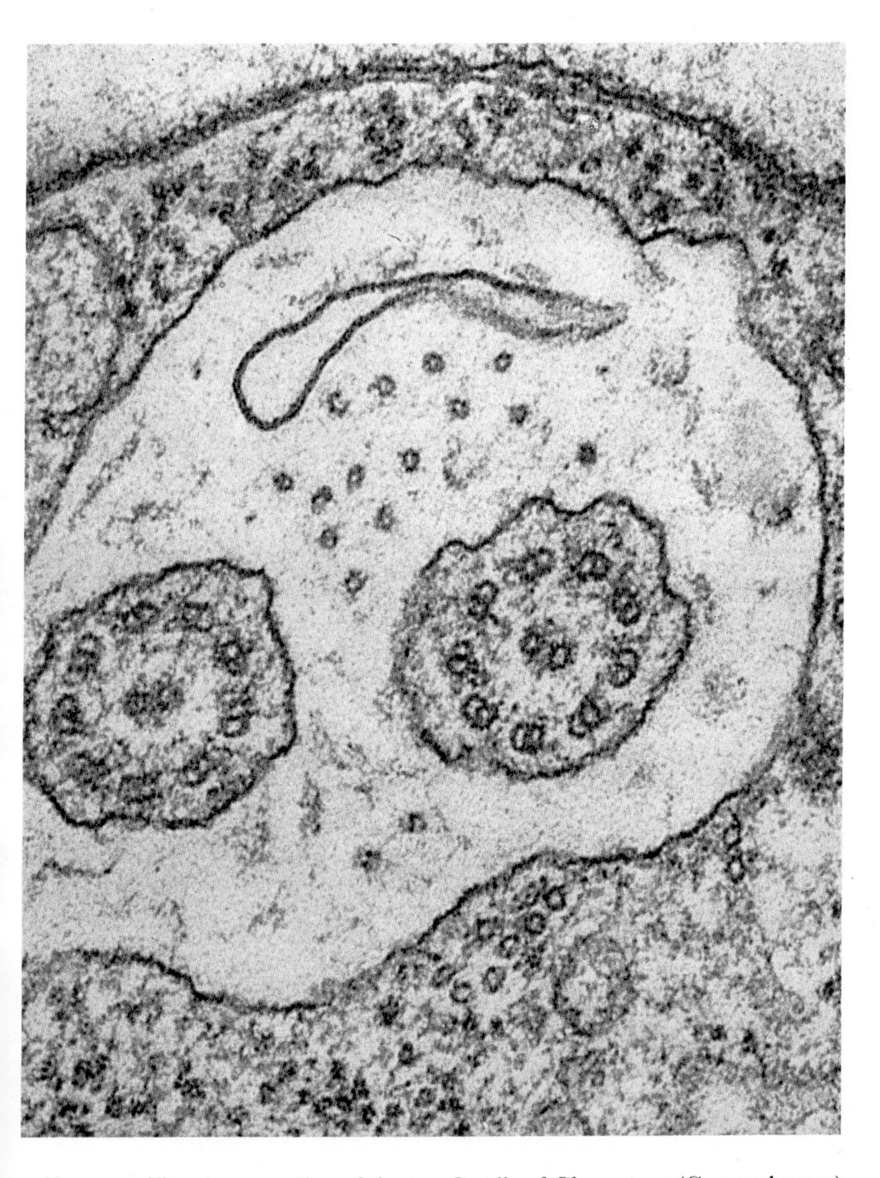

FIG. 3.10. Transverse section of the two flagella of *Chroomonas* (Cryptophyceae) showing their typical construction. Above the right hand flagellum are seen a number of sections of flagellar hairs which are obviously tubular structures. In the cell can be seen sections of microtubules (×118,000)

FIG. (3.11) Longitudinal section across one 'wave' of a flagellum from *Oxyrrhis* (Dinophyceae), showing the axoneme sectioned obliquely, at left and right, and the para-flagellar strand in the centre (×36,000); (3.12) a shadowed preparation of the transverse flagellum of a dinoflagellate showing the single row of long fine hairs and the striated strand which takes a shorter course than the axoneme. (×9600) a complex structure termed the paraflagellar rod which lies adjacent to and parallel with the axoneme (Mignot, 1966; Leedale, 1967). This rod runs almost the whole length of the flagellum. A rather similar structure, although one which is not solid, is seen in one of the flagella of the dinoflagellate *Oxyrrhis* (Fig. 3.11) (Dodge and Crawford, 1972). This consists of four or five helically arranged strands each 10–15 nm thick which are linked together by short cross-pieces. The rod is hollow and although it seems closely appressed to the axoneme it appears to take a slightly straighter course along the flagellum.

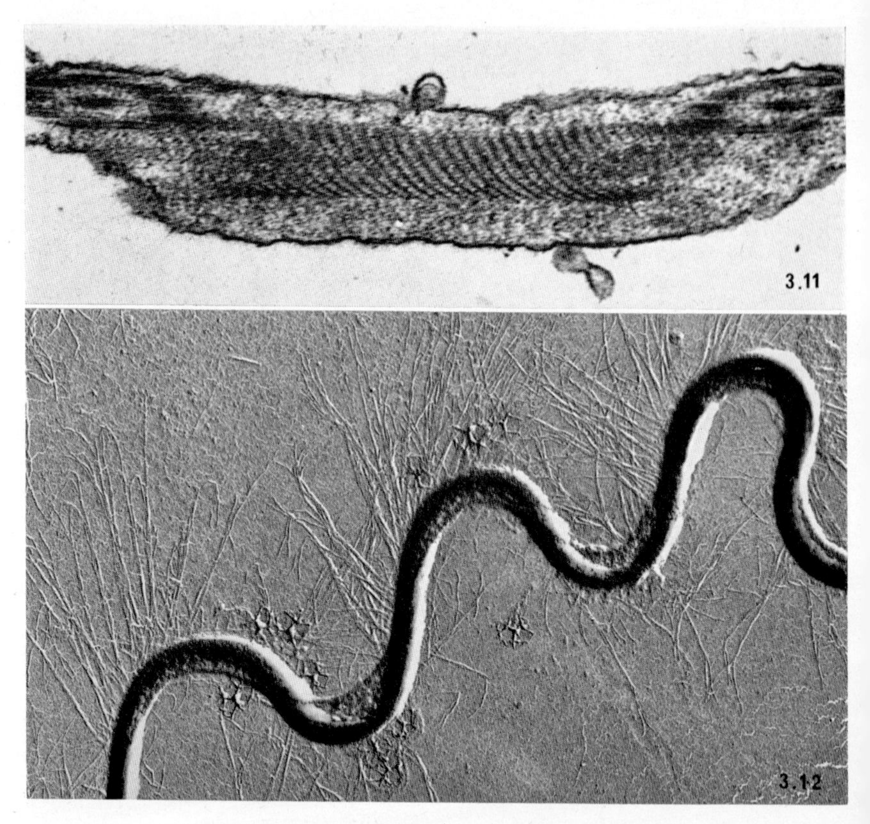

FIGS. 3.11–3.12. (See caption on p. 67.)

The majority of members of the Dinophyceae have a transversely orientated flagellum which has a helical form. Studies using shadowed (Fig. 3.12), replicated and sectioned material (Leadbeater and Dodge, 1967a) showed that the flagellar sheath was rather extended and that it also contained a narrow striated strand which had regular transverse banding with a major periodicity of 66 nm. This strand is apparently under tension and the axoneme, which is much longer than the strand, follows a spiral course around the strand. The two structures are kept separate by packing material but are of course held together by the flagellar membrane. A somewhat similar type of flagellum has been reported from a small member of the Chrysophyceae, *Pedinella hexacostata* (Swale, 1969), although here the extra band appears to run at one side of the axoneme (Fig. 3.3) thus throwing it into a simple planar wave form rather than the helical wave found in dinoflagellates.

Various flagellar mutants of *Chlamydomonas* have been described (McVittie, 1972). These invariably have a normal basal body but some have virtually no axoneme or just a short length. Often electron-dense material was present between the axoneme and the membrane of short flagella. Long-flagellum mutants were also observed where the structure was normal but the control of flagellar development was defective.

The free part of the flagellum is always bounded by a membrane sheath (Fig. 3.10) which is a continuation of the cell plasmalemma. Thus the flagellum is an extension of the cell and not an extra-cellular structure. The membrane usually lies fairly close to the doublets but may be slightly wrinkled.

B. THE TRANSITION ZONE

At the junction between the free part of the flagellum and the basal body there is a short transition zone which often has a number of unique features, first reported by Lang (1963b) working with *Polytoma*. In most members of the Chlorophyceae which have been investigated (e.g. *Stigeoclonium*, Manton 1964b; *Volvox*, Olson and Kochert, 1970; Pickett-Heaps, 1970a; *Chlorogonium*, Manton, 1964c; *Chlamydomonas*, Ringo, 1967b) *Eudorina* (Hobbs, 1971); some members of the Prasinophyceae (e.g. *Heteromastix*, Manton, 1964c) and *Prymnesium parvum* a member of the Haptophyceae (Manton and Leedale, 1963a), a complex stellate structure is found either side of the basal disc which marks the proximal end of the external portion of the flagellum. The central two tubules terminate at the distal end of the stellate structure and peripheral to it the doublets appear to be fused to the flagellar membrane. In *Chlaymdomonas* (Ringo, 1967b), where this area has been most carefully investigated (Fig. 3.13), the stellate structure is about 150 nm long, of which 100 nm is outside the disc and 50 nm inside it. In part it appears to consist of a central cylinder, 80–90 nm in diameter, to which are attached triangular structures which make contact with the 'A' tubules of the doublets (Fig. 3.13C). However, this may be an over-simplification for sometimes it is clear that the structure consists of two intermeshed stars, one with five points and the other with four. The points of any one star make contact with alternate doublets and the two stars connect together, as first deduced by Manton (1964c), so that a continuous repeating pattern is obtained. It is tempting to speculate that this structure is responsible for the control of contraction in the flagellar tubules. Unfortunately for this hypothesis, there are many organisms which do not have this structure in their otherwise rather similarly constructed flagella.

In *Uroglena* (Chrysophyceae) Caspar (1972) has described a spiral structure which is found between the doublets and the central pair of tubules immediately

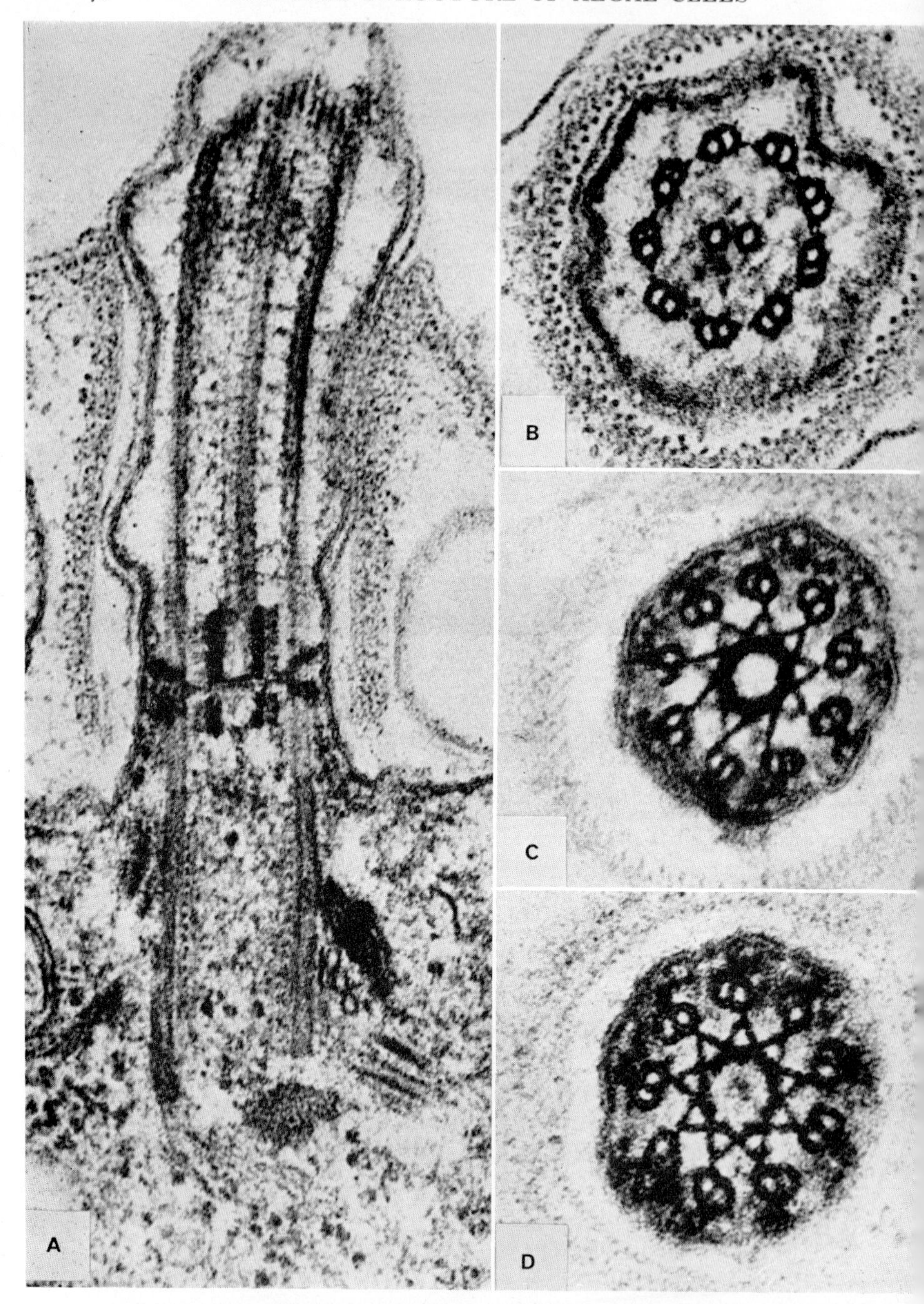

FIG. 3.13. The flagellum of *Chlamydomonas* (Chlorophyceae). A, Longitudinal section through the base and transition zone. Parts of the root system can be seen to the right of the basal body and the complex stellate structure (appearing tubular in this section) can be seen either side of the basal disc (×92,000); B, transverse section of the flagellum where it is still within the collar of the cell (×120,000); C, transverse section of the transition region showing the stellate structure connecting with the doublets (×120,000); D, another, and probably lower, section of the transition region. (×120,000) Reproduced, with permission, from Ringo (1967b).

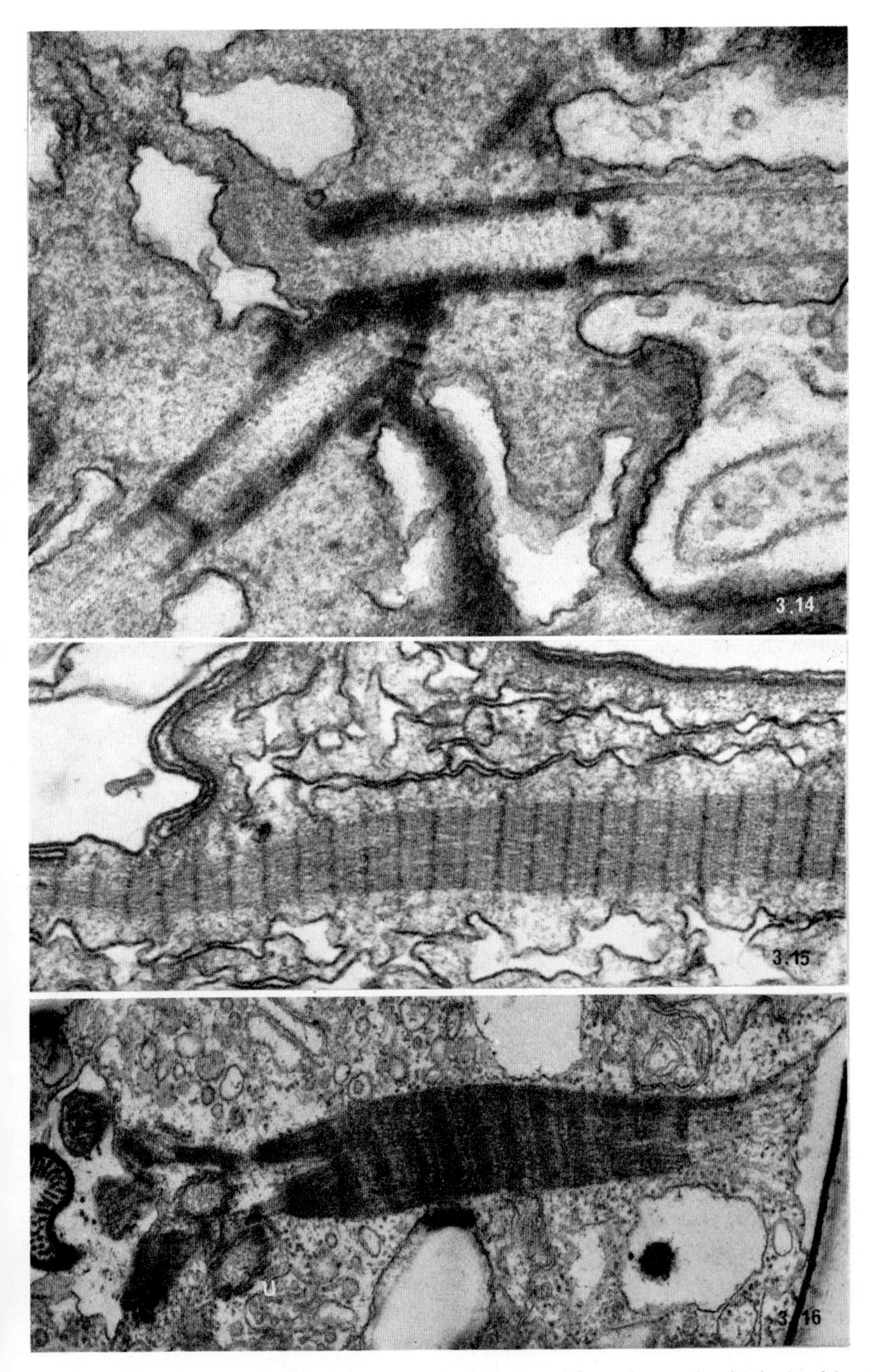

FIG. 3.14. The two flagellar bases in *Amphidinium* (Dinophyceae) which, in this organism, meet at an oblique angle. Note the relatively simple transition region and the regular pattern within the basal bodies (×63,000); (3.15) a striated flagellar root from the dinoflagellate *Ceratium* (×59,000) (After Dodge and Crawford, 1970b); (3.16) a root of striated fibrils in *Platymonas* (Prasinophyceae), (×32,400)

above the basal disc. This spiral body consists of 4–5 turns of highly electron dense material and it appears to lie in the space between the tubules without connecting to any of them. It is suggested that this structure might be present in members of the Chrysophyceae, Eustigmatophyceae and Xanthophyceae.

In the dinoflagellates the transition zone is more simply constructed. As with many other algae the central tubules terminate at the basal disc. There is then a short zone with only doublets and, in *Gymnodinium* (Leadbeater and Dodge, 1967a), there is a second disc. A short distance below this are found two diaphragms. In all probability there is a connection between the inner disc and the outer diaphragm such that the flagellar base is sealed off when the flagellum is shed or damaged. Flagella are often lost during preparation for electron microscopy and they always appear to be detached at this point. In *Amphidinium* where there is only one disc and one diaphragm (Fig. 3.14) the connection between the two is very clear.

C. THE FLAGELLAR BASE, BASAL BODY OR KINETOPLAST

The basal part of the flagellum appears to have a fairly consistent structure in all algae which have been investigated. Its size does not vary much and it is normally about 200 nm in diameter and about 400 nm long. Shortly beneath the basal disc additional tubules are added to the peripheral doublets to convert them into 'triplets'. These triplets soon become reorientated so that instead of being placed around the circumference of a circle (at 90° to the radii) they come to lie at an angle of about 130° to the radii. Adjacent triplets are connected together by thin filaments and towards the proximal end of the basal body they are each connected to a single central strand (Fig. 3.17). Thus we have the characteristic 'cart-wheel' appearance in transverse sections. Near the distal end of the basal body there are often fibrils which project out from the triplets at an angle which is almost the tangent to the circle. In *Amphidinium* (Dodge and Crawford, 1968) these fibrils are about 100 nm long and each appears to terminate in an electron-dense knob. Similar structures have been observed in *Colacium* (Leedale, 1967) and *Volvox* (Olson & Kochert, 1970).

III. Flagellar Roots

All algal flagella appear to have 'roots' attached to their basal bodies. Whether these are for anchorage or for the conduction of stimuli or material to the flagella has yet to be determined. Generally the roots arc of two kinds, bands of striated fibrils and groups of microtubules. The orientation of the roots with respect to the basal bodies and the cell is extremely variable and, for example, of four dinoflagellates which have been studied in detail the arrangement is quite different in each case (Dodge, 1971b; Dodge and Crawford, 1972). As an example of these the situation in *Amphidinium* will be described (Fig. 3.17) (Dodge and

Crawford, 1968). Here the two basal bodies lie more or less in line but at an orientation of 180° to each other. One composite root structure joins the two bases together. In part this consists of a row of 6–11 microtubules which run from the foot of one basal body to some distance past the base of the other.

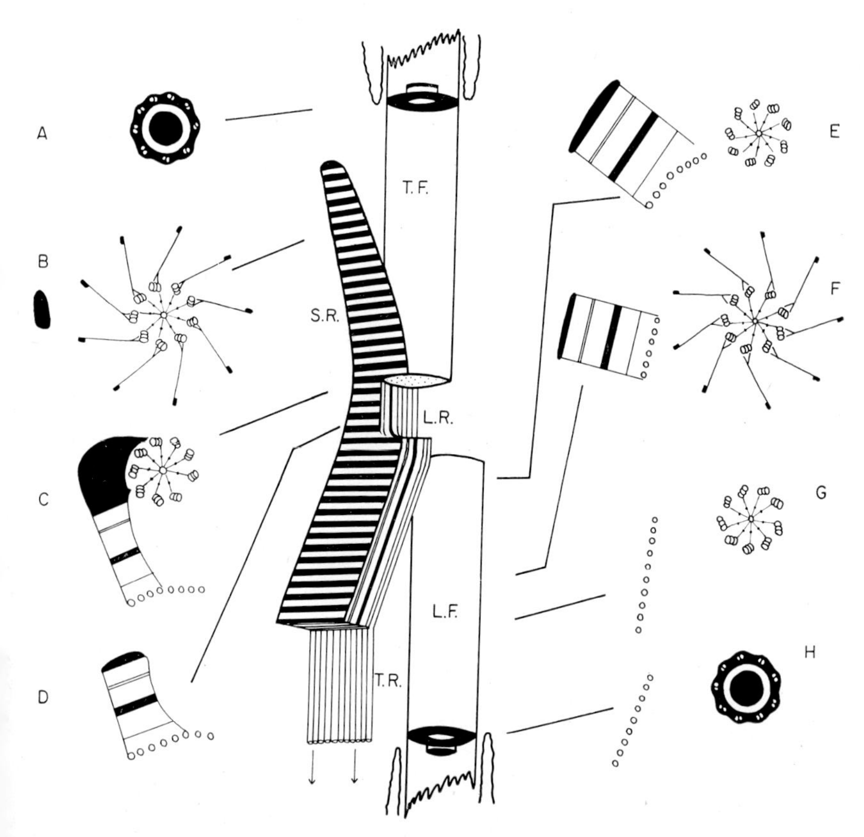

FIG. 3.17. A diagram to show the arrangement of the two flagellar bases and the associated root system in *Amphidinium carterae*. On the left are shown transverse sections (A–D) of the base of the transverse flagellum (T.F.) and on the right similar sections (E–H) of the longitudinal flagellum (L.F.). The parts of the root complex are labelled striped root (S.R.), layered root (L.R.) and tubular root (T.R.). (After Dodge and Crawford, 1968.)

It probably then goes to join the sub-thecal microtubule system. Adjacent to the tubules is a thick banded root which has longitudinal striations (the layered root). This runs from the base of one flagellum to about half-way along the other. Lastly there is attached to this a thin root with transverse banding (the striped root) which runs from the distal end of the basal body of one flagellum and may end in narrow branches when it is adjacent to the other basal body. The striped root of another dinoflagellate, *Ceratium hirundinella*, is shown in Fig. 3.15 where the transverse banding and longitudinal arrangement of the fibrils can be clearly discerned.

By way of contrast, in *Chylamydomonas*, where the two flagella are anterior and orientated at an angle of about 90° to each other, Ringo (1967b) found that they were connected together by a broad striated root. Arising from this were four microtubular roots which spread out around the cell, lying just beneath the plasma-membrane. In the rather complex motile cells of *Oedogonium* a wide striated band connects together the anterior ring of 100 or so flagella (Hoffman and Manton, 1962, 1963; Hoffman, 1970). In addition, between adjacent bases both striated and tubular roots run at right angles to the striated band and are probably connected to it. This combination of striated and tubular roots is quite common and frequently, as in *Chilomonas* (Cryptophyceae) (Mignot *et al.*, 1968) the striated root runs into the cell whilst the microtubular root is directed towards the cell periphery where it probably joins the microtubular system which normally exists beneath the cell membrane.

A somewhat different type of root system has been found in members of the Prasinophyceae (Parke and Manton, 1965), *Olithsodiscus* a member of the Chrysophyceae (Leadbeater, 1969), and members of the Chloromonadophyceae (Mignot, 1967). In these organisms there is a broad striated band (Fig. 3.16) which runs directly from the basal bodies to the nucleus where it appears to be attached to the nuclear envelope. This type of root is clearly the structure termed the 'rhizoplast' by many earlier workers and which was thought to provide an essential connection with the nucleus in all flagellated organisms. As will be seen from the above account, only in some groups of algae is this connection made.

IV. The Origin of Flagellar Hairs

Until 1969 the origin of the hairs found attached to the flagella of many algae was a mystery, however, workers had frequently reported the presence, in cells, of vesicles containing 'fibrous' material. It was Bouck (1969) who, working with the developing spermatozoids of the brown algae *Fucus* and *Ascophyllum*, solved the problem when he noticed the similarity in size and appearance of the structures in the vesicles and flagellar hairs. Further investigations revealed that the vesicles originate as swellings of the perinuclear space. As these enlarge fibrillar material forms in them and they are eventually budded off from the nuclear envelope. The vesicles containing distinct 'hairs' migrate towards the base of the flagella. This process has now been clearly demonstrated in members of the Chrysophyceae, Xanthophyceae, Cryptophyceae and Chloromonadophyceae where the hairs are all of the stiff variety (Leadbeater, 1969; Heath *et al.*, 1970; Leedale *et al.*, 1970; Deason, 1971b; Bouck, 1971; Heywood, 1972). It also appears to be identical in the Dinophyceae where the hairs are very delicate (Leadbeater and Dodge, 1967a; Leadbeater, 1971b) and vesicles containing

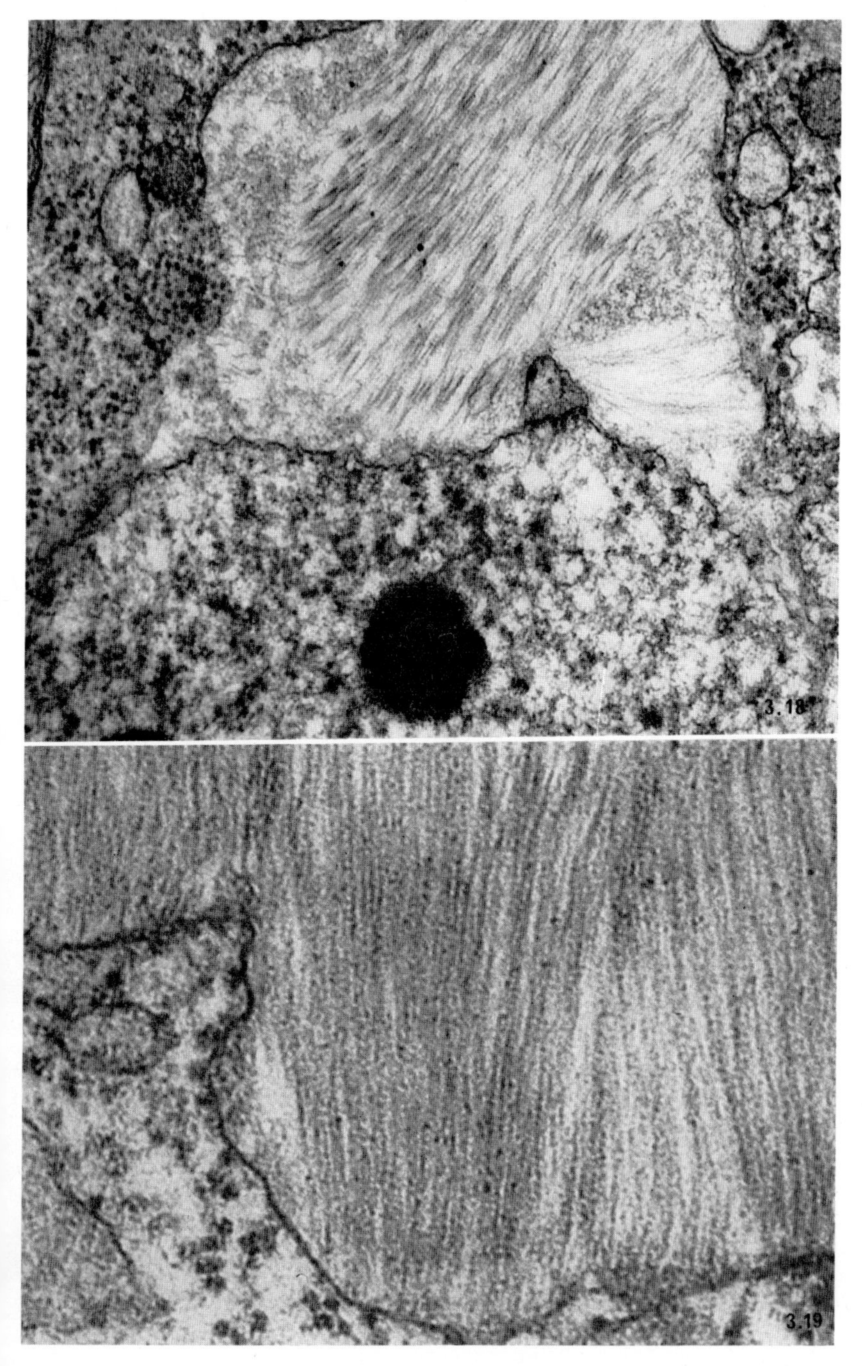

FIGS 3.18–3.19. Flagellar hair production in *Amphidinium* (Dinophyceae). (3.18) The hairs are seen to be forming in the expanded perinuclear cavity. Part of the nucleus is at the bottom of the picture (×84,000); (3.19) a more highly magnified portion of a flagellar hair vesicle showing the apparently 'beaded' nature of the hairs and ribosomes on the membrane bounding the vesicle. (×113,000)

presumptive hairs are frequently seen in sections of these organisms (Figs 3.18 and 3.19).

Mignot *et al.* (1972) working with various members of the Euglenophyceae and *Ochromonas* (Chrysophyceae) do not entirely agree with the findings of the other workers noted above. They think that the hairs begin to form in the endoplasmic reticulum but are then transferred to the Golgi cisternae. They are not certain that the microtubules seen in the perinuclear space are flagellar hair precursors.

In *Vacuolaria* (Chloromonadophyceae) Heywood (1972) found a vesicle, containing complete flagellar hairs, which opens into the groove at the foot of the flagella. This appears to be in position to supply hairs for a newly forming flagellum. It is of interest that internal vesicles in the cell only appear to contain the main shafts of the hairs, the basal and tip portions presumably being synthesized just before the vesicle fuses with the plasma-membrane, to open to the exterior.

V. Pseudocilia

In the green alga *Tetraspora* there are immobile projections of the cells which have long been known as pseudocilia. Electron microscope studies (Lembi and Herndon, 1966; Wujek, 1968; Lembi and Walne, 1971) have shown that these structures are derived from flagella and have basal bodies and root systems which are almost identical to those of *Chlamydomonas*. The transition zone also appears to have the typical stellate pattern but then, immediately above this, abnormalities appear. The free part of the organelle has no central tubules and the 'B' tubules of the doublets are mostly very short. Thus, most of the pseudocilium consists of a number of single tubules, often arranged in a 9+0 or 8+1 pattern, with the sheath contracted around them. There is no sign of the short projections normally to be found on 'A' tubules of flagella (cf. Fig. 3.9).

VI. The Haptonema

It was thought that some of the small yellow-brown flagellates, then placed in the Chrysophyceae, possessed three flagella although it was clear that one of them behaved differently from the other two (cf. Fritsch, 1935). Once electron microscopical techniques were applied to these organisms it became clear (Fig. 3.20) that the 'third flagellum' was a unique structure and it was then named the *haptonema* (Parke *et al.*, 1955). Later work involving sections has confirmed the distinctive structure of the haptonema. Basically it consists of three concentric membranes which enclose about 7 microtubules (Fig. 3.21). As with flagella, the outer membrane is continuous with the cell boundary membrane. The two inner membranes enclose a cylindrical cavity which is continuous over the tip of the haptonema (Manton, 1964d) and connects with a layer of superficial endoplasmic reticulum where the organelle enters the cell (Manton, 1968a). The tubules, which are of similar dimensions to cell microtubules (*c.* 20 nm), are 6 or 7 in number in the free part of the haptonema and then increase to 8 or even 9 within the cell. The basal portion of the haptonema is about the same length as a flagellar base but there are no roots or other structures associated with the tubules (Fig. 3.22).

FIG. 3.20. A shadowed preparation of a *Chrysochromulina* cell showing the characteristic features of a member of the Haptophyceae: two smooth flagella; scales, here both flattened oval scales and elongated tubular structures are present; the haptonema which is mainly extended but still partly coiled at the tip. (× 5000)

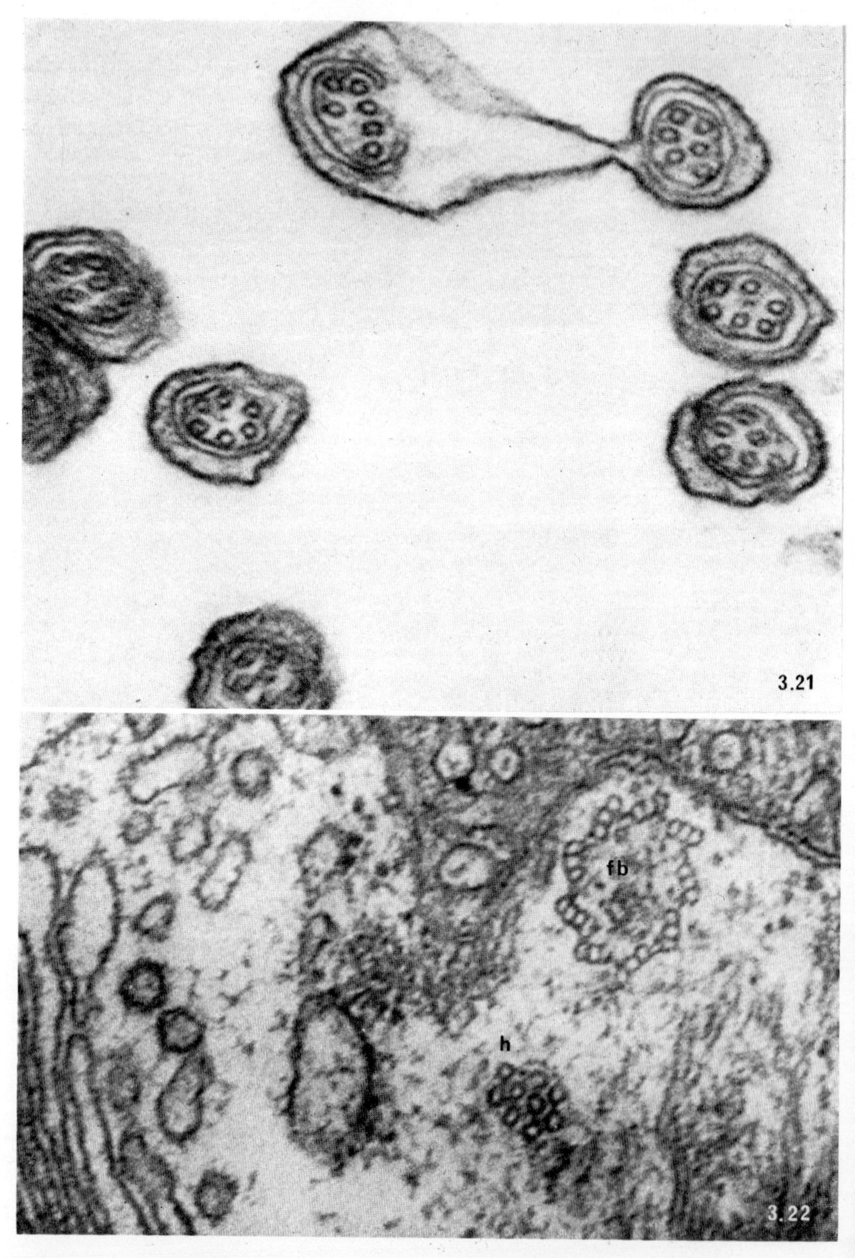

FIG. 3.21. A section through part of a coiled haptonema showing the outer membrane, the two membranes of the E.R. jacket and the six tubules which they enclose. (×98,000); (3.22) section of *Chrysochromulina* through the anterior end of the cell. A flagellar base (fb) shows the characteristic triplets and the base of the haptonema (h) shows sections of nine microtubules. (×98,000)

The haptonema can vary considerably in length from up to 80 μm when fully extended in *Chrysochromulina parva* (Parke *et al.*, 1962) to the short stump of *Prymnesium parvum* (Manton and Leedale, 1963a) and there are some organisms in which the structure of the haptonema is modified. In the coccolithophorid *Hymenomonas* Manton and Peterfi (1969) found that the short bulbous appendage, which has been termed a 'stummel'; contains flattened vesicles related to the endoplasmic reticulum, in addition to tubules which run into the cell for some distance. It clearly is a type of haptonema although almost certainly has lost the power of movement. A somewhat similar situation exists in *Pavlova* (Green and Manton, 1970) where the basal and transition portions of the short haptonema have the normal construction but the free portion is much altered. This consists of a crescentic profile of ER together with three tubules at the proximal end, reducing to two and finally to one or none at the extreme tip. In part of the haptonema the tubules are embedded in electron opaque material. No doubt further types of haptonema will be discovered as more organisms are examined by electron microscopy.

4. *Chloroplasts (Plastids)*

In this chapter we will consider the membranes, envelope and stroma of algal chloroplasts. The pyrenoids and eyespots which, when present, normally form an integral part of these chloroplasts will be separately described in later chapters.

I. Chloroplast Structure

The fact that algae of various groups have different colours, due to the presence in the chloroplasts of a variety of pigments, might lead to the supposition that the structure of the chloroplasts should also be variable. In fact it has been found that nearly every class of algae has its own distinctive and consistent chloroplast structure. The first clear intimation that this was so came from the pioneering survey of Gibbs (1962c) who examined the fine structure of chloroplasts in 13 species belonging to nine different classes. This revealed that the groups of membranes running across the chloroplast, and which had been first observed by Wolken and Palade in 1952, were variously arranged in the different algal classes. The main units were termed *bands* by Gibbs and *stacks* by other workers but are now generally referred to as *lamellae*. The basic membrane units of which they are composed are termed thylakoids after Menke (1962) although the terms *disc* and *partition* have also been used to describe them. The thylakoid is a flattened sack, composed of a single membrane.

Chloroplast structure has been reviewed by various authors: Ueda (1961), Greenwood (1964), Manton (1966d), Kirk and Tilney-Bassett (1967), Dodge (1970), and most recently the subject has been brought up-to-date again by Gibbs (1970).

A. STRUCTURE OF THE THYLAKOID

Although the thylakoids of higher plants have been the subject of much research, and also controversy resulting from differing interpretations of their basic structure, those of algae have not been studied very extensively. However, such evidence as we have suggests that in general thylakoids from algae or higher plants are constructed in a similar manner.

The first detailed description of the structure of algal thylakoids was provided by Weier *et al.* (1966). They found that the *c.* 11 nm thick thylakoid membranes in *Scenedesmus* revealed, after conventional fixation and sectioning, a single row of more or less spherical subunits having light cores and dark rims. Similar structures had also been observed in the thylakoids of higher plants (Weier *et al.*, 1965) when similar techniques were used.

Most recent work on thylakoid structure has involved the use of freeze-etching techniques. Neushul (1971) compared the structure of the thylakoids of red, brown and blue-green algae. He found no striking differences and in all cases one fracture face (after the thylakoid has been split along the centre of the membrane) has rows consisting of a few large 10 nm particles, whilst the other face has many small particles (about 5 nm) which are irregularly arranged (Fig. 4.3B). In a more detailed study of *Chlamydomonas* Goodenough and Stahelin (1971) found four distinct fracture faces. The Bs and Cs faces are complementary and are found where the thylakoids are stacked together. The Bs face has a dense population of large particles including many which are about 16 nm in diameter. In places where the thylakoids are not stacked Bu and Cu faces were found and these could be continuous with Bs and Cs faces, as at the edge of a stack. The Bu faces are almost completely lacking in 16 nm particles but have many at about 10 nm diameter. When mutants which lack stacking were examined only Bu and Cu faces were found although mutant ac-31 could be induced to stack by isolating the chloroplasts in a high-salt medium and then typical Bs and Cs faces were found. The authors conclude that the type of particle distribution in the thylakoid membranes can be modified by the stacking process and the ability to stack can be affected both by gene mutation and the environment (see Chapter 13).

Working with *Chaetomorpha* Robinson (1972) has found the two basic fracture faces described above but, in addition, he finds a face bearing both large and small particles and a completely smooth face which cannot yet be explained. Holt and Stern (1970), using *Euglena*, find the thylakoids to be particulate and similar to those found in higher plants by Branton and Park (1967), Park and Pfeifhofer (1969) and other workers.

The form of the thylakoids has been studied in the dinoflagellate *Gymnodinium micra* (Dodge, 1968). Cells were smashed and the chloroplast fragments separated by density gradient centrifugation. In shadow-case (Fig. 4.2) and negative stained preparations the thylakoids were mostly circular in outline but with varying diameters of between 0·15–3·6 μm. The surface of the thylakoids showed some evidence of the presence of large granules partially embedded in the membrane. Sections of these thylakoids showed that the membrane was about 6 nm thick and the central space about 12 nm (although very subject to variation due to fixation techniques). In *Batrachospermum* (Brown and Weier, 1970) the thylakoids are about 20 nm wide, the membrane of 6·0–6·5 nm enclosing a space of 6·5–8·0 nm in sectioned material and 9·0 nm after freeze-etching.

Normally, in section, thylakoids do not contain material that can be resolved in the electron microscope but, in the Cryptophyceae, the intrathylakoidal space contains dense material (Fig. 4.4). The loculus of the thylakoid is 20–30 nm thick and filled with a finely granular substance which was suspected of containing the phycobilin pigments (Dodge, 1969b). Selective extraction procedures (Gantt *et al.*, 1971) have now shown conclusively that this is this case, for when

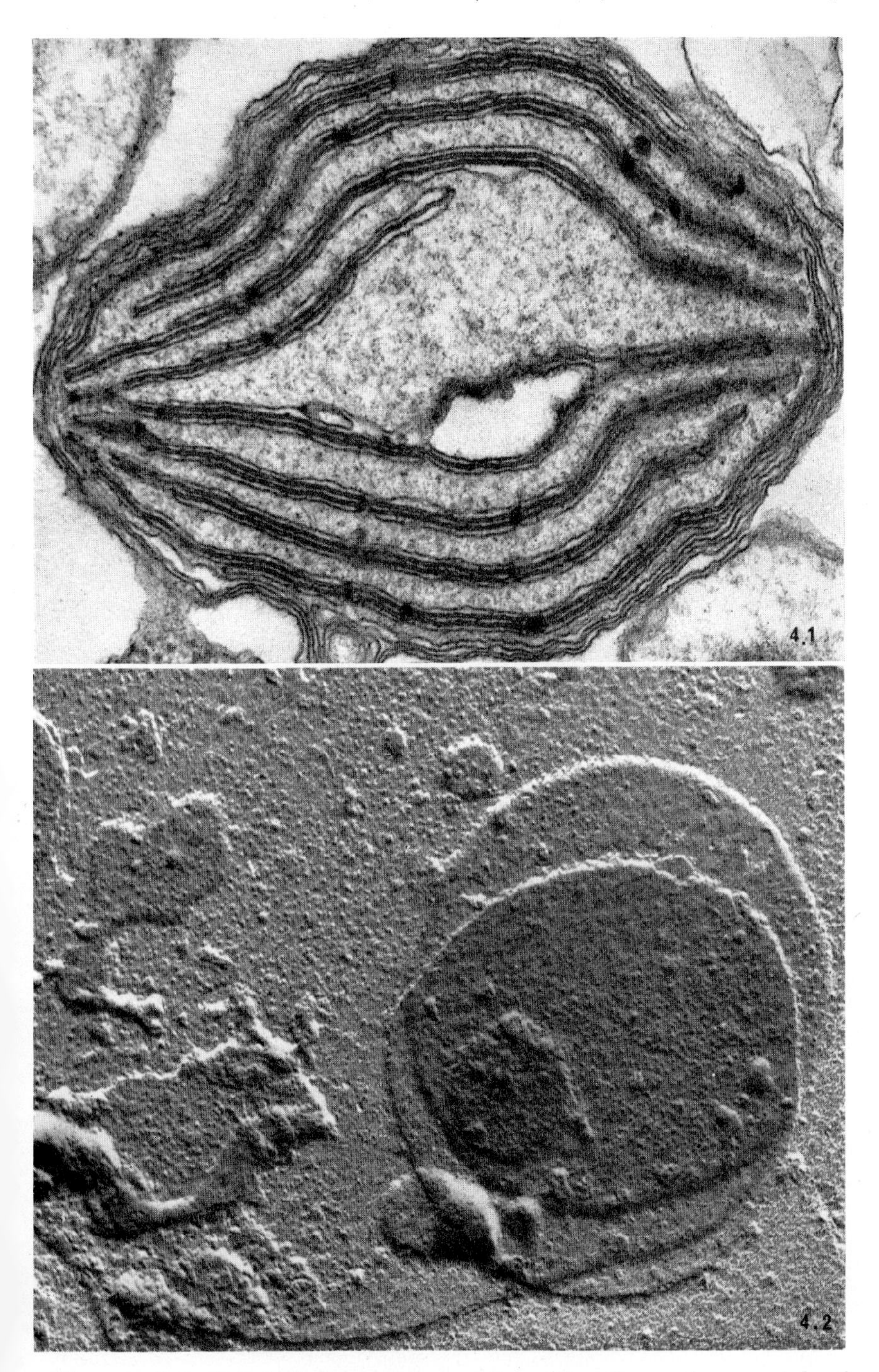

FIG. 4.1. A typical algal chloroplast consisting of lamellae made up mainly of stacks of three thylakoids. *Amphidinium britanicum* (Dinophyceae) (×60,000); (4.2) thylakoids separated from the chloroplast of *Gymnodinium micra* (Dinophyceae) and shadowed with platinum. Note their circular profiles and evidence of granularity in the structure of the membranes. (×50,400) (After Dodge, 1968.)

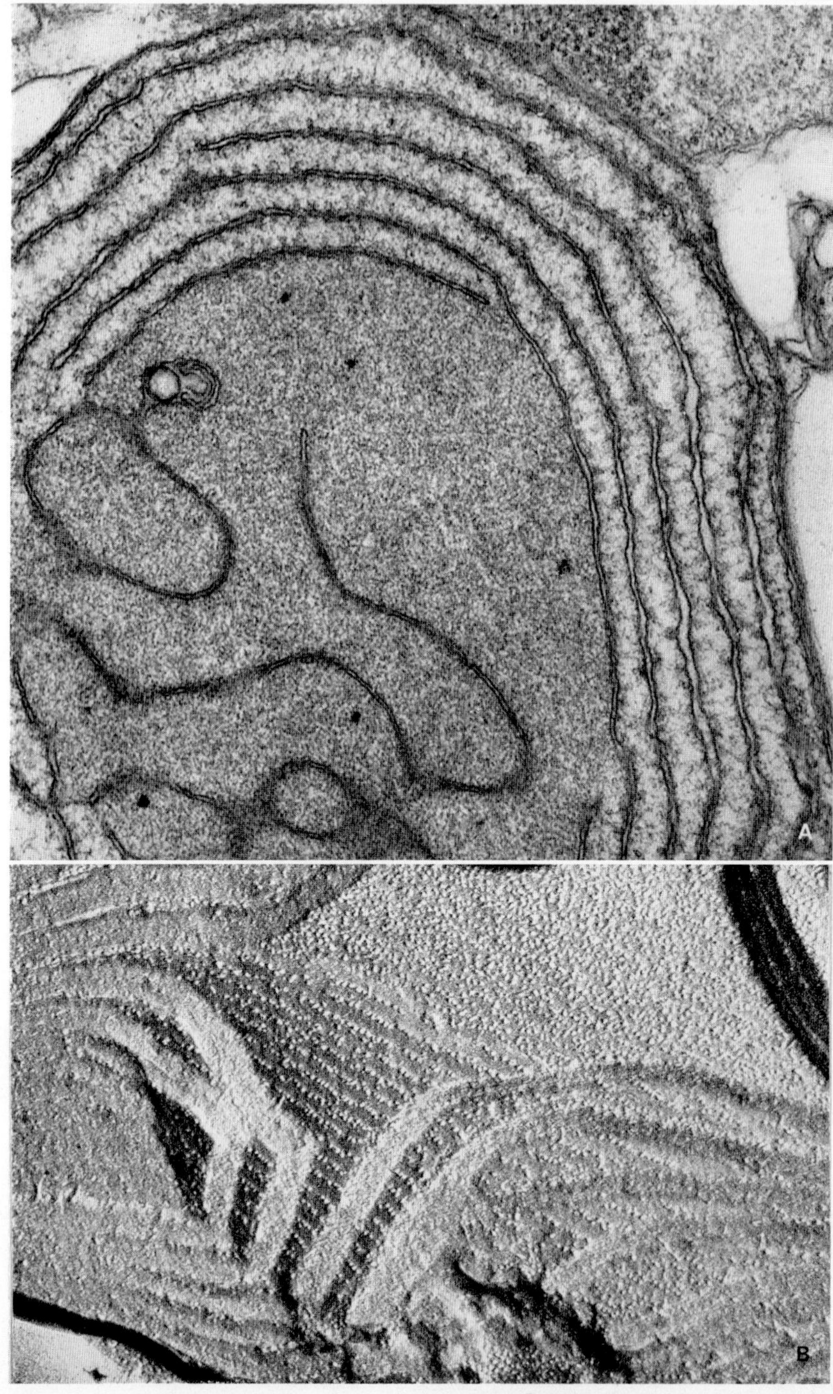

FIG. 4.3. A, The chloroplast, and enclosed pyrenoid, of *Porphyridium cruentum* (Rhodophyceae). Note the single thylakoids and the rather poorly preserved phycobilisomes in the stroma (×46,000); B, a freeze-etch preparation of the chloroplast of *Porphyridium*. The steplike fracture reveals the regular arrangement of particles on the surfaces of the thylakoid membranes. (×49,000) (Micrograph provided by M. Neushul, after Neushul, 1971.)

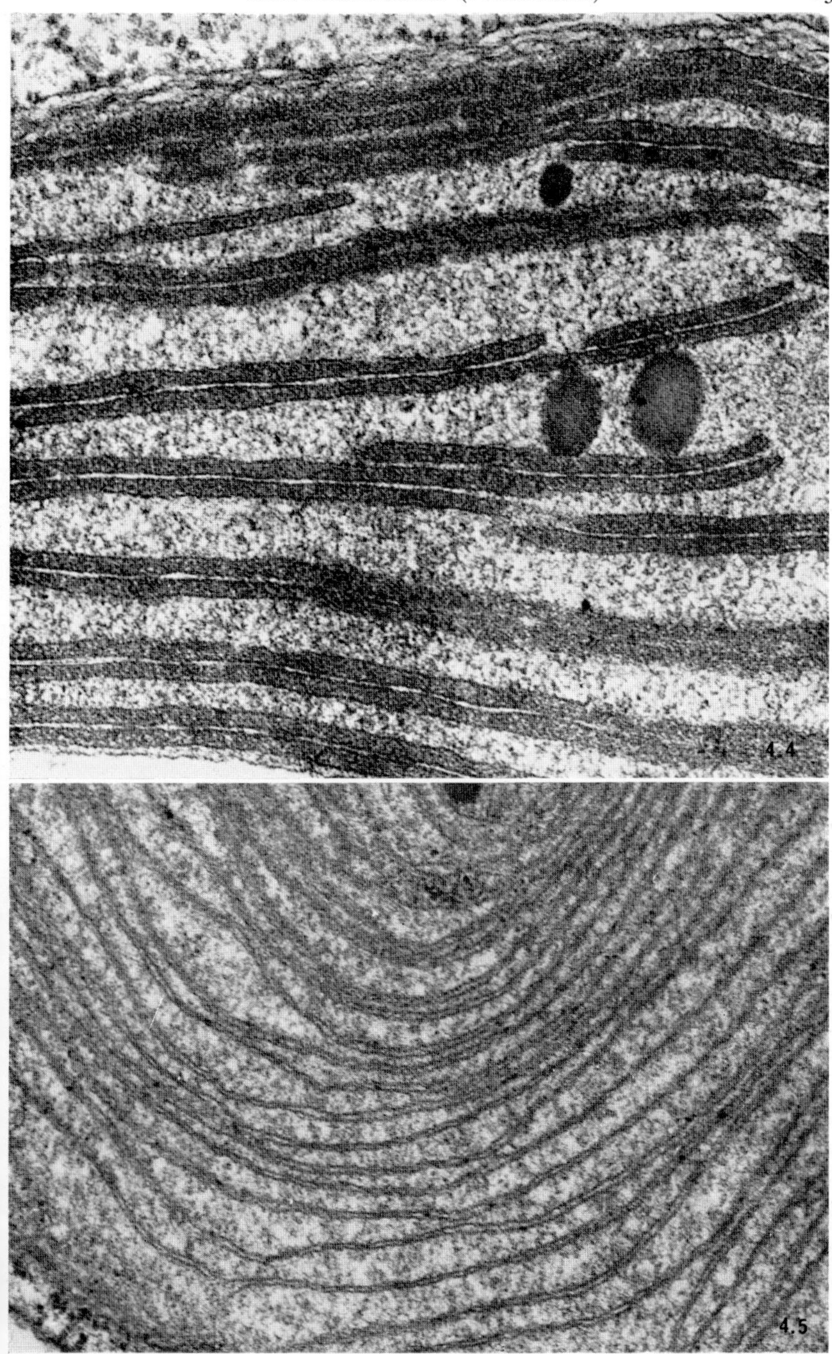

FIGS 4.4–4.5. Chloroplasts in the Cryptophyceae. (4.4) Part of the chloroplast of *Chromonas mesostigmatica* in which the thylakoids are mainly arranged in pairs. The thylakoid lumen is dense owing to the presence of phycobilin pigments (×81,000) (After Dodge, 1969b); (4.5) a chloroplast from *Cryptomonas salina* in which the thylakoids are single and not stacked together. This is probably a result of the conditions under which the cells had been grown. (×69,000)

the phycobilins had been extracted from the cells the thylakoids appear empty. This location of the biliproteins within thylakoids is in marked contrast with the situation in the red and blue-green algae where they are located outside the thylakoids (see later). Gantt *et al.* (1971) also noted a marked periodicity in the thylakoid membranes of *Chroomonas*. This took the form of parallel arrays of subunits with a repeat distance of 14–16 nm. The significance of this finding is not apparent and as yet no freeze-etching has been carried out to determine whether the structures are a part of the membrane or simply a reflection of the arrangement of the biliproteins on the membrane.

B. ARRANGEMENT OF THE THYLAKOIDS

The most simple arrangement is found in the Rhodophyceae where the thylakoids lie singly in the stroma (Fig. 4.3) and virtually never come into contact with other thylakoids except for occasional inter-connections. They are normally evenly spread through those parts of the chloroplast not occupied by DNA areas or pyrenoid and are only rarely perforated. In some red algae such as *Laurencia* (Bisalputra and Bisalputra, 1967a) the thylakoids lie in more or less parallel lines across the chloroplast but often there is a single thylakoid which lies parallel to the chloroplast envelope and thus encloses the others. Many red algae have a concentric arrangement of thylakoids and this is particularly well seen in *Porphyridium cruentum* (Fig. 4.3) (Gantt and Conti, 1965), a spherical unicellular alga, but it also occurs in the filamentous *Compsopogon* (Nichols *et al.*, 1966) and to some extent in *Batrachospermum* (Brown and Weier, 1970).

In the Cryptophyceae the arrangement is a little more complex. Here the thylakoids tend to be stacked in pairs (Fig. 4.4) (Gibbs, 1962c; Dodge, 1969b; Lucas, 1970a,b) but in some species the arrangement is more variable and single thylakoids (Fig. 4.5) and stacks of variable numbers have been reported, particularly in the genus *Cryptomonas* (Gantt *et al.*, 1971; Glazer *et al.*, 1971; Taylor and Lee, 1971); the thylakoids never appear to be very closely appressed together and there is often a gap of 3–8 nm between the members of a pair. Wehrmeyer (1970b) has shown that there may be occasional connections between the thylakoids and adjacent lamellae.

In several algal classes (see Table III) the thylakoids are predominantly arranged in threes (Figs 4.1, 4.6–4.8). In the Dinophyceae and Euglenophyceae adjacent thylakoids appear to be firmly attached together, to form partitions which are about 10 nm thick compared to the 5–6 nm of single unfused membranes. In some cases, especially the Phaeophyceae, the thylakoids are not closely appressed together and there may be a distinct space of about 4 nm (Gibbs, 1970) although this probably varies with the fixation procedure employed. Also in the Phaeophyceae, which were first studied by Leyon and Wettstein (1954), it is usual to find occasional thylakoids passing from one lamella to another (Bouck, 1965; Evans, 1966; Bisalputra and Bisalputra, 1967b). These connections tend to be mainly located towards the ends of the lamellae.

TABLE III

A summary of the main characteristics of chloroplast structure in the algal classes

	Chloroplast lamellae					Envelope		Reserve products	
	Number of thylakoids per lamella	Girdle lamella present	Inter-connections between lamella	Grana formed	Main chloro-phylls	Number of membranes around chloroplast	ER-connections	Main food reserve	Starch within chloroplast
Chlorophyceae	2–many	−	+	+	*a b*	2	−	Starch	+
Prasinophyceae	2–4	−	+	+?	*a b*	2	−	Starch	+
Euglenophyceae	3	−	?	−	*a b*	3	−	Paramylon	−
Chloromonadophyceae	3	+	+	−	*a*	4	−	?	−
Eustigmatophyceae	3	−	−	−	*a*	4	±	?	−
Xanthophyceae	3	+	+	−	*a*	4	+	Leucosin Oil	−
Chrysophyceae	3	+	+	−	*a c*	4	+	Leucosin	−
Bacillariophyceae	3	+?	+	−	*a c*	4	+	Leucosin Oil	−
Phaeophyceae	3	+	+	−	*a c*	4	+	Mannitol Laminarin	−
Haptophyceae	3	−	+	−	*a c*	4	+	?	−
Dinophyceae	3	−	−	−	*a c*	3	−	Starch Oil	−
Cryptophyceae	2	−	−	−	*a c*	4	+	Starch	−
Rhodophyceae	1	−	−	−	*a*	2	−	Starch	−

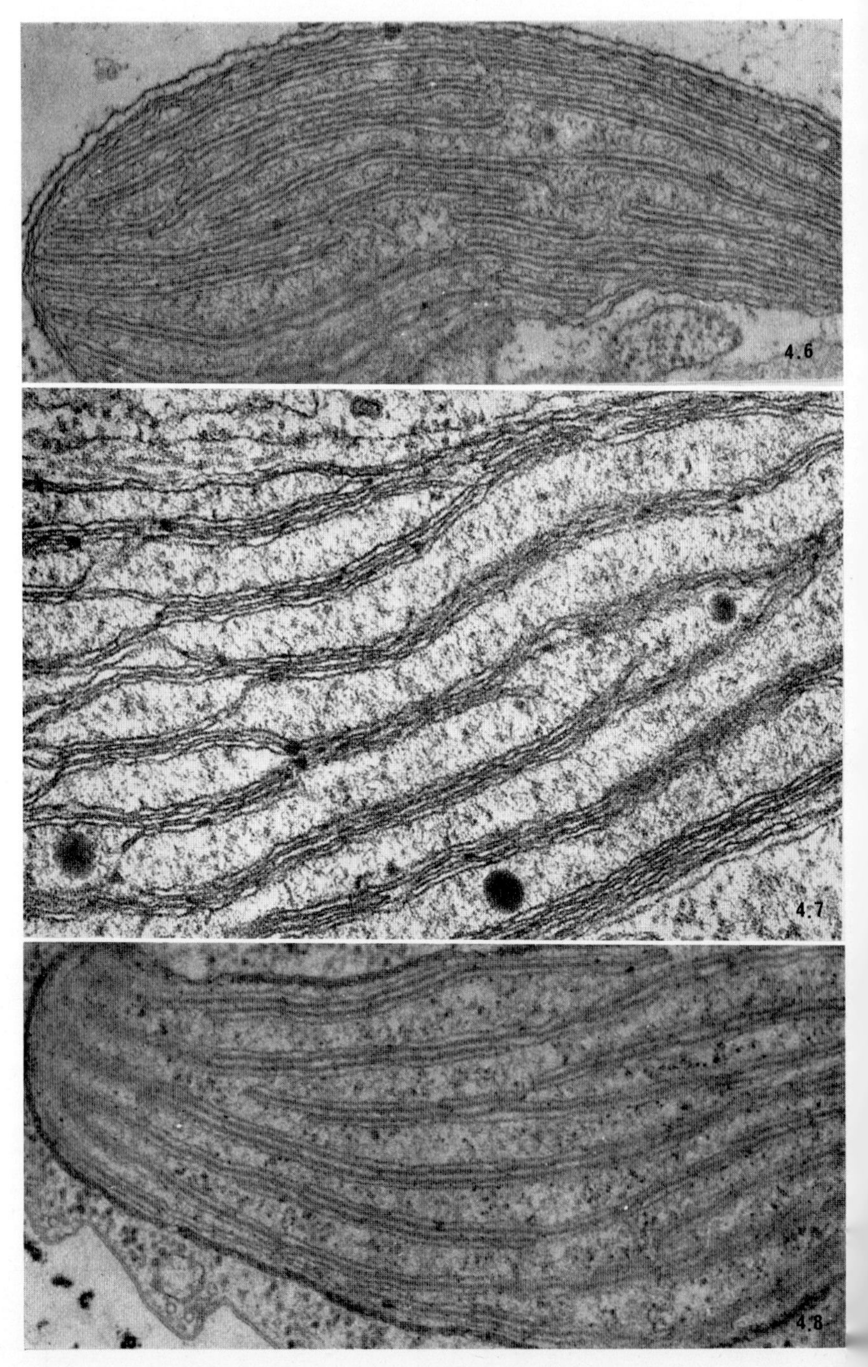

FIGS 4.6–4.8. Chloroplasts with mainly 3-thylakoid lamellae. (4.6) *Chrysochromulina* (Haptophyceae). (×49,000); (4.7) *Dinobryon* (Chrysophyceae) note interconnections between lamellae here (×59,000); (4.8) *Trachelomonas* (Euglenophyceae) where the chloroplast envelope consists of three membranes. (×59,000)

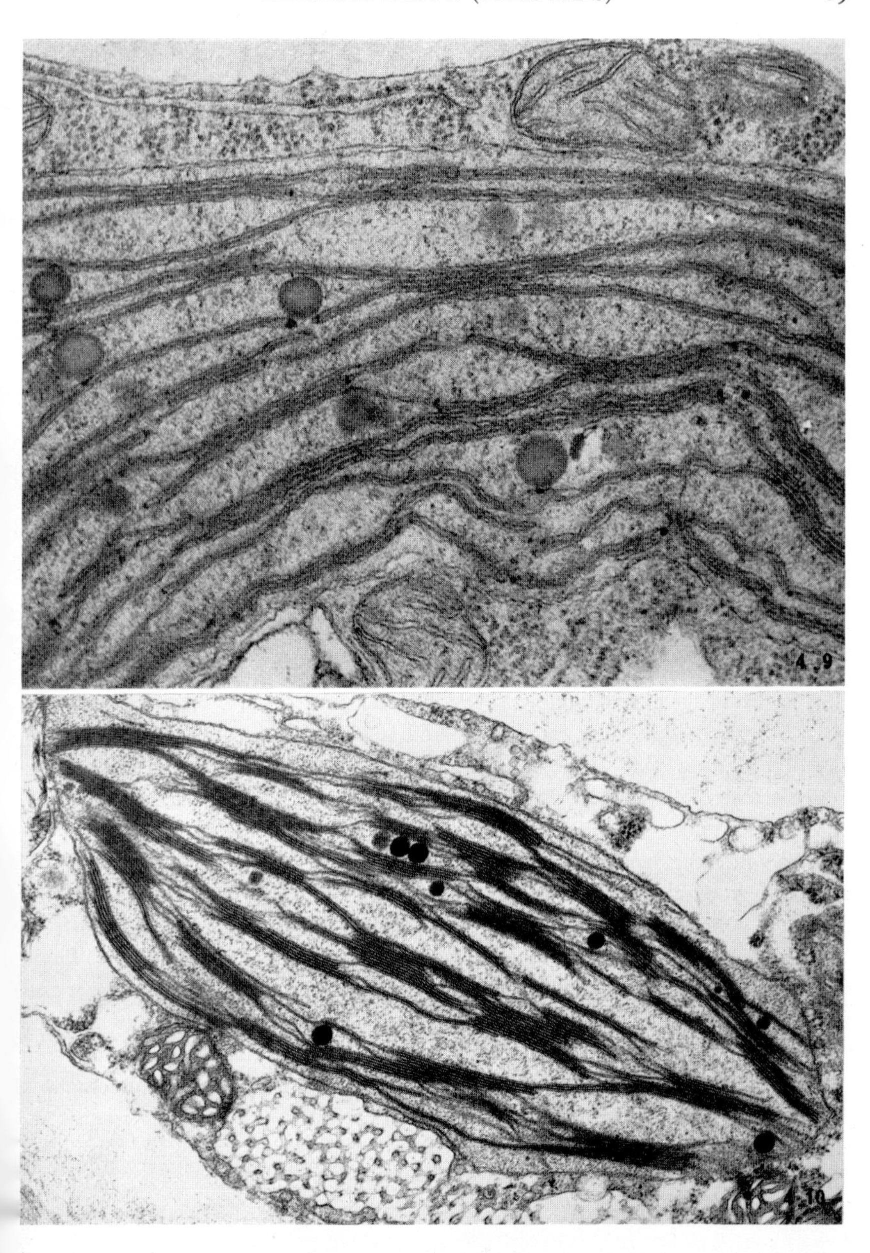

FIG. 4.9. Part of the reticulate chloroplast of *Cladophora* showing the two-membrane envelope and the thylakoids stacked in a rather random manner (×75,000); (4.10) a cross-section of a chloroplast of the stonewort *Chara* showing the thylakoids arranged into grana and inter-grana regions. This chloroplast might easily be mistaken for one from a higher plant. (×26,000) (Micrograph provided by R. M. Crawford.)

Similar connections have also been reported from a number of other organisms with 3-thylakoid lamellae (e.g. Chysophyceae, Fig. 4.7) but they do not appear to be found in the Euglenophyceae and Dinophyceae.

Many of the 3-thylakoid algae have a girdle lamella. This is a lamella which encircles the rim of the chloroplast and generally lies parallel to the chloroplast envelope. The phenomenon was first noted by Greenwood (1959) in *Vaucheria* (Xanthophyceae) and has since been seen in most other members of that class which have been examined, and also in the Chrysophyceae, Phaeophyceae and Chloromonadophyceae and some diatoms (Fig. 5.1). In *Dinobryon* (Chrysophyceae) the girdle lamella has been seen to be composed of an aggregation of single thylakoids from several other lamellae. The Haptophyceae lack girdle lamellae but here occasional single-thylakoids have been seen to run around the edge of the chloroplast and enclose the ends of the lamellae. (Gibbs, 1970 re: Manton, 1966a, 1967c) Apart from these girdle lamellae the other lamellae normally run more or less straight from side to side of the chloroplast. However, some anomalies have been noted. In *Glenodinium* (Dinophyceae) and in *Melosira* (Bacillariophyceae) (R. M. Crawford, pers. comm.) occasional chloroplasts are encountered which have short stacks of thylakoids, like grana, and in some dinoflagellates the lamellae from time to time appear to be radially arranged. The influence of environmental conditions on the structure of the chloroplast is a field that remains virtually unexplored, although Gergis (1972) has shown that this can have a considerable effect on chloroplast structure (see Chapter 13).

In the green algae—Prasinophyceae and Chlorophyceae—the arrangement of the thylakoids is very variable and much more like that seen in higher plants. Short stacks of thylakoids termed grana are often formed and at the edges of these there are frequent connections with other grana or lamellae (Wygash, 1963; Lembi and Lang, 1965; Ohad *et al.*, 1967b). However, owing to the irregular shape of green algal chloroplasts (e.g. cup-shaped, girdle-like, spiral) it is not normally possible for thylakoids to run in more or less straight lines from side to side as they tend to do in the typically discoid chloroplasts of angiosperms. Thus, in a plant like *Cladophora*, the arrangement is irregular (Fig. 4.9) (Strugger and Peveling, 1961) and the numbers of thylakoids adhering together in any one place normally varies from two to about six, but up to 20 have been found in *Chlamydomonas* (Ohad *et al.*, 1967a). The situation is often made more complex by the interpolation of starch grains between the lamellae. Where thylakoids are appressed together they are very clearly fused and a dense 2·5 nm layer has been detected between such appressed thylakoids (Ohad *et al.*, 1967a) following staining with uranyl acetate.

C. THE CHLOROPLAST STROMA OR MATRIX

1. *Ribosomes*

In all algal classes the chloroplast stroma contains ribosomes which look smaller than those of the cytoplasm. In *Ochromonas* (Gibbs, 1968) cytoplasmic ribosomes are 21–23 nm diameter whilst those of the chloroplast are only 17–20 nm. In *Chlamydomonas*, after glutaraldehyde-osmic fixation they are $26 \cdot 6 \pm 3 \cdot 2$ nm long by $20 \cdot 8 \pm 3 \cdot 2$ nm wide in the chloroplast compared with $31 \cdot 5 \pm 2 \cdot 5$ by $23 \cdot 2 \pm 3 \cdot 3$ nm in the cytoplasm (Ohad *et al.*, 1967a). The chloroplast ribosomes have been shown to have a sedimentation coefficient of 70S (Sager and Hamilton, 1967) which is similar to that of ribosomes from angiosperm chloroplasts and from the cells of blue-green algae and bacteria. The ribosomes are fairly evenly dispersed through the stroma and do not appear to form polyribosomes. Gibbs (1968) has shown, by use of labelled uridine, that RNA synthesis takes place in chloroplasts and mitochondria of *Ochromonas* and at a faster rate than in the cytoplasm. She suggests that most of the RNA found in these organelles is synthesized *in situ*. During chloroplast development in the light the number of ribosomes increases 10-fold. Bourque *et al.* (1971) working with *Chlamydomonas* found that the relative amounts of chloroplast ribosomes did not seem to be correlated with chlorophyll synthesis and the formation of the lamellar system. This result is somewhat at variance with that of Goodenough and Levine (1970) working with the ac-20 mutant of *Chlamydomonas reinhardi*. They found that proper membrane organization, photosystem II activity, etc., were dependent on the presence of a more or less normal amount of chloroplast ribosomes.

2. *Deoxyribonucleic acid*

Chloroplasts of most algae contain DNA and, in electron micrographs, this is generally seen as relatively electron-translucent areas containing irregularly arranged fibrils about 2·5 nm thick. The first alga in which these fibrils were clearly shown to be composed of DNA was *Chylamydomonas* which Ris and Plaut (1962) studied, using cytochemical tests and enzymic digestion. A survey of the chloroplasts of many red, brown and green algae (Yokomura, 1967a) revealed the presence of DNA fibrils in every organism examined. In many organisms it has been shown that similar fibrils disappear after treatment with the enzyme deoxyribose nuclease (Bisalputra and Bisalputra, 1967a,b; Leadbeater, 1967; Puiseux-Dao *et al.*, 1967). In the Chrysophyceae it has been shown, by use of autoradiography, that these areas take up tritium labelled thymidine (Slankis and Gibbs, 1968) indicating that DNA synthesis takes place there. In the Euglenophyceae DNA has been located in the chloroplasts both by autoradiography (Sagan *et al.*, 1965) and by extraction procedures (Brawerman and Eisenstadt, 1964; Ray and Hanawalt, 1964).

In organisms which have girdle lamellae the DNA always seems to be located iu a ring around the periphery of the chloroplast, but enclosed by the girdle lamella. This has been particularly clearly demonstrated in *Ochromonas* (Chrysophyceae) (Gibbs, 1968) and *Sphacelaria* (Phaeophyceae) (Bisalputra and Bisalputra, 1969) (Fig. 4.11) where sections were obtained parallel to the lamellar plane. This confirmed the suggestion from transverse sections that the DNA zone or 'genophore' does run right around the chloroplast. Also working with *Sphacelaria*

Bisalputra and Burton (1969) used osmotic shock and spreading techniques to release DNA fibrils from the chloroplasts. The DNA was in the form of straight filaments, single, and double loops and was always attached to a membrane which was thought to be from the edge of a thylakoid. It would seem likely that a similar situation exists in the Xanthophyceae, Chloromonadophyceae and Bacillariophyceae where sections of chloroplasts invariably reveal the characteristic DNA areas beneath the girdle lamellae (e.g. Falk, 1967; Falk and Kleinig, 1968; Hibberd and Leedale, 1971; Manton and Von Stosch, 1966; Mignot, 1967; Taylor, 1972).

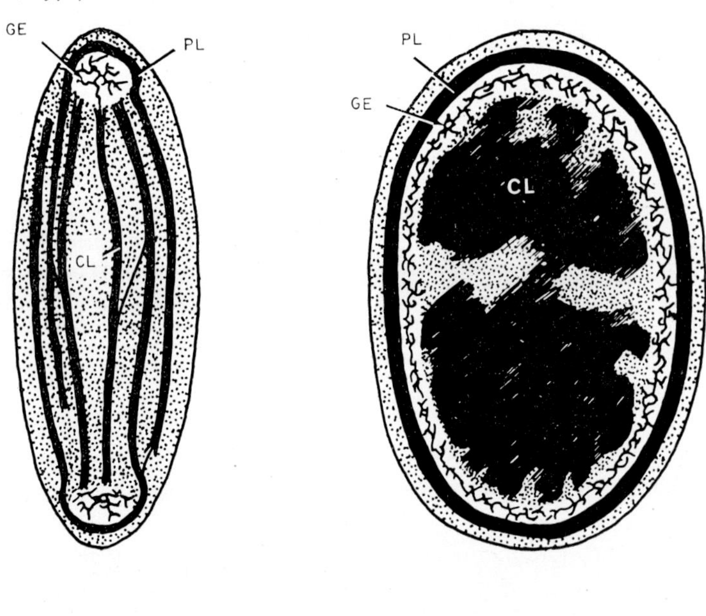

FIG. 4.11. Diagrams illustrating the arrangement of the chloroplast DNA in *Sphacelaria* (Phaeophyceae). A, A vertical section in which the DNA genophore (GE) is seen as patches at each end of the chloroplast, enclosed by the girdle lamella (PL); B, a horizontal section (parallel with the largest face of the chloroplast) which shows the DNA (GE) in the form of a ring within the girdle lamella (PL). CL represents a tangentially sectioned lamella. Reproduced, with permission, from Bisalputra and Bisalputra (1969).

In the remaining classes of algae (Rhodophyceae, Cryptophyceae, Dinophyceae, Englenophyceae, Haptophyceae, Prasinophyceae, Chlorophyceae) the DNA is generally scattered through the chloroplast matrix in a rather irregular manner. Yokomura (1967a) has suggested that the DNA areas in red algae may be interconnected to form a reticulum. In the dinoflagellate *Prorocentrum* Kowallik (1971) and Kowallik and Haberkorn (1971) used serial section techniques and found that each of the two multilobed plastids contained 80–100 discrete DNA regions. These regions differed in size (and probably in DNA content) and were usually in the form of flattened irregular discs. In some cases the DNA fibrils seemed to be attached to a tube or tongue-like evagination of a thylakoid and the fibrils had a complex possibly spiral organization similar to that seen in chromosomes of dinoflagellates. In *Euglena*, chloroplast DNA was extracted and spread out using the techniques normally employed for the examination of viral and bacterial DNA (Manning *et al.*, 1971). It was found to be in the form of circular threads each some 40 μm long.

In the giant unicellular green alga *Acetabularia* Woodcock and Bogorad (1970) have shown that some of the thousands of chloroplasts appear to have no DNA in them. Using serial sections, acridine orange fluorescence and osmotic bursting of chloroplasts to reveal the DNA strands, they found DNA in only 20–35% of the plastids and even in these the amount was rather variable.

3. *Lipid globules*

Osmiophilic lipid globules are found in the chloroplasts of most algal classes. They very rarely form the dominant component of the chloroplast in the way that they often do in angiosperms during senescence or during conversion into a chromoplast. Sprey (1970) has shown that in *Haematococcus* lipid globules increase in size as the thylakoids break down. Lipid globules also appear in the specialized areas of chloroplasts which are called eyespots. These are discussed in Chapter 6.

4. *Phycobilisomes*

In members of the Rhodophyceae dark-staining granules or bodies are frequently seen between the thylakoids or attached to them. These are *phycobilisomes*, the site of the biliprotein pigments phycoerythrin and phycocyanin. The first organism in which these were studied in detail was the unicellular *Porphyridium cruentum* (Gantt and Conti, 1966). Here they are more or less spherical with a diameter of 32 nm and are regularly spaced in rows on the outer surface of the thylakoids (Fig. 4.3A). They are definitely attached to the thylakoids, as Gantt and Conti found that they adhere to chloroplast fragments. Because of the ready solubility of the pigments they may often appear wispy and then appear not to be attached to the membranes unless careful fixation is carried out (Cohen-Bazire and Lefort-Tran, 1970). The composition of the phycobilisomes has been characterized by various techniques (Gantt and Conti, 1967). When cells were extracted with an 80% methanol: 20% acetone mixture the chlorophylls and many carotenoids were removed but the phycobilisomes were still present in the cells. However, after extraction with 0·2% deoxycholate the red pigments were removed from the cells and electron microscopy showed no trace of the phycobilisomes. The arrangement of these granules in *Porphyridium* has been studied in freeze-etch preparations (Neushul, 1970). A model of the chloroplast structure is proposed in which the phycobilisomes join together adjacent thylakoids.

In a number of other red algae the phycobilisomes appear to have different forms. Thus in *Porphyridium aeruginosum* (Gantt and Conti, 1967) and *Rhodella violaceum* (Wehrmeyer, 1971) they are said to be disc shaped. In the latter case they are arranged in rows or stacked piles. In the filamentous alga *Batrachospermum virgatum* (Lichtlé and Giraud, 1970) the phycobilisomes appear as long cylinders which are attached to the outer surface of the thylakoids. On any one thylakoid they are always parallel but the orientation may change from one thylakoid to the next. Strangely, a freeze-etch study of a related species, *Batrachospermum moniliforme* (Brown and Weier, 1970) did not reveal the presence of any phycobilisomes but presumably these were dissolved out by the glycerol in which the material was suspended prior to freezing. By way of contrast a red alga which lacks phycobilin pigments—*Rytiphlea*—was found to have 35 nm particles on the lamellae which look like phycobilisomes (Peyrière, 1968).

5. *Starch grains and amyloplasts*

In the green algae—Chlorophyceae and Prasinophyceae—frequently large areas of the chloroplasts are occupied by starch grains. In these classes starch only occurs in plastids whereas in all the other algal classes reserve polysaccharide is found in the cytoplasm (see Chapter 10). Occasionally it may lie just outside the chloroplast adjacent to the pyrenoid, and in the Cryptophyceae it is found between the chloroplast envelope and the endoplasmic reticulum sheath (Gibbs, 1962d; Dodge, 1969b), but generally it lies completely free in the cytoplasm (Fig. 10.14). In the green algae starch is particularly abundant in chloroplasts when cells have been grown under high light intensity or on a carbohydrate rich medium (Fig. 13.3) The starch grains normally lie between the lamellae and may cause them to become very distorted. In chloroplasts with a central pyrenoid it is normal to find most of the starch in hemispherical or segmental grains around the pyrenoid (Figs 5.14 and 5.15), but additional grains may still form between the lamellae.

In a number of algae which lack chlorophyll, rudimentary chloroplasts remain and these often function as leucoplasts or amyloplasts for the storage of starch. They have been found in the 'green' algae *Polytoma* (Lang, 1963c), *Prototheca* and *Polytomella* (Menke and Fricke, 1962; Webster *et al.*, 1968); the diatom *Nitzschia alba* (Schnepf, 1969); trichoblasts of the red alga *Laurencia* (Godin, 1970).

6. *Miscellaneous components*

The chloroplast stroma must contain a fairly large amount of protein and what is presumed to be this appears in micrographs as finely granular amorphous material. There have been occasional reports of other structures in the stroma such as crystals in *Chlamydomonas* (Stein and Bisalputra, 1969) and tubules in *Oedogonium* (Hoffman, 1967), *Chara* and *Volvox* (Pickett-Heaps, 1968a). These latter were about 32 nm in diameter, which is considerably larger than normal cell microtubules, and appeared to have a helical or banded construction. They seemed to be more conspicuous when chloroplasts were elongating or enlarging.

D. THE CHLOROPLAST ENVELOPE AND ENDOPLASMIC RETICULUM SHEATH

In all algae except the Dinophyceae and Euglenophyceae the basic chloroplast envelope consists of two parallel membranes separated by a space of up to 10 nm, the whole structure normally being 12–15 nm wide (Gibbs, 1962d). In *Chlamydomonas* (Ohad *et al.*, 1967a) the envelope has been seen very clearly and here it has an overall thickness of about 20 nm. The two membranes are markedly dissimilar, the outer being 7·5 nm thick and having the normal (dark-light-dark) tri-layer appearance of a unit membrane whilst the thinner (5·5 nm) inner membrane normally appears as a single band and looks very similar to the thylakoid membranes in the same chloroplasts. The basic chloroplast envelope has no obvious pores and no connections with other organelles and in this is similar to the envelope of higher plant chloroplasts. The chloroplasts of the Rhodophyceae, Prasinophyceae and Chlorophyceae are enclosed only in such an envelope.

In the two exceptions to the structure outlined above, the dinoflagellates and euglenids, the chloroplast envelope consists of three membranes (Leedale, 1967; Dodge, 1968). In the Dinophyceae this was found to have a total width of 23 nm and in *Euglena* it appears to be rather thicker, 35–45 nm (as measured from the photographs of Ploaie, 1971). No connections have been observed between the envelope and other organelles, in dinoflagellates, and ribosomes never appear to be attached to the outer surface. In the Euglenophyceae connections have been noted to what is said to be tubular (smooth) ER (Leedale, 1968).

In the Cryptophyceae, Chrysophyceae, Phaeophyceae, Xanthophyceae, Haptophyceae, Bacillariophyceae and Chloromonadophyceae, besides the basic two membrane envelope described above, the chloroplast is surrounded by a sheath of endoplasmic reticulum which is usually continuous with the nuclear envelope (Fig. 4.12). Thus it can be said that in these organisms the chloroplasts and nucleus are in one compartment whilst the remaining cell organelles are in another. This arrangement was first noted by Gibbs (1962b) in *Ochromonas* and two members of the Cryptophyceae. It appears to be present in all members of the classes listed above. Frequently the outer surface of the ER sheath is covered with ribosomes and these may occasionally be linked into polysomes (Gibbs, 1970). Where the nucleus lies adjacent to a chloroplast the space between the nuclear envelope and the true chloroplast envelope may contain tubular structures which form a branched network. It is not entirely clear whether these are associated with the ER or the chloroplast. They have been clearly observed in *Ochromonas* (Gibbs, 1970), various members of the Xanthophyceae (Descomps, 1963b; Falk, 1967; Falk and Kleinig, 1968; Massalski and Leedale, 1969;

Hibberd and Leedale, 1971) and in certain diatoms (Stoermer *et al.*, 1965). The structure of this reticulum, as seen in *Tribonema* (Falk and Kleinig, 1968), is shown in Fig. 4.13. In *Ophiocytium* and *Bumilleria* dense spherules are also present in this region (Hibberd and Leedale, 1971).

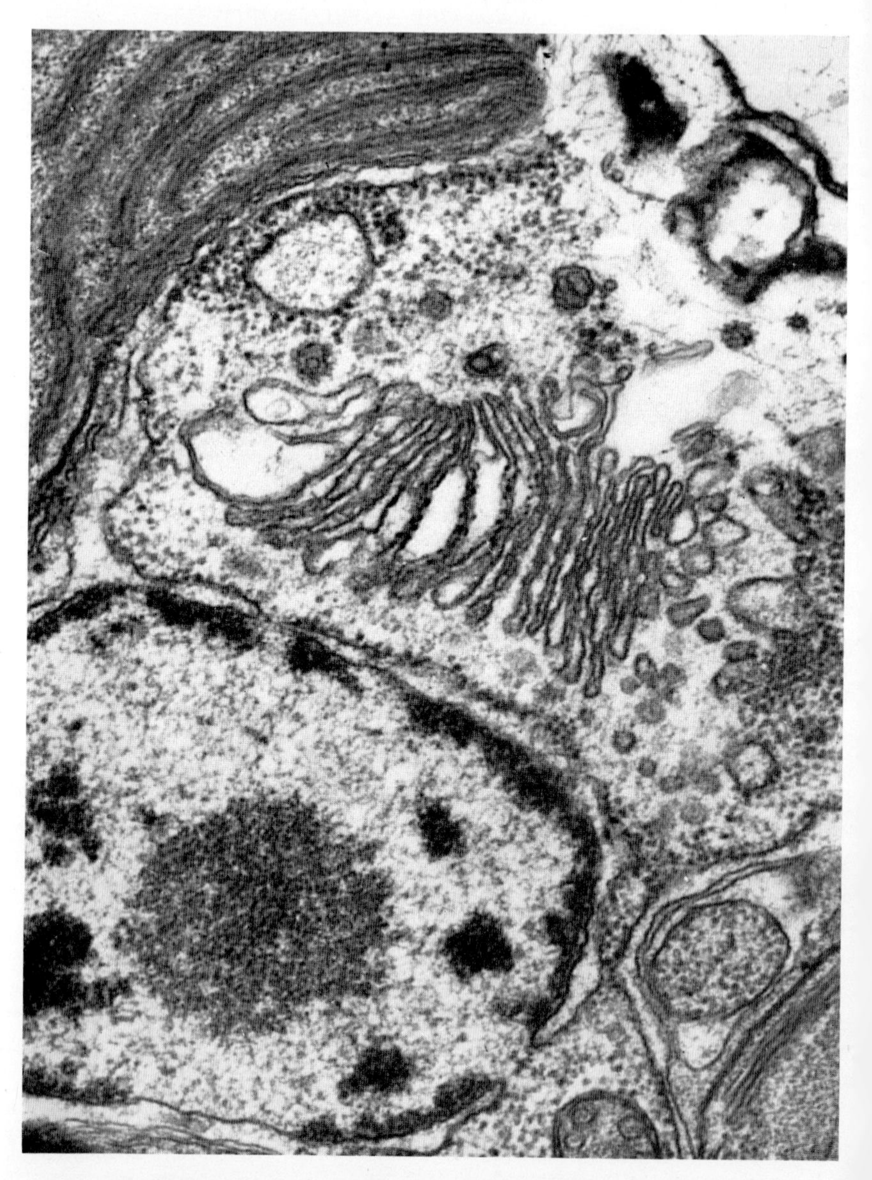

FIG. 4.12. The continuation of nuclear envelope to form the chloroplast endoplasmic reticulum in *Chrysochromulina* (Haptophyceae). Note also the large scale-producing Golgi body above the nucleus. (× 53,000)

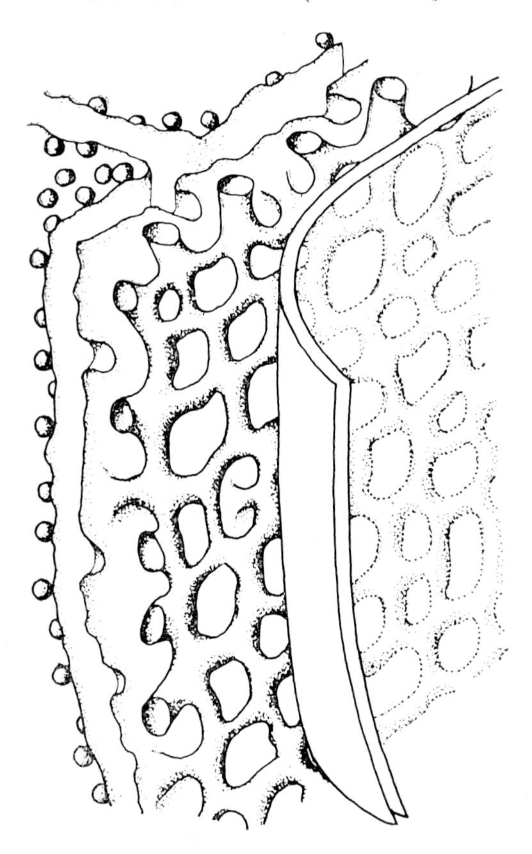

FIG. 4.13. A diagrammatic representation of the periplastidial reticulum in *Tribonema* (Xanthophyceae) situated between the chloroplast envelope (right) and chloroplast–ER (left). Reproduced, with permission, from Falk and Kleinig (1968).

II. Unusual Chloroplast Structure in Mutants, etc.

Mutation studies in the algae have been virtually confined to the Chlorophyceae and especially to *Chlamydomonas*. A very early study by Sager and Palade (1954) compared the chloroplast structure of green and yellow strains. In this genus numerous mutants have been isolated which are deficient in some aspect of the photosynthetic process. In many cases they also show structural changes in their chloroplasts. Thus, Goodenough and Levine (1969) found that mutant *ac-206* (which lacks cytochrome 553) could only produce two-thylakoid stacks, as compared to the 2–5 thylakoids per stack of the wild type. Mutants *ac-115* and *ac-148*, which lack cytochrome 559 and are incapable of any 'Hill' activity, had chloroplasts with mainly long single thylakoids. However, another mutant *ac-31* is capable of most photosynthetic activities in spite of the fact that its thylakoids are mainly single (Goodenough *et al.*, 1969). The only functional difference noted between this mutant and the wild type is that a higher light intensity is required for light saturation of the photo-systems. This would suggest that stacking of the thylakoids is not an absolute necessity for photosynthetic activity but merely confers a functional advantage. It is not clear how the red algae, for example, overcome the disability of only having single thylakoids.

A colourless mutant of *Chlorella vulgaris* which was studied by Gergis (1969) was obviously incapable of photosynthesis. Its chloroplasts were of normal size and contained long single thylakoids.

A photosynthetic mutant of *Euglena* has been studied by Schwelitz *et al.* (1972). This strain, in which photosynthesis was blocked in photosystem II, had a complete lack of thylakoid pairing and the thylakoids were more varied in size and fewer in number than in the wild-type. No fundamental differences were seen in the structure of the thylakoid membranes by use of freeze-etching. Comparisons have been made between chloroplast structure in normal and 'bleached' strains of *Euglena* (Gibor and Granick, 1962; Rogers *et al.*, 1972).

In the chloroplast of the filamentous green alga *Zygnema* McLean & Pessoney (1970) have reported the presence of 'A large scale quasi-crystalline lamellar lattice . . .'. These complex structures only appeared in cultures at the stationary phase of growth and could therefore be brought about by a nutritional deficiency. The peripheral parts of the stellate chloroplast remained normal but the central area adjacent to the large pyrenoid became filled with more or less parallel lamellae, each apparently consisting of several membranes which are in part wavy and in part forming tubular structures which join adjacent lamellae together. There are some slight similarities between this structure and the prolamellar body of etiolated higher plant plastids.

Changes which can be brought about in the structure of chloroplasts, by experimental treatment or nutritional conditions are discussed in Chapter 13.

III. Chloroplast Development

This has been extensively studied in two organisms, *Chlamydomonas reinhardi* and *Euglena gracilis* var. *bacillaris*. The first work on Chlamydomonas was carried out by Sager and Palade (1957). More recently Ohad *et al.* (1967a,b) used a strain (*y-l*) of *C. reinhardi* which cannot synthesize chlorophyll in the dark. This was first grown in the dark on an acetate carbon source for 5–6 generations (about 7 days), during which time the photosynthetic ability was almost completely lost and the chloroplasts came to consist of a few scattered thylakoids interspersed between numerous starch grains. The pyrenoid and eyespot remained unaffected. When the cells were brought into the light a rapid synthesis of membranes took place and at the same time chlorophyll, lipids and proteins were formed, whilst the starch was used up. Within 6–8 h the cells contained normal chloroplasts which were fully able to carry out photosynthesis (Fig. 4.14).

Study of the kinetics of the development suggested that the chloroplast membranes were assembled from their components by a single-step process. However, later studies with the same mutant, but employing fractionated cells (Petrocelis *et al.*, 1970) gave a rather different result and suggested that the process of membrane assembly during greening probably follows a multistep pattern. In a radioautographic experiment with *Chlamydomonas* Goldberg and Ohad (1970)

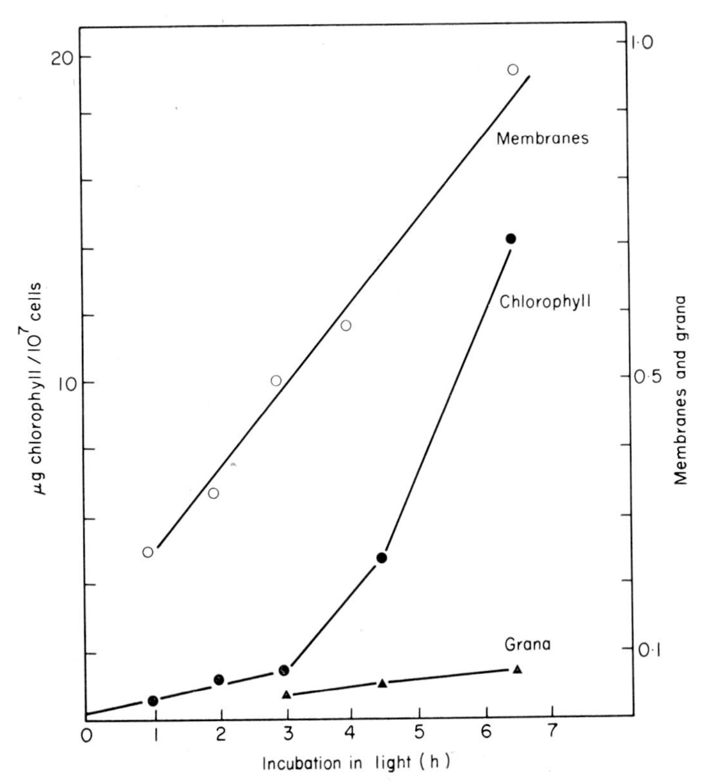

FIG. 4.14. Synthesis of chlorophyll and the associated increase in chloroplast membranes in *Chlamydomonas*. The alga was grown in the dark, resuspended in fresh medium and then exposed to light. Samples were removed at intervals for chlorophyll estimation and electron microscopy. Reproduced, with permission, from Ohad Siekevitz and Palade (1967b).

found that the chloroplast lamellae develop by incorporation of new material and different types grow at differing rates. Single thylakoids were found to grow twice as fast as those which are paired.

With *Euglena* most experiments have been concerned with the development of chloroplasts in strains which have been artificially 'bleached'. This bleaching can be readily carried out by growing cells in the dark in complete medium or by treatment with agents such as Streptomycin or ultra-violet light (Gibor and

Granick, 1962; Moriber *et al.*, 1963). Over about eight generations (144 h) the cells lose their chlorophyll, the chloroplasts decrease in size and within them the lamellae come apart and the discs are gradually lost. Finally the chloroplasts become small proplastids (about 1 μm diameter) which consist of little more than an envelope with small membranous inclusions (Lefort, 1964; Ben-Shaul *et al.*, 1965). The early studies with this system showed that when such cells were placed in light of adequate intensity the thylakoids (= discs) began to appear as

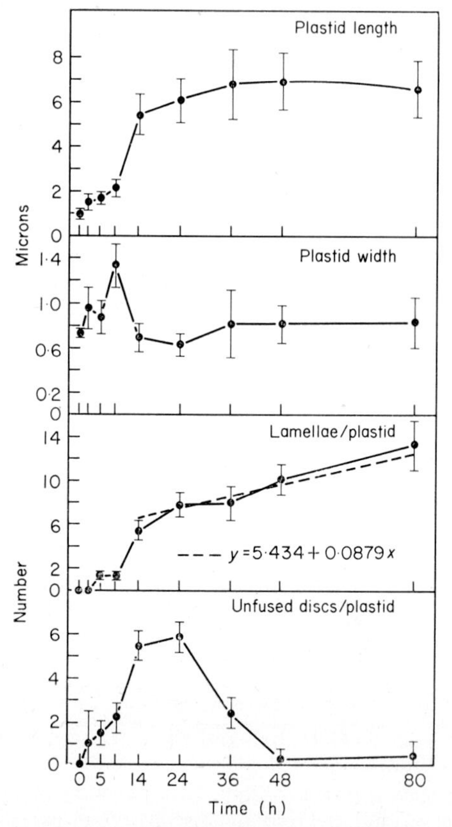

FIG. 4.15. Data on the development of the plastids after bringing bleached *Euglena* into the light. The bars across the lines represent 95% confidence intervals for the mean points. Reproduced, with permission, from Ben-Shaul, Schiff and Epstein (1964).

ingrowths of the inner membrane of the envelope (Ben-Shaul *et al.*, 1964). After about 6 h the first lamella was formed by fusion of two thylakoids. From 4–8 h after the start of illumination chlorophyll began to be synthesized and photosynthesis started up. The number of thylakoids continued to increase and every 6 h or so a new lamella was formed. This continued in linear fashion from

14 h to about 72 h by which time the chloroplast was complete with 13–14 lamellae and had reached its normal mature size (Fig. 4.15). This sequence correlated well with the development of photosynthetic oxygen evolution and carbon dioxide fixation. The pyrenoids appeared in the plastid after 18–20 h.

More recent work using improved fixation techniques (Klein *et al.*, 1972) has given a slightly different picture of chloroplast development. The proplastids in the dark-grown cell are now known to have a number of girdle-like thylakoids, which are located close to the chloroplast envelope, and may be connected to one or more small prolamellar bodies. On illumination no new membranes are formed during the first few hours but, after about 12 h, the girdle thylakoids fuse, and convert into straight thylakoids. New thylakoids form and the prolamellar body disappears. This is the stage of rapid increase in chlorophyll content. One major difference from the earlier work is that it is now thought that there is no connection between the thylakoids and the plastid envelope. The presence of a prolamellar body in plastids of dark-grown cells brings *Euglena* into line with the higher plants. As yet such structures have not been observed in any other algae except *Zygnema* (cf. p. 98).

In another study Schiff *et al.* (1967) investigated chloroplast development in the presence of the photosynthetic inhibitor DCMU [3, (3, 4-dichlorophenyl) 1, 1-dimethyl urea]. When cells were brought into the light in the presence of DCMU photosynthetic carbon fixation was completely inhibited and yet the cells were able to develop 70% of their normal chlorophyll. They also developed more or less normal chloroplasts, the only significant difference from normal being the appearance of more thylakoids per lamella and the absence of a pyrenoid. It was concluded that in *Euglena* chloroplasts can develop in the absence of photosynthetic competence. This contrasts with what was found in higher plants (Klein and Neuman, 1966) where DCMU inhibited chlorophyll formation during greening although, as in *Euglena*, it did allow the formation of fewer and larger grana than is usual.

In *Ochromonas* (Chrysophyceae) Gibbs (1962e) studied the development of chloroplasts in cells which were brought into the light after being grown for a long period in the dark. The small proplastid of the dark-grown cells produced vesicles which became aligned into rows and then fused to form thylakoids. Multiple-thylakoid lamellae arose, either from simultaneous fusion of adjacent plates, or by addition of vesicles to thylakoids already formed. After 24 h most cells contained small chloroplasts with 3-thylakoid lamellae and by 48 h, the chloroplasts had reached normal size. It took several more days for the pigment to reach its normal amount for mature light-grown cells.

IV. Chloroplast Division

This process has been very little studied, perhaps because it is rapid and is not always linked with cell division. In organisms which have only one chloroplast the division clearly is linked with that of the cell. In *Chlamydomonas* (Goodenough, 1970) the chloroplast divides by vertical fission after mitosis is complete and cytokinesis has begun. A similar process takes place in cells of the multicellular alga *Ulva* (Løvlie and Bråten, 1968) where the chloroplast is bisected by the advancing division furrow.

In algae with numerous chloroplasts division can take place at any time and it is consequently less easy to observe. In *Euglena* Schiff and Epstein (1965) show a picture of a dividing chloroplast in which it appears that longitudinal fission has taken place, perhaps starting from both ends of the long chloroplast. In *Acetabularia* Puiseux-Dao (1966) reports that before the chloroplasts divide they are seen to contain two starch grains. The chloroplast membranes appear to become arranged in two groups around the starch grains which are now located near to opposite ends of the chloroplast. Rapid constriction takes place across the centre of the chloroplast and it divides into two by a transverse division.

V. Chloroplasts as Symbionts*

Recent work has shown that a number of herbivorous opisthobranchs, which normally feed on siphonaceous green algae, can have a symbiotic relationship with chloroplasts which they extract from the algae. Thus, Taylor (1968a) found that in *Elysia viridis*, which normally feeds upon *Codium tomentosum*, morphologically intact chloroplasts could be observed in certain cells of the animal's digestive gland. These chloroplasts had an identical fine structure to those in normal *Codium* plants and experiments showed that whilst in the host they were capable of photosynthesis for an indefinite period (Taylor, 1970). Similar associations of different animals with chloroplasts of other species of *Codium* and *Caulerpa* have been reported (Kawaguti and Yamasu, 1965; Trench *et al.*, 1969) and this suggests that it may be a relatively common phenomenon amongst marine animals.

There are a number of flagellates, generally of uncertain systematic position, which appear to utilize whole blue-green algae in place of chloroplasts. The fine structure of these associations has been extensively investigated, particularly as they appear to support the theory of the cyanophytic origin of chloroplasts first propounded nearly 100 years ago and recently revived (see Margulis, 1970; Taylor, 1970; Lee, 1972; see also Chapter 12). The main organisms involved are *Geosiphon pyriforme* (Schnepf, 1964), *Glaucocystis nostochinearum* (Schnepf *et al.*, 1966; Hall and Claus, 1967; Bourdu and Lefort, 1967; Echlin, 1967) and *Glaucosphaera vacuolata* (Richardson and Brown, 1970), which have been thought to be apochlorotic green algae, and *Cyanophora paradoxa* (Hall and Claus, 1963; Mignot *et al.*, 1969) which has been variously ascribed to the Cryptophyceae and Dinophyceae. In each case the cyanelles consist of little more than

* See also Chapter 12.

a membrane surrounding a number of concentric thylakoids. A central DNA-containing area may be present. The evolutionary scheme worked out by Lee (1972) would place *Cyanophora*-type organisms as representing the initial type of aquisition of cyanophytic 'chloroplasts'. From these he thinks that the 'green', 'yellow-brown' and 'red' types of chloroplast would have evolved along more or less separate lines. Clearly very much more biochemical and structural work is necessary before this interesting theory can be accepted as a fact.

5. *The Pyrenoid*

This term was first used by Schmitz (1882) to describe the dense and often refractive bodies associated with the chloroplasts of many algae. Pyrenoids are now known to be present in some representatives of most algal classes and in probably all the members of some, such as the Chlorophyceae. Although they are extremely varied in morphology they have certain basic structural similarities. The first common feature is that pyrenoids are always part of a chloroplast. They may be entirely embedded in the plastid or they may protrude from it but, even then, the pyrenoid is bounded by a continuation of the chloroplast envelope. The second common feature concerns the matrix or ground substance of the pyrenoid. In all cases this consists of proteinaceous material which has a rather uniform granular appearance. In a few cases the matrix is para-crystalline. Thus we can define a pyrenoid as a distinct part of a chloroplast which is mainly occupied by a proteinaceous matrix. Pyrenoids have been previously reviewed by Leyon (1954), Gibbs (1962a,b), Manton (1966d) and more recently by Griffiths (1970).

For convenience of description in the account which follows, the pyrenoids have been divided into a number of somewhat arbitrary categories. There are probably intergradations between many of the types as will almost certainly be revealed by work on the numerous algae which have not yet been studied by electron microscopy.

I. Types of Pyrenoid

A. SIMPLE INTERNAL PYRENOID

1. *No thylakoids entering the pyrenoid*

This appears to be the most simple type of pyrenoid. It consists of a fusiform or flattened structure which generally lies across the centre of the plastid (Figs 5.1–5.3). In transverse section it is often oval. Where an organism has several chloroplasts each will have its own pyrenoid and thus each cell can contain many pyrenoids. The granular pyrenoid matrix is often bounded by a narrow dark layer which may be extended to form a tail or projection at the ends of the pyrenoid. This has been described as a membrane but it is clearly thinner than a unit membrane (Parke *et al.*, 1971). In some organisms such as *Glenodinium foliaceum* (Dodge and Crawford, 1971a) tubular structures may be seen in the centre of the pyrenoid. These do not appear to be connected to the lamellar

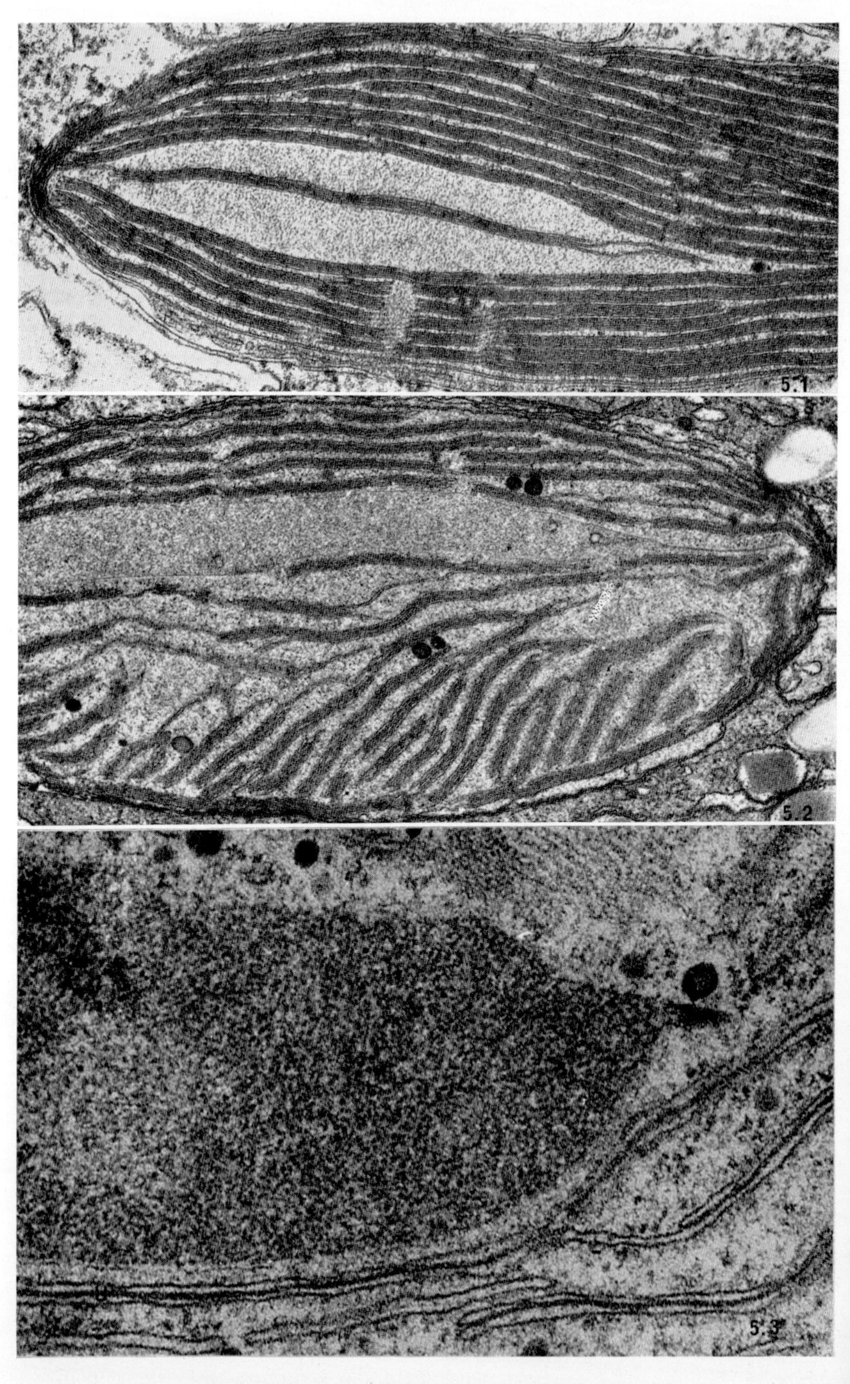

FIGS 5.1–5.3. Simple internal pyrenoids. (5.1) *Melosira* (Bacillariophyceae) showing a single lamella passing through the centre of the pyrenoid (×42,000) (Micrograph provided by R. M. Crawford); (5.2) *Glenodinium foliaceum* (Dinophyceae) showing the 'tail' at the end of the simple pyrenoid (×29,400); (5.3) *Gymnodinium* pyrenoid at much higher magnification to show the granules of which it is made up. (×78,000) (After Dodge and Crawford, 1971a.)

system of the chloroplast. The matrix consists of evenly dispersed granular material (Fig. 5.3) and only in one organism, *Chrysochromulina acantha* (Haptophyceae) (Leadbeater and Manton, 1971) has a regular para-crystalline packing of the matrix been observed. The rows of granules here were arranged with a repeating pattern at 12 nm centres. There is no starch or other reserve material associated with these pyrenoids and the chloroplast lamellae normally lie close to the sides of the pyrenoid.

This type of pyrenoid has been found in several members of the Dinophyceae (Dodge and Crawford, 1971a) and *Phaeocystis* (Parke *et al.*, 1971) and *C. acantha* (Leadbeater and Manton, 1971) of the Haptophyceae. In *Dictyocha fibula* (Van Valkenburg, 1971b) a member of the Chrysophyceae, the pyrenoid is often situated at one side of the plastid.

2. *Thylakoids entering the pyrenoid*

This type of pyrenoid is similar to that described above but with one difference. There is a somewhat undulating membranous structure, probably consisting of one or two thylakoids, running through the centre of the matrix. Pyrenoids like this are common in the diatoms (Drum and Pankratz, 1964; Manton and von Stosch, 1966; Esser, 1967; Drum, 1969; Taylor, 1972). In one diatom, *Cymbella affinis* (Drum and Pankratz, 1964) the fusiform pyrenoid is shown in the unusual position of lying at right angles to the chloroplast lamellae. This type of pyrenoid is quite common in the Haptophyceae where it is found in *Chrysochromulina parva* (Parke *et al.*, 1962), *C. strobilus* (Manton, 1966d) and *Prymnesium parvum* (Manton and Leedale, 1963a). It also appears to be present in the red alga *Kylinia* (Gibbs, .962a).

B. COMPOUND INTERNAL PYRENOID

1. *With random arrangement of thylakoids*

Here the pyrenoid is much larger than in the previous types and is penetrated by a number of chloroplast lamellae. As before, no starch is found near the pyrenoid. There are normally no more than one or two of these pyrenoids per cell.

In *Porphyridium cruentum* (Gantt and Conti, 1965) and *Nemalion multifidum* (Gibbs, 1962a), both red algae, the pyrenoid is situated at the centre of the rather large chloroplast. A number of single-thylakoid lamellae enter the pyrenoid and these generally follow a sinuous course through the matrix (Fig. 5.4) and may branch and rejoin. Osmiophilic droplets are often seen around the periphery of the matrix. A very similar type of pyrenoid has recently been described from the chloroplasts of the coenocytic alga *Vaucheria sphaerospora* (Descomps, 1972) and from germinating monospores of the red alga *Smithora* (McBride and Cole, 1972).

Recently Belcher and Swale (1972a) have reported the presence of 'pyrenoid areas' at the ends of the chloroplast in *Chrysococcus cordiformis* (Chrysophyceae).

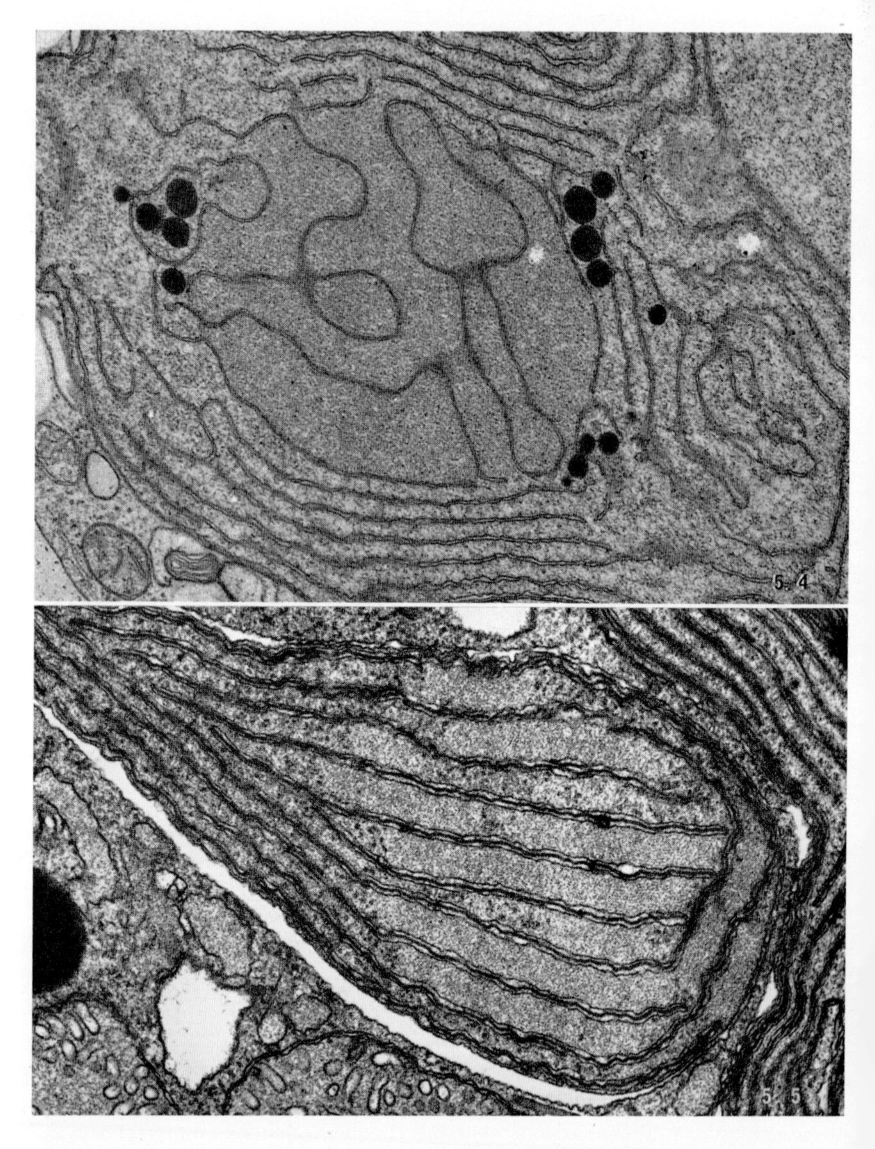

FIGS 5.4–5.5. Compound internal pyrenoids. (5.4) *Porphyridium* (Rhodophyceae). The single pyrenoid in the centre of the chloroplast which is penetrated by a number of thylakoids. Lipid droplets lie around the pyrenoid matrix (×26,000); (5.5) *Prorocentrum* (= *Exuviaella*) *mariae-lebouriae* (Dinophyceae). Here the compound pyrenoid forms between a number of 2-thylakoid lamellae. Much of the pyrenoid matrix is regularly arranged. (×39,600) (After Dodge and Crawford, 1971a.)

These areas are surrounded by the girdle lamella and are penetrated by a number of single, sinuously convoluted thylakoids which issue from the ends of the 3-thylakoid lamellae of the chloroplast. In that the granular matrix of the 'pyrenoid area' is said to be no different from that of the chloroplast stroma there must be some doubt as to whether these are really pyrenoids. In any case they appear to be unique structures.

2. *Compound pyrenoid with more or less parallel lamellae*

In many organisms with compound pyrenoids the chloroplast lamellae run in a more or less parallel course straight through the pyrenoid. Sometimes there appears to be a regular reduction in the number of thylakoids from three to two per lamella in the pyrenoid, but in the Xanthophyceae there is no such reduction.

A good example of this type is found in the dinoflagellates *Prorocentrum micans* (Kowallik, 1969) and *P.* (= *Exuviaella*) *mariae-lebouriae* (Dodge and Crawford, 1971a). Here, each cell has two pyrenoids, one situated each side of the cell, and their bulk causes the chloroplast to protrude towards the cell centre. In the pyrenoid (Fig. 5.5) the lamellae consist of pairs of thylakoids and the enlarged space between the lamellae contain regularly packed material (see later section). This type of pyrenoid is very common in the Xanthophyceae where it has been reported in the genera *Botrydium* (Falk, 1967) *Bumilleria* (Massalski and Leedale, 1969) *Mischococcus*, *Pleurochloris*, *Bumilleriopsis* (Hibberd and Leedale, 1971), *Pseudobumileriopsis* (Deason, 1971a). It is perhaps the standard type of pyrenoid in this class. Rather similar pyrenoids have been found in *Euglena granulata* (Walne and Arnott, 1967), where each plastid appears to have a pyrenoid; in the diatom *Gomphonema parvulum* (Drum and Pankratz, 1964); in the haptophyceans *Hymenomonas roseola* (Manton and Peterfi, 1969) and *Apistonema* (Leadbeater, 1970) and in the chrysophyte *Olisthodiscus* (Gibbs, 1962a; Leadbeater, 1969). Thus, this type is present in members of at least six different algal classes.

C. STALKED PYRENOIDS

1. *With single stalk*

This type of pyrenoid clearly protrudes from the chloroplast, generally towards the centre of the cell as the chloroplasts tend to occupy a peripheral position (Figs 5.6 and 5.7). In some brown algae (Evans, 1968) the pyrenoids at first project laterally from the ends of the plastids but may then bend through 180 degrees to lie beside the chloroplasts. The connection between pyrenoid and plastid is normally narrowed to form a stalk or neck and the pyrenoid matrix is bounded by chloroplast envelope plus endoplasmic reticulum, when this occurs around the chloroplast. The one exception to this rule is the Cryptophyceae where the ER does not immediately surround the pyrenoid but is found outside the polysaccharide cap. In the representatives of the Cryptophyceae,

Dinophyceae and Phaeophyceae which have this type of pyrenoid no thylakoids enter the matrix but in *Colacium* (Leedale, 1967) *Trachelomonas* (Euglenophyceae), and several species of *Chrysochromulina* (Haptophyceae) (Manton, 1966a, 1972a) a number of two-thylakoid lamellae enter the pyrenoid and terminate there.

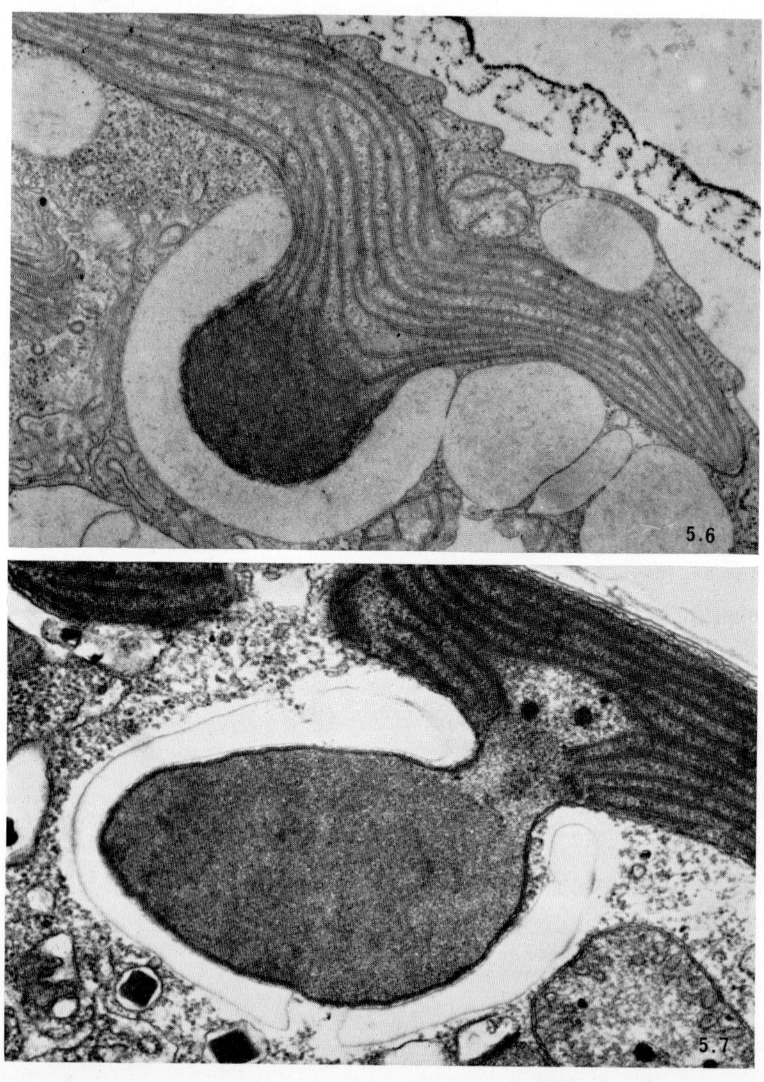

FIGS 5.6–5.7. Single-stalked pyrenoids. (5.6) *Trachelomonas* (Euglenophyceae). The pyrenoid projects into the cell and is covered by a cap of paramylon; (5.7) *Glenodinium* (Dinophyceae). In this no lamellae enter the pyrenoid and the cap consists of starch plates. (×30,000) (After Dodge and Crawford, 1971a.)

Single-stalked pyrenoids are normally closely surrounded by a cap or sheath of polysaccharide reserve material (Figs 5.6 and 5.7) which lies outside the bounding membranes. In *Chrysochromulina chiton* Manton (1966a,d) found that the cap, which is bounded by its own membrane, becomes conspicuous in the early hours of darkness (when grown in light: dark cycle) although in full daylight it may not be formed. Both in *Colacium* (Leedale, 1967) and in the brown algae (Bouck, 1965) the cap also appears to be surrounded by a membrane but in no case is the origin of the membrane clear. In the Cryptophyceae the cap forms between the pyrenoid envelope and the ER sheath and other starch grains also occur in this cavity. In the Dinophyceae (Dodge and Crawford, 1971a) no membrane of any type is found outside the cap and in some species, such as *Glenodinium hallii* no regular cap is found but several flat starch plates lie near the pyrenoid. This also seems to be the case in some members of the Eustigmatophyceae (Hibberd and Leedale, 1972) where the flattened plates give the pyrenoid an angular appearance. In *Cryptomonas cryophila* (Taylor and Lee, 1971) the pyrenoid is surrounded by two distinct hemispherical starch grains and in *Cryptomonas* sp. (Glazer *et al.*, 1971) there are several small grains around the pyrenoid.

In *Vaucheria woroniniana* (Xanthophyceae) Marchant (1972) has reported the presence of single-stalked pyrenoids in the short filaments of recently germinated aplanospores. There was no polysaccharide cap and no thylakoids entered the matrix which was described as being similar to chloroplast stroma. No pyrenoids were found in the mature filaments so the pyrenoids would appear to be associated only with this particular growth phase. This may also be the case in some brown algae (Bourne and Cole, 1968) where stalked pyrenoids have been found in sporelings but not in the cells of the mature plant. The reverse seems to occur in *Macrocystis* (Chi, 1971) where pyrenoids are absent from the zoospores but present in vegetative cells and this is also the situation in members of the Eustigmatophyceae (Hibberd and Leedale, 1972).

It will be seen from the above that single-stalked pyrenoids are found in many of the algal classes. In the Phaeophyceae (Bouck, 1965; Evans, 1966, 1968; Cole and Lin, 1968; Cole, 1969, 1970; Chi, 1971) they are perhaps the most important type. They are also found in a few members of the Cryptophyceae (Wehrmeyer, 1970b; Taylor and Lee, 1971; Glazer *et al.*, 1971) Dinophyceae (Dodge, 1968; Dodge and Crawford, 1971a) Haptophyceae (Manton and Leedale, 1961a; Manton, 1966a, 1972a) Euglenophyceae (Leedale, 1967) the Xanthophyceae (Marchant, 1972) and the Eustigmatophyceae (Hibberd and Leedale, 1972).

2. *Multiple-stalked pyrenoid*

A development of the stalked pyrenoid is seen in some members of the Euglenophyceae (Mignot, 1966; 1967), Dinophyceae (Dodge and Crawford, 1971a), Cryptophyceae (Gibbs, 1962a; Dodge, 1969b) and Rhodophyceae (Evans, 1970; Wehrmeyer, 1971) where the pyrenoid is supported or subtended by two or more branches of the chloroplast reticulum. The variations of structure

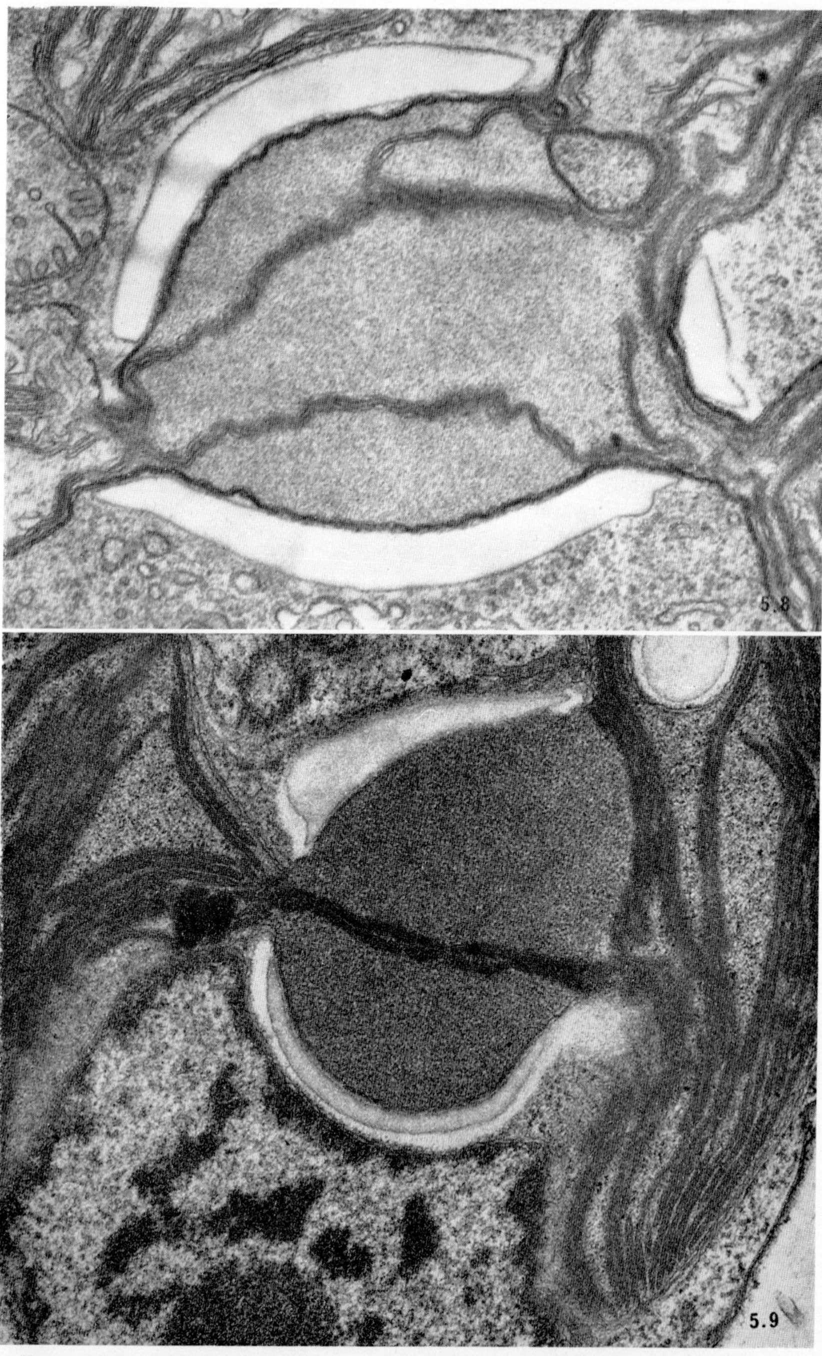

FIGS 5.8–5.9. Multiple-stalked pyrenoids. (5.8) *Amphidinium carterae* (Dinophyceae). The single pyrenoid can be seen to be suspended by at least three projections from the chloroplast (×30,000) (After Dodge and Crawford, 1971a); (5.9) *Chroomonas* (Cryptophyceae). The starch plates here are situated outside the pyrenoid and chloroplast envelope but underneath the chloroplast endoplasmic reticulum sheath. (×39,600)

found here are similar to those found in single-stalked pyrenoids. Thus, in the red alga *Rhodella maculata* (Evans, 1970) and *R. violaceum* (= *Porphyridium violaceum*) (Wehrmeyer, 1971) the pyrenoid is not penetrated by any thylakoids and has a number of discrete starch grains lying around it. In *R. maculata* occasional protrusions of the nuclear envelope penetrate into the pyrenoid. In these examples each cell contains only one pyrenoid. This is also the case in the dinoflagellate *Amphidinium carterae* (Dodge and Crawford, 1968) where the large pyrenoid (Fig. 5.8) is suspended above the nucleus by two or three branches of the peripheral chloroplast. A number of two-thylakoid lamellae pass completely through the pyrenoid and around it there is a thick starch sheath which usually appears to be in two or more segments. In another member of this class, *Peridinium trochoideum* (Dodge and Crawford, 1971a), there are usually several multiple-stalked pyrenoids but these contain only occasional chloroplast lamellae and polysaccharide is present as somewhat detached grains.

In the Cryptophyceae pyrenoids of this type have been seen in *Rhodomonas lens*, *Cryptomonas sp.* (Gibbs, 1962a) and in *Chroomonas mesostigmatica* (Dodge, 1969b) (Fig. 5.9). In the latter case the pyrenoid was normally crossed by a single 2-thylakoid lamella and a number of curved starch plates were present around the pyrenoid but beneath the ER sheath.

Several species of *Euglena* have pyrenoids which perhaps come into this category because a polysaccharide sheath (Paramylon) is always present. However, the stalks are very short and thick and the pyrenoids bear a strong resemblance to those placed in Section B1. In *Euglena gracilis* Gibbs (1960) and Ploaie (1971) showed that the pyrenoid was a differentiated region which is narrower than the rest of the chloroplast. It contained a dense homogeneous matrix which was crossed by numerous 2-thylakoid lamellae, contrasting with the 3-thylakoid lamellae of the typical parts of the chloroplast. Mignot (1966; 1967) reported a similar arrangement in *E. stellata* and *E. splendens*.

D. PYRENOID WITH NUCLEAR OR CYTOPLASMIC INVAGINATIONS

In some of the types of pyrenoid already described, occasional cells are discovered in which the pyrenoid is penetrated by invaginations of the nucleus or cytoplasm. Thus, as already noted in the red alga *Rhodella maculata* (Evans, 1970), the nucleus may extend into the pyrenoid. In *Cryptomonas* sp. (Glazer *et al.*, 1971) and *C. reticulata* (Lucas, 1970a) finger-like cytoplasmic invaginations are found in the pyrenoid. In these examples invaginations into the pyrenoid appear to be an irregular feature. However, there are a number of algae in which they are always present and these are included in this section.

1. *In algae with no starch in the chloroplasts*

In the dinoflagellate *Heterocapsa triquetra* (Dodge and Crawford, 1971a) there is a large multiple-stalked pyrenoid. This never contains chloroplast lamellae but it is perforated by numerous invaginations of its outer envelope (Figs 5.12 and 5.13). These tubes are about 60 nm in diameter, are circular in section, and appear to be bounded by only two membranes whereas the chloroplast is bounded by three. The number of invaginations appear very variable and from 12 to 60 profiles per pyrenoid have been counted. Although the tubes appear to open into the cytoplasm they contain a rather sparsely granular material and no ribosomes or other cytoplasmic organelles have been definitely identified within them.

In *Chrysamoeba radians* (Chrysophyceae) (Hibberd, 1971) a large pyrenoid is situated in the hollow of the deeply curved chloroplast. Immediately above it lies the nucleus. The pyrenoid is penetrated by a number of channels, 125–250 nm in diameter, containing vesicular inclusions which are probably part of the periplastidial network. The channels are lined by the two membranes of the chloroplast envelope and, as the ER sheath remains outside, they are not in direct communication with the cytoplasm.

2. *In chloroplasts which contain starch*

This type of pyrenoid appears to be absolutely characteristic of the Oedogoniales (Chlorophyceae) and is also found in *Ankyra* (Chlorophyceae) and many members of the Prasinophyceae. In *Oedogonium* (Hoffman, 1968a,b) and *Bulbochaete* (Retallack and Butler, 1970a) the pyrenoids are more or less spherical structures which are penetrated by a system of ramified cytoplasmic channels lined by two membranes continuous with those of the chloroplast envelope. At times the pyrenoid appears to be almost dissected into pieces by the channels which often contain ribosomes and membranous structures. In *Ankyra*, a chlorococcalean green alga, the large pyrenoid is penetrated by wide cytoplasmic channels (Swale and Belcher, 1971).

In flagellate members of the Prasinophyceae, such as *Platymonas* (Gibbs, 1962a; Manton and Parke, 1965; Manton, 1966d) a single pyrenoid is situated at the centre of the cup-shaped chloroplast. There is usually a single large branched cytoplasmic invagination into this pyrenoid (Figs 5.10 and 5.11) but in *P. convolutae* (Parke and Manton, 1967) and *P. impellucida* (McLachlan and Parke, 1967) there are a number of cytoplasmic invaginations which, as in *Oedogonium*, enter from various directions. These channels often contain ribosomes.

A slightly different situation exists in another member of the Prasinophyceae, *Prasinocladus* (Parke and Manton, 1965). Here, the pyrenoid lies immediately beneath the nucleus and a wide invagination, bounded by chloroplast envelope and nuclear envelope, enters the centre of the pyrenoid and ends as a few short branches. Granular nucleoplasm fills the channels so produced. No thylakoids enter the pyrenoid and the matrix is bounded by a discontinuous double membrane (a single thylakoid?) outside which lie the curved starch grains making up the cup-shaped pyrenoid sheath.

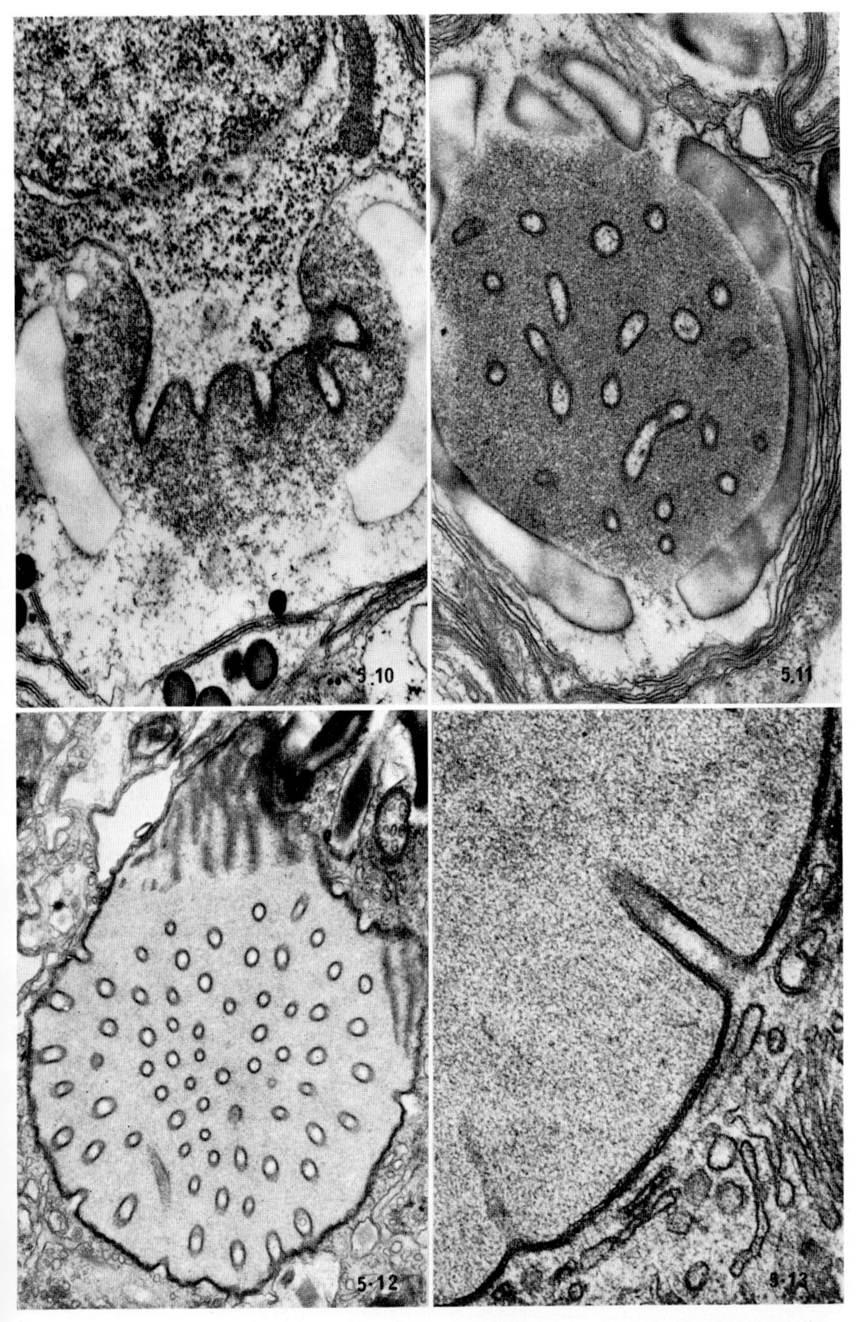

FIGS 5.10–5.13. Invaginated pyrenoids. 5.10 shows a vertical section and 5.11 a transverse section of the central pyrenoid of *Platymonas* (Prasinophyceae). Cytoplasmic invaginations penetrate deep into the pyrenoid from the side adjacent to the nucleus. (5.10, ×25,200); (5.11, ×26,400). Figures 5.12 and 5.13 show a general view and a higher magnification of the pyrenoid of *Heterocapsa* (Dinophyceae) which is penetrated by numerous cytoplasmic invaginations. (5.12, × 20,000); (5.13, ×59,000)

E. PYRENOID ENTIRELY EMBEDDED IN A STARCH-CONTAINING CHLOROPLAST

1. *Pyrenoid normally with starch sheath*

This is the usual type of pyrenoid found in most members of the Chlorophyceae. It has now been described from some 30 species belonging to the orders Volvocales, Chlorococcales, Ulotrichales, Chaetophorales, Cladophorales, Caulerpales, Siphonocladiales and Conjugales. A few members of the Prasinophyceae also have pyrenoids of this type.

In members of the Volvocales and Chlorococcales each cell usually possesses a more or less cup-shaped chloroplast which occupies the posterior end of the cell. The pyrenoid is embedded in a central position in this plastid (Fig. 5.15) and is usually penetrated by lamellae on all sides. These lamellae normally consist of pairs of thylakoids and they terminate before reaching the centre of the pyrenoid. At times they appear to be tubular rather than lamellar structures. Between the pyrenoid matrix and the chloroplast proper is found the starch sheath which may either consist of a few large curved plates or a number of smaller discoidal grains.

In cells of certain other orders of green algae there is either a plastid reticulum or the chloroplasts are large and variously shaped. In *Spirogyra* (Dawes, 1965; Fowke and Pickett-Heaps, 1971) the spiral chloroplast contains numerous pyrenoids which fit the type already described above but with one difference, the lamellae appear to run right through the pyrenoid, often along a sinuous course. In the desmid *Micrasterias* the large pyrenoids are traversed by numerous single-thylakoid lamellae (Drawert and Mix, 1962), and a similar pyrenoid is also found in *Closterium* (Chardard, 1971). In these two desmids it appears that the 3-thylakoid chloroplast lamellae splay out into single thylakoids at the edge of the pyrenoid matrix. In *Cladophora* the chloroplast reticulum contains many simple pyrenoids (Fig. 5.14). No obvious lamellae enter the pyrenoid although small tubular structures have been observed. No thylakoids were observed in the similar pyrenoids of *Urospora*, *Caulerpa* (Hori and Ueda, 1967) and *Scenedesmus* (Gibbs, 1962b; Bisalputra and Weier, 1964b; Dodge and Crawford, unpub. obs.).

In two genuine members of the Prasinophyceae, *Halosphaera* (Manton, 1966d) and *Mesostigma* (Manton and Ettl, 1965), and *Pedinomonas* (Belcher, 1968b) which is placed by some authors in the related class Loxophyceae, there are pyrenoids of this type. In *Halosphaera* the pyrenoid is elongated rather than globular, but in all cases the pyrenoid is embedded in the plastid, is surrounded by a starch sheath, and is traversed by a number of 2-thylakoid lamellae.

2. *Pyrenoids not surrounded by starch*

In some green algae a compound pyrenoid is found which consists of a large area of plastid modified in various ways, and swollen, but not very distinct from the rest of the chloroplast. In *Carteria acidicola* (Joyon and Fott, 1964) and

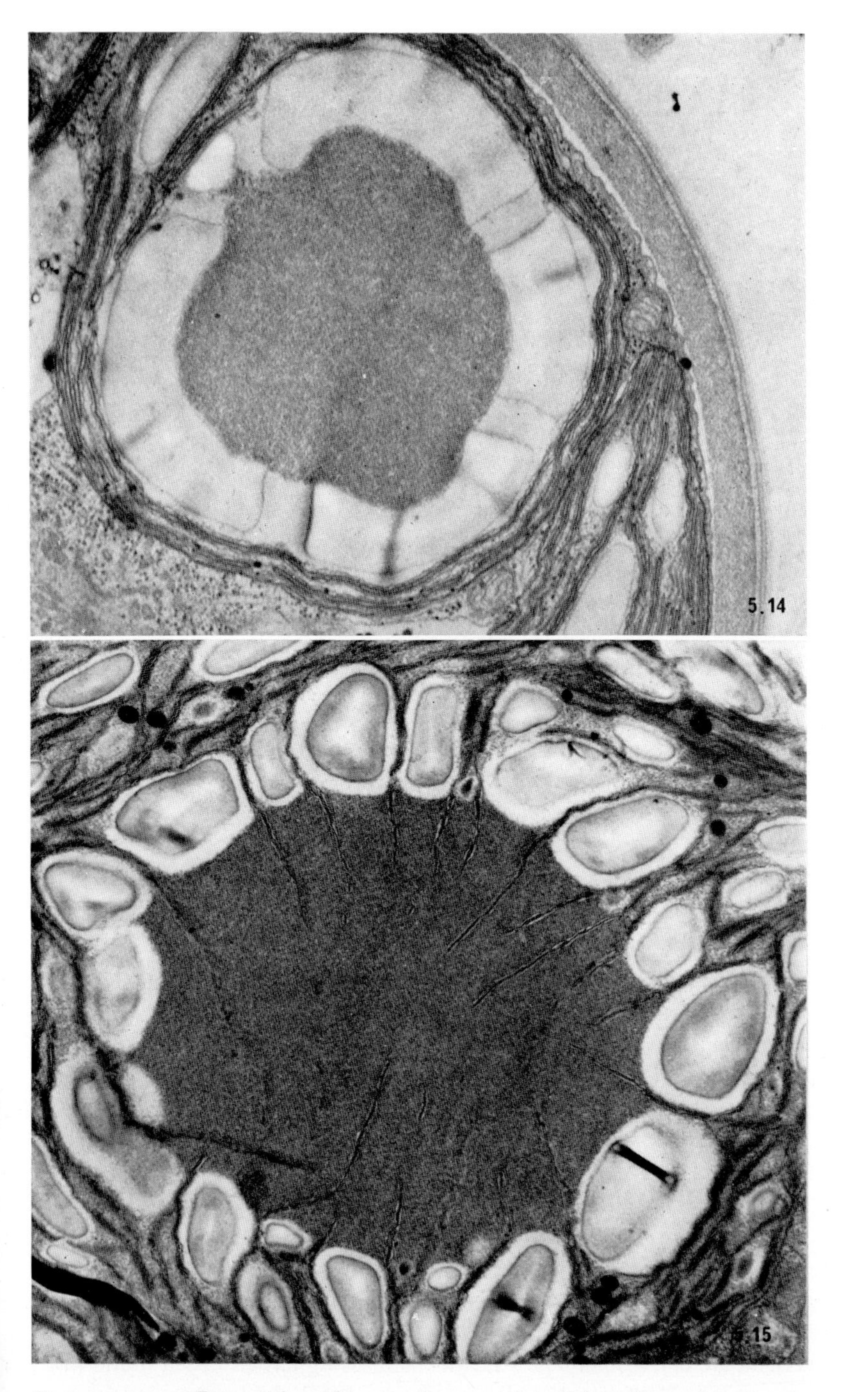

FIGS 5.14–5.15. Pyrenoids within starch-containing chloroplasts of green algae. (5.14) *Scenedesmus* (?) No lamellae enter the pyrenoid and the starch sheath is composed of a few large segments. (× 24,000); (5.15) *Chlamydomonas*. Numerous chloroplast lamellae enter the pyrenoid matrix and the starch sheath consequently consists of very small grains. (× 13,200)

C. quadrilobata (Fott, 1967) the pyrenoid consists of a region of the chloroplast in which the matrix between adjacent lamellae is much thicker than normal and is composed of regularly packed material. The lamellae are parallel to each other thus giving the pyrenoid a very uniform appearance. Discoid starch grains are mainly found between the lamellae in the normal part of the chloroplast. Apart from this the pyrenoid looks very similar to those reported in type B1 above.

Somewhat similar pyrenoids are found in the green alga *Trebouxia* (Fisher and Lang, 1971a,b; Jacobs and Ahmadjian, 1971) which is normally only found as the phycobiont in a number of lichens. Here, the large pyrenoid is situated at the centre of the stellate chloroplast. Numerous, mainly single, thylakoids run through the pyrenoid, along a sinuous course in *Trebouxia* from the lichen *Cladonia* but in the alga from *Ramalina* the single thylakoids were mainly swollen to form vesicular structures. Large numbers of lipid globules (pyrenoglobuli) were either found around the periphery of the pyrenoid or associated with the intrapyrenoidal thylakoids. It is not clear whether the differences here are due to cultural conditions or are species characteristic. A similar pyrenoid has been found in the green alga *Leptosira* (Wujek, 1971) which, also, is usually a lichen phycobiont.

It is of interest that the only group of non-algae which possess pyrenoids, the bryophyte order Anthocerotales (Burr, 1970), have multiple pyrenoids which fit into this type. In the members of that order the pyrenoid consists of a central portion of plastid in which the 2- or 3-thylakoid lamellae are kept apart by a dense matrix. Occasional starch grains may appear at the edge of the pyrenoid but mostly they are found in the normal part of the chloroplast.

II. The Pyrenoid Matrix

Typically the matrix or ground substance of the pyrenoid consists of rather uniformly dense granular material. Most authors agree that it is composed of protein, and ribosomes are not normally seen in it. One report (Esser, 1967) suggests that DNA fibrils are attached to the single thylakoid which runs through the pyrenoid of *Streptotheca*, (Bacillariophyceae) but normally DNA patches are not observed in pyrenoids.

In a few algae the matrix shows a regular, paracrystalline arrangement of the subunits. In the diatom *Acnanthes* (Holdsworth, 1968) the lattice is composed of spherical units *c.* 5 nm diameter. At times these are seen in straight rows in which the subunits are packed closely in one direction, or linked by short bridges. Sometimes they show hexagonal packing, when the general appearance is as of two intersecting sets of lines. It is suggested that only parts of the pyrenoid are crystalline. The straight-line packing has also been found in parts of the pyrenoid of another diatom *Cocconeis* (Taylor, 1972) and crystalline areas appeared in *Navicula peliculosa* after treatment with colchicine (Coombs *et al.*, 1968b). A regular lattice has been found in a member of the Haptophyceae,

Chrysochromulina acantha (Leadbeater and Manton, 1971) which, like the diatoms described above, also has a simple internal pyrenoid. Here, the subunits are arranged in straight lines, parallel to the long axis of the pyrenoid. The periodicity of adjacent lines is 12 nm.

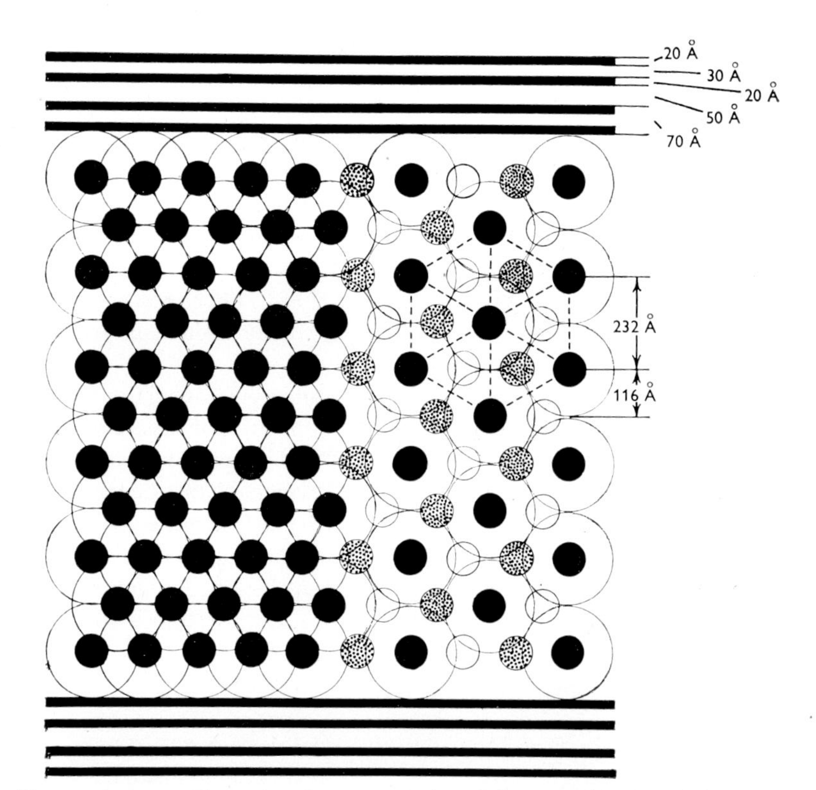

FIG. 5.16. A two-dimensional representation of the crystal lattice of the pyrenoid of *Prorocentrum micans* (Dinophyceae). The left hand side shows projections of three superimposed close-packed layers which are shown singly, and differently shaded, on the right. The lines top and bottom represent chloroplast lamellae which traverse the pyrenoid (cf. Fig. 5.5). Reproduced, with permission, from Kowallik (1969).

In the dinoflagellates *Prorocentrum micans* (Kowallik, 1969) and *P.* (= *Exuviaella*) *mariae-lebouriae* (Dodge and Crawford, 1971a; Dodge and Bibby, 1973) the matrix between the numerous lamellae which traverse the pyrenoid has an even thickness and a very regular structure which Kowallik has analysed in detail. He concludes that between adjacent lamellae there appear to be eleven rows of *c.* 23 nm diameter globular units. The units are not arranged in parallel layers, however, this appearance being the result of superimposition of several images in the thickness of the section as explained in Fig. 5.16. The basic arrangement of the units is thought to be in the form of a cubic face-centred lattice.

A regular pyrenoid matrix has been observed in some species of *Carteria* (Joyon and Fott, 1964) and this is said to consist of five layers of 15 nm granules between adjacent lamellae. At times there appear to be disc-like structures at right angles to the lamellae but this is probably a fixation artifact. Crystalline bodies have been observed in the pyrenoid of *Chlorella*, (Bertagnolli and Nadakavukaren, 1970b) but these are so distinct from the usual form of the matrix that it would seem likely that they have formed as a result of abnormal growth conditions. Evans (1966) reported the presence of crystalline areas in the pyrenoids of some brown algae but these were not illustrated or described in detail.

Recently the pyrenoid of the green alga *Eremosphaera* has been isolated and partially characterized by physical and enzymic techniques (Holdsworth, 1971). In the electron microscope this pyrenoid has a homogeneous granular matrix with no obvious crystalline areas. It is penetrated by one or two thylakoids and separated from the chloroplast by a wide starch sheath. Analysis of the extracted pyrenoid material showed the presence of considerable amounts of protein and the probable absence of nucleic acids. Two of the possibly 16 proteins present accounted for 90% of total protein and these had molecular weights estimated as 59,000 and 12,300. The physical and enzymic properties of these two proteins were found to closely resemble the properties of chloroplast fraction I protein which also has two protein components. Enzymes of the Calvin-Bassham pathway were found in the pyrenoid protein extract and these are also present in fraction I protein. Thus it is suggested that the pyrenoid might serve as a store of enzymes particularly for the use of newly formed cells. In higher plants a storage region is probably provided by the stroma centre which is packed with fraction I protein.

III. The Development of Pyrenoids

This takes place in several ways. In many organisms the pyrenoid divides either before or during cell division so that daughter cells share out the pyrenoid material of their parent. This process has been described for *Chlamydomonas* (Goodenough, 1970) where both the chloroplast and the enclosed pyrenoid divide following nuclear division. In *Chrysochromulina chiton* (Manton, 1966a) the stalked pyrenoid becomes bilobed and then divides just prior to nuclear and cell division. A slightly different situation is found in some brown algae (Evans, 1966) where the projecting pyrenoids do not appear to divide but new pyrenoids are budded off from the base of existing ones just before cell division.

In many green algae the pyrenoids disappear at cell division and are then reformed in the daughter cells. The first detailed study of this was carried out on *Scenedesmus* (Bisalputra and Weier, 1964b). Just before division the pyrenoid is very large with a distinct starch sheath. The whole structure disappears as the cell divides. In the daughter cells the new pyrenoid is first recognized as an accumulation of dense material between two broad bands of stacked chloroplast lamellae and it is flanked on either side by two small round starch grains. As the pyrenoid enlarges so the starch sheath extends to keep pace with it and also thickens. No thylakoids enter the pyrenoid matrix. A somewhat similar situation exists in *Tetracystis* (Brown and Arnott, 1970). Here, the pyrenoid breaks down during the formation of zoospores. The ground substance appears to be dissolved and is replaced by reticulate fibrillar structures. The starch plates disappear but much starch appears elsewhere in the plastid. New pyrenoids arise late in the division cycle after most other structures have been formed and they develop in a clear area of chloroplast which looks rather like a DNA containing zone. The new pyrenoid core arises in close proximity to stroma starch grains but then two new hemispherical starch plates are developed as the pyrenoid enlarges. By the time the zoospore has settled and begun to enlarge into a vegetative cell two convoluted thylakoids, which are typical of this organism, have formed between the matrix and the starch sheath. Much the same process appears to take place in *Oedogonium* (Hoffman, 1968a) and *Bulbochaetae* (Retallack and Butler, 1970a) where new pyrenoids also arise during zoospore formation. They first appear as small areas (*c.* 0·3 μm diameter) containing dense granules. These areas increase in size and in so doing force aside the chloroplast lamellae. By the time they have reached their normal size of about 5 μm diameter starch grains are present around the pyrenoid. No information was obtained about the development of the cytoplasmic channels in these organisms although it is suggested that this may be by 'active' invagination of the pyrenoid by the chloroplast envelope.

From the foregoing it will appear that we have as yet no information about the origin or development of many of the pyrenoid types and such information as we have is mainly confined to zoospore formation or binary fission.

IV. Pyrenoids and Taxonomy

Reference to Table IV will show that several algal classes possess many types of pyrenoid. Thus we find that the Dinophyceae have five fairly distinct types, the Haptophyceae have four, and the Rhodophyceae have three. In the Chrysophyceae most genera have no pyrenoids but in three which do have them, the pyrenoids are all of different types. Clearly, in these classes there is no typical pyrenoid type. The Phaeophyceae, however, with their single-stalked pyrenoids and the Xanthophyceae with mainly multilamellar internal pyrenoids do both appear to have pyrenoids which are of taxonomic importance. Similarly, amongst the green algae the Oedogoniales, which some authors (cf. Round, 1971b) would raise to independent class rank, have a distinctive type of invaginated pyrenoid whilst the Chlorophyceae as a whole are typified by large starch-sheathed pyrenoids. The taxonomic implications of absence of pyrenoid lamellae, and the two types of pyrenoid development as described in the previous section are not clear as yet. Only one class, the Chloromonadophyceae, completely lacks pyrenoids although there are representatives of most classes which also appear to

TABLE IV

The distribution of pyrenoid types amongst the algal classes

Pyrenoid types:	Internal Simple		Internal Compound		Stalked		Invaginated		In starch-containing plastid	
	No lam.	1 lam.	Random lam.	Parallel lam.	Single	Multiple	In non-starchy chloroplast	In starchy chloroplast	Single central	Multiple or compound
Section:	A1	A2	B1	B2	C1	C2	D1	D2	E1	E2
Chlorophyceae								+++	+++	+++
Prasinophyceae								+++	+	
Euglenophyceae				+++	+	+				
Chloromonadophyceae										
Eustigmatophyceae					+++					
Xanthophyceae			+	+++	+					
Chrysophyceae	+		+	+			+			
Haptophyceae	+	+		+	+					
Bacillariophyceae		+++		+						
Phaeophyceae					+++					
Dinophyceae	++			+	+	++	+			
Cryptophyceae					+	++	+			
Rhodophyceae		+	++			+				

+ = known from one species; ++ etc. = known from several species.

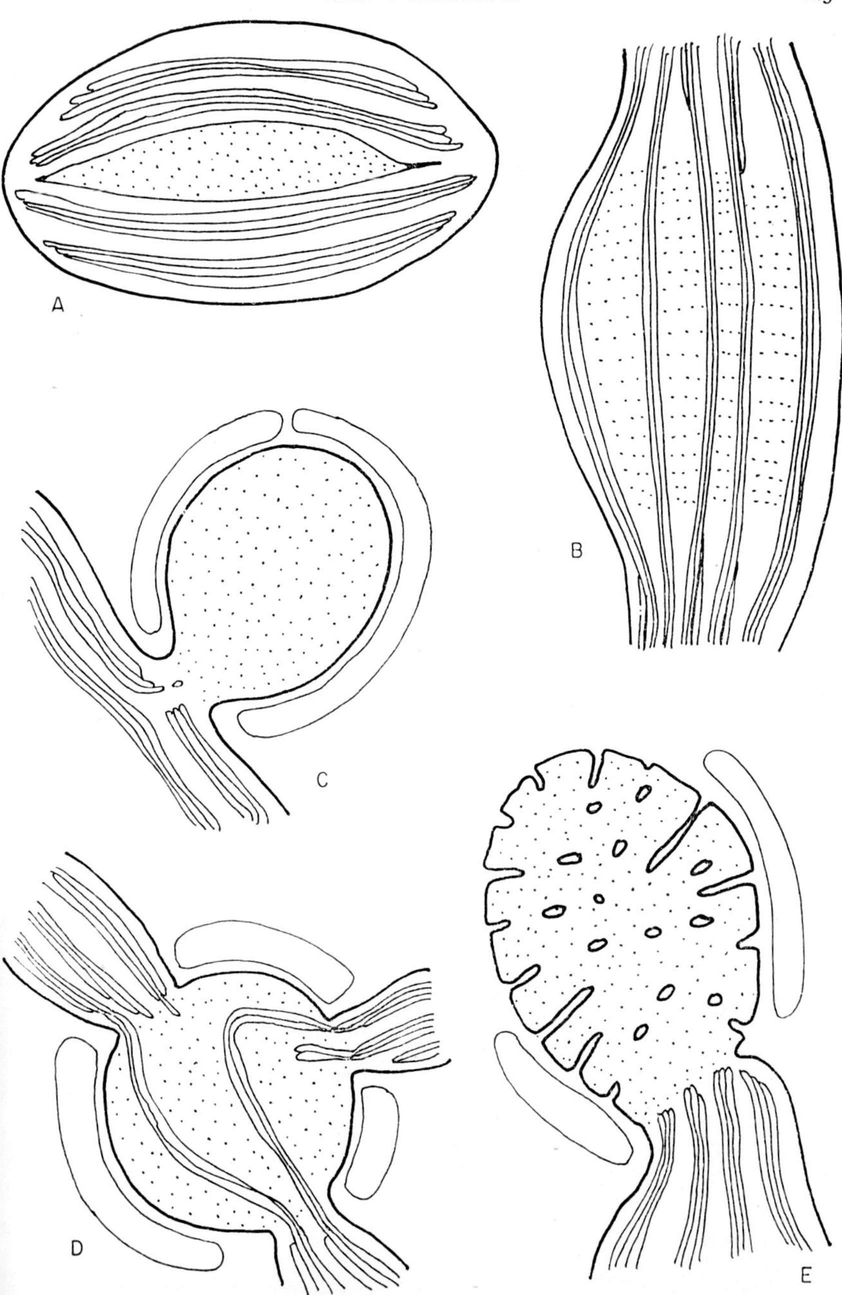

FIG. 5.17. Diagrams to illustrate the various types of pyrenoid which have been found in the Dinophyceae. Similar pyrenoids are also present in various other algal classes. A, Simple internal type; B, compound internal pyrenoid; C, single-stalked pyrenoid; D, multiple-stalked type; E, invaginated pyrenoid. (After Dodge and Crawford, 1971a.)

have none. The problem of the definition of pyrenoids often makes it difficult to be certain when a pyrenoid is absent for it is not easy to say when a rather uniformly granular area of stroma has become a simple pyrenoid. Also it appears that in some organisms pyrenoids are present for some stage of the life history but absent from the main vegetative phase. This brings problems when the life histories of so many algae are still not clearly established. One thing seems certain, as more algae are examined by electron microscopy further types of pyrenoid will surely be discovered.

6. *The Eyespot*

Many motile algae and motile stages (gametes and zoospores) of non-motile algae possess a structure termed the eyespot or stigma. This is an organelle which consists of a number of osmiophilic globules (often called granules) containing orange or red carotenoid pigment. It is generally located in a definite position in the cell in any particular species and, as will be seen, each class of algae in which they are found tends to have the eyespot in a characteristic postion. Electron microscopical studies have shown that there are a number of different types of eyespot and the 20 or so papers describing algal eyespots, which have been published since the subject was last reviewed (Dodge, 1969a) have tended to confirm the classification of eyespots which was then proposed. The present chapter follows a similar plan to the previous review with the exception of the section dealing with the Dinophyceae.

For the sake of completeness it should be mentioned that in two classes which mainly consist of motile unicells, Chloromonadophyceae and Haptophyceae, eyespots have not been found, with the exception of one possible member of the latter class. Eyespots are also relatively uncommon in the Dinophyceae and Cryptophyceae whereas virtually every flagellated cell in the Chlorophyceae, Prasinophyceae, Euglenophyceae, Xanthophyceae, Chrysophyceae and Phaeophyceae has an eyespot.

I. The Types of Eyespot

TYPE A. EYESPOT PART OF A CHLOROPLAST BUT NOT OBVIOUSLY ASSOCIATED WITH THE FLAGELLA

This is the typical form found in two classes of green algae, the Chlorophyceae and Prasinophyceae. It is also found in the few stigmatic members of the Cryptophyceae which have been sectioned (Dodge, 1969b and unpub.; Lucas, 1970a,b).

In the green algae the eyespot is generally situated at one side of the cell, often in line with the plane of beat of the flagella. It is anything from one-quarter to one-half the length of the cell distant from the insertion of the flagella. In general this type of eyespot occupies part of the edge of the chloroplast which here tends to be closely pressed up against the plasmalemma (Fig. 6.1). Sometimes, as in *Tetracystis* (Arnott and Brown, 1967) the eyespot causes a slight bulge

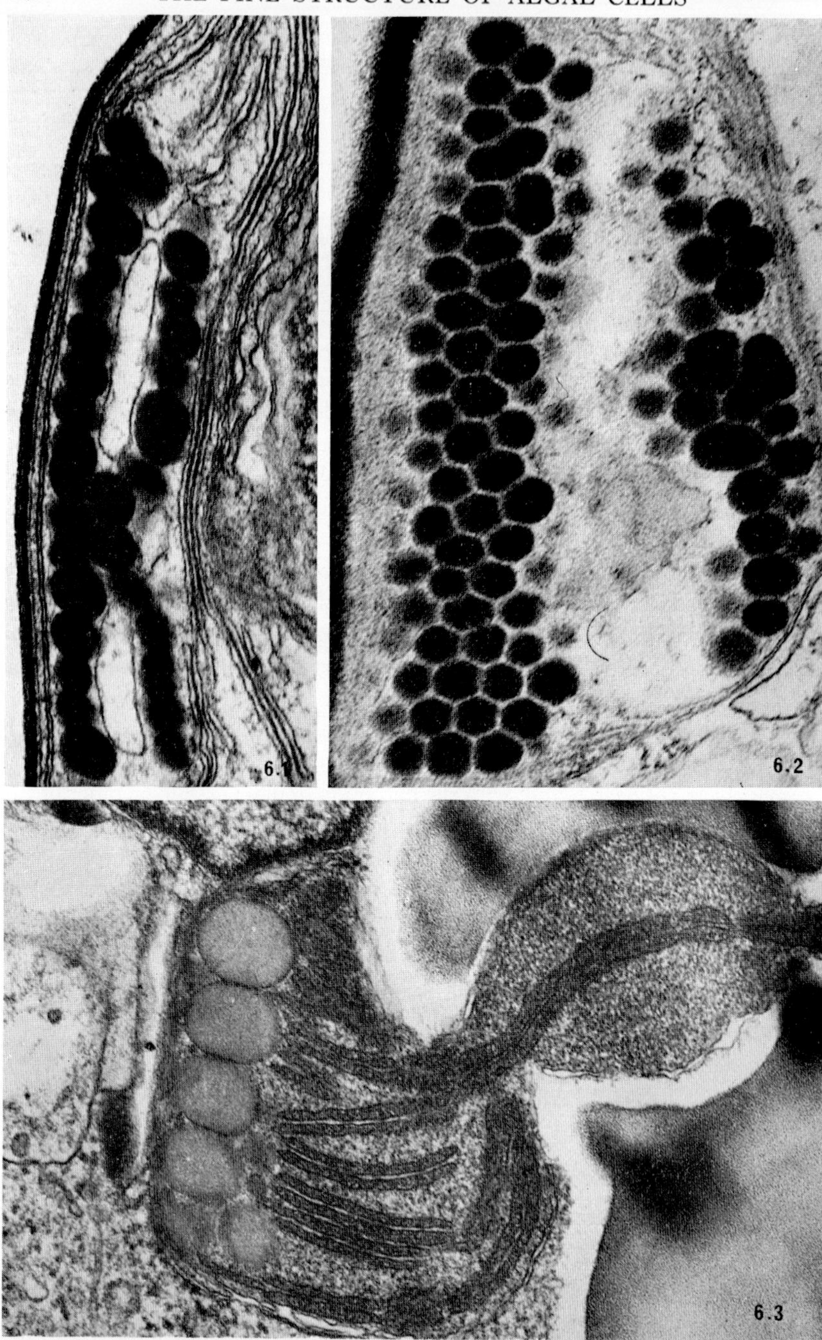

FIGS 6.1–6.2. Two views of the type A eyespot of *Platymonas* (Prasinophyceae). (6.1) In this vertical section the two layers of eyespot globules are seen to be separated by one swollen thylakoid. The eyespot lies very close to the cell membrane and wall (×59,000); (6.2) in this oblique section the two layers of globules are sectioned tangentially and their hexagonal close-packing can be seen (×59,000) (After Dodge, 1969a); (6.3) the eyespot of *Chroomonas mesostigmatica* (Cryptophyceae). This is situated at the end of a short spur of chloroplast which projects beyond the pyrenoid seen at the right of the photograph. The eyespot consists of a single layer of globules. (×49,000)

at the edge of the chloroplast but it may also, in *Pyramimonas* (Maiwald, 1971) and sometimes in *Volvox* (Pickett-Heaps, 1970a), produce a concave depression in the plastid. It has been suggested that the plasma-membrane above the eyespot is unusually thickened (Arnott and Brown, 1967) and may be specially constructed to act as a photoreceptor. Occasionally, as in *Chlorogonium* (de Haller and Rouiller, 1961) and *Volvulina* (Lang, 1963a), the eyespot is deep-seated.

Eyespots of this type consist of a variable number of layers of osmiophilic globules. By far the most usual number is one or two, but in the colonial members of the Volvocales up to nine layers have been recorded (Lang, 1963a; Hobbs, 1972). The higher numbers tend to be found in the anterior members of colonies and the number usually decreases towards the posterior of the colony where the eyespots are very small. Where there is only one layer of globules these are situated immediately under the double membrane chloroplast envelope but when more are present adjacent layers are separated by single swollen thylakoids which appear to lack any contents (Fig. 6.1).

The layers of the eyespot each contain a large number of osmiophilic globules which tend to be packed close together and arranged in hexagonal close packing (Fig. 6.2) (Arnott and Brown, 1967; Dodge, 1969a). In two organisms with only single-layer eyespots tangential sections of the cell enable the number of globules to be counted accurately and there are said to be 50–60 in *Chlamydomonas eugametos* (Walne and Arnott, 1967) and 110 in *Tetracystis excentrica* (Arnott and Brown, 1967). In the motile female gamete of *Bryopsis* the eyespot is particularly extensive (Fig. 11.3) being about 4 μ in diameter and 57 globules of varying sizes can be counted in a single section (Burr and West, 1970). This suggests that in the complete eyespot there are something like 3000 globules. By way of contrast the male gamete contains no eyespot (Burr and West, 1970). In the multilayered eyespot of *Volvox aureus* (Dolzmann and Dolzmann, 1964) 100 granules can be counted in a single vertical section, indicating that there may be about 1000 granules in the complete eyespot. Probably a slightly higher number is to be found in *Pyramimonas montana* for here a section through the five layers reveals about 120 globules (Maiwald, 1971). The size of globules does vary considerably even within a single eyespot. Arnott and Brown (1967) collected data from a variety of organisms and found the range to be 75–300 nm with the most common size being of the order of 100 nm.

A few members of the Cryptophyceae possess eyespots which are situated near the centre of the cell. In *Chroomonas mesostigmatica* (Dodge, 1969b) a single flat layer of eyespot granules is situated at the end of a short square spur of the chloroplast which extends beyond the pyrenoid (Fig. 6.3). There are about 35 of these granules, which can be up to 260 nm in diameter. On its outer side the eyespot is surrounded by the two membranes of the chloroplast envelope and the additional two of the endoplasmic reticulum sheath. The eyespot often lies adjacent to the nucleus and at least one-quarter of the cell length away from the flagellar bases. In *Cryptomonas rostrella*, Lucas (1970a), found a similar arrangement but here there was always an empty vesicle over the eyespot. Central eyespots have also been found in *Chroomonas collegionis* (Dodge, unpub. obs.) where the granules were not very regularly arranged and the eyespot was often dome-shaped. In *Cryptomonas* sp. (Wehrmeyer, 1970b) the eyespot is said to be situated in a groove of the pyrenoid but no clear pictures are provided to explain this unusual situation.

TYPE B. EYESPOT PART OF A CHLOROPLAST AND CLOSELY ASSOCIATED WITH A FLAGELLUM

This type of eyespot is characteristic of the Chrysophyceae, Xanthophyceae and male gametes of the Phaeophyceae. It has also been found in one or two members of other classes. In the Chrysophyceae and Xanthophyceae most motile cells have eyespots and in the typical arrangement the eyespot occupies the anterior end of one of the chloroplasts. It consists of a single layer of about 40 osmiophilic granules, each about 500 nm diameter, which are packed between the ends of the thylakoids and the chloroplast envelope (Fig. 6.4). Often the plasmalemma and chloroplast envelope are closely associated over the eyespot. In most of these organisms the shorter posterior flagellum emerges adjacent to the eyespot and immediately enlarges to form a swelling which lies above the eyespot, often in a groove or depression of the cell surface. The shape of the swelling varies from rounded, as in *Ophiocytium* (Hibberd and Leedale, 1971) where it is 1 μm long and 600 nm in diameter, to almost anchor-shaped in transverse section and wedge shape in longitudinal section in *Ochromonas tuberculatus* (Hibberd, 1970). In the region of the swelling the axoneme passes along the side of the flagellum away from the eyespot and the other side is occupied by what is generally described as granular material. In fact the appearance of this material varies from fine dense granules to large irregular granules, but a distinctly packed or para-crystalline arrangement has not been observed, with the possible exception of *Sphaleromantis* (Manton and Harris, 1966). Here, the wedge-shaped swelling was said to contain a layered structure orientated in the same plane as the tubules of the axoneme.

In some members of the Chrysophyceae the eyespot is not situated in a superficial position but is to some extent embedded within the cell. Thus in *Chromulina psammobia* (Rouiller and Fauré-Fremiet, 1958) although the eyespot is situated at the anterior end of a chloroplast there is an extension of the cell above this which forms a pouch, completely enclosing the short flagellum with its swelling. This process is taken a stage further in *Chromulina placentula* (Belcher and Swale, 1967) where the eyespot is situated in a short diverticulum of the chloroplast near the centre of the flattened cell. The short swollen flagellum lies in a pouch immediately above the eyespot. In two other members of the Chrysophyceae, *Mallomonas papillosa* (Belcher, 1969b) and *Chrysamoeba radians* (Hibberd, 1971) there is no eyespot but nevertheless the short flagellum still bears a swollen region, which in section looks similar to that described from other members of the class, and which appears to be pressed against a concavity on the cell surface. *Chrysococcus cordiformis* (Belcher and Swale, 1972a) also has no eyespot but here the swelling on the short flagellum contains layers of unusually electron-opaque material and dark granules. The swelling lies on the side of the flagellum away from the cell. In a non-pigmented member of the Chrysophyceae, *Anthophysa*, Belcher and Swale (1972b) found both a flagellar swelling and a simple eyespot. This latter consisted of a number of pigmented lipid globules situated in a leucoplast which was closely linked with the nucleus by the ER envelope.

In the Phaeophyceae the spermatozoids normally have a laterally placed eyespot

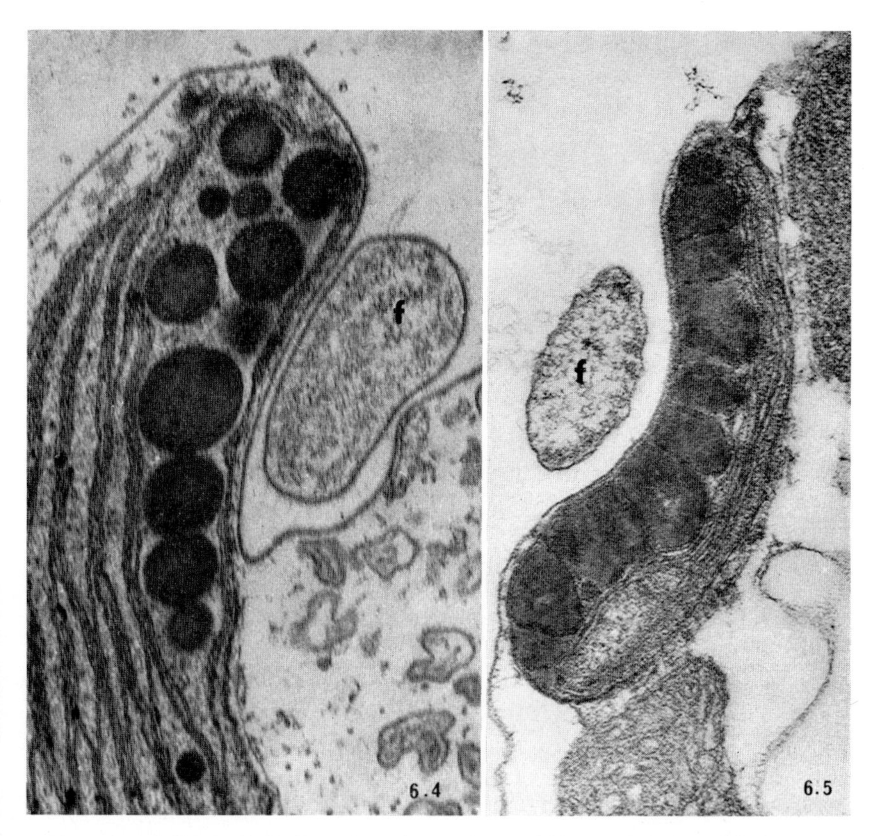

FIG. 6.4. The type B Eyespot of *Dinobryon* (Chrysophyceae). The eyespot globules are situated at the anterior end of a chloroplast and adjacent to a groove through which passes the posterior flagellum (f) with its swelling (×49,000) (After Dodge, 1969a); (6.5) the type B eyespot in the male gamete of *Fucus serratus* (Phaeophyceae). In this transverse section the eyespot is seen to form a curved plate which almost fills the reduced plastid. The swollen part of the posterior flagellum (f) lies above the eyespot (×49,000) (see also Fig. 11.1 for a longitudinal section).

which is situated below the lateral insertion of the flagella. Here, the eyespot consists of a reduced chloroplast which contains pigment granules and a few degenerate thylakoids. In the early work (Manton and Clarke, 1956) it was suggested that there were a number of hexagonal pigment chambers. However, as with the similar observation on *Chromulina* (Rouiller and Fauré-Fremiet, 1958) it would appear that this is an artifact caused by the removal of most of the lipid during methacrylate embedding. More recent studies of *Fucus* gametes using other techniques (Dodge, 1969a) show the presence of about 60 spherical granules (Figs 6.5 and 11.1) lying above two or three short thylakoids. The eyespot is concave and projects slightly from the surface of the cell. The eyespot envelope (4 membranes) and the plasmalemma are closely appressed together. The posterior flagellum bears a rounded swelling containing granular material and this appears to fit into the concave outer face of the eyespot although, as in all organisms with this type of eyespot, no direct connection has been observed between eyespot and flagellum.

Until recently it was thought that no members of the Haptophyceae possessed eyespots. However, *Pavlova gyrans* has now been provisionally transferred to that class from the Chrysophyceae (Green and Manton, 1970) and this has a small eyespot. Basically, this is constructed in the typical manner with a curved plate of about 15 osmiophilic globules situated at the anterior end of one lobe of the chloroplast but not quite at the anterior end of the cell. Adjacent to the eyespot there is a narrow pit which opens between the insertion of the two flagella and follows a curving path from there to the eyespot. Dense material envelops the tip of the pit. Neither flagellum bears a swelling, in fact the basal body of the long flagellum is situated nearer to the eyespot than any of the free parts of the flagella. This is a most unusual arrangement.

TYPE C. EYESPOT INDEPENDENT OF CHLOROPLASTS BUT ADJACENT TO THE FLAGELLA

This is the characteristic condition in the Euglenophyceae where a very large, brightly pigmented, eyespot is situated adjacent to the anterior invagination or reservoir from which the flagella emerge. In many organisms the eyespot moves about as the organism changes its shape (Walne, 1971). Recently, a number of organisms which were classed in the Xanthophyceae have been found to have eyespots which are essentially of this type and these have been placed in a new class, Eustigmatophyceae (Hibberd and Leedale, 1970, 1972).

This type of eyespot consists of a variable number of different sized granules (Fig. 6.6) lying adjacent to each other but not in any way associated with the chloroplasts—they are extrachloroplastidic. It is interesting that when chloroplasts of euglenids are experimentally 'bleached' the eyespots remain (Kivic and Vesk, 1973). When *Euglena* is grown heterotrophically for long periods in the dark the globules appear much less electron dense and optically they are seen to be pale yellow instead of red (Kivic and Vesk, 1972). Within 12 h after exposure to

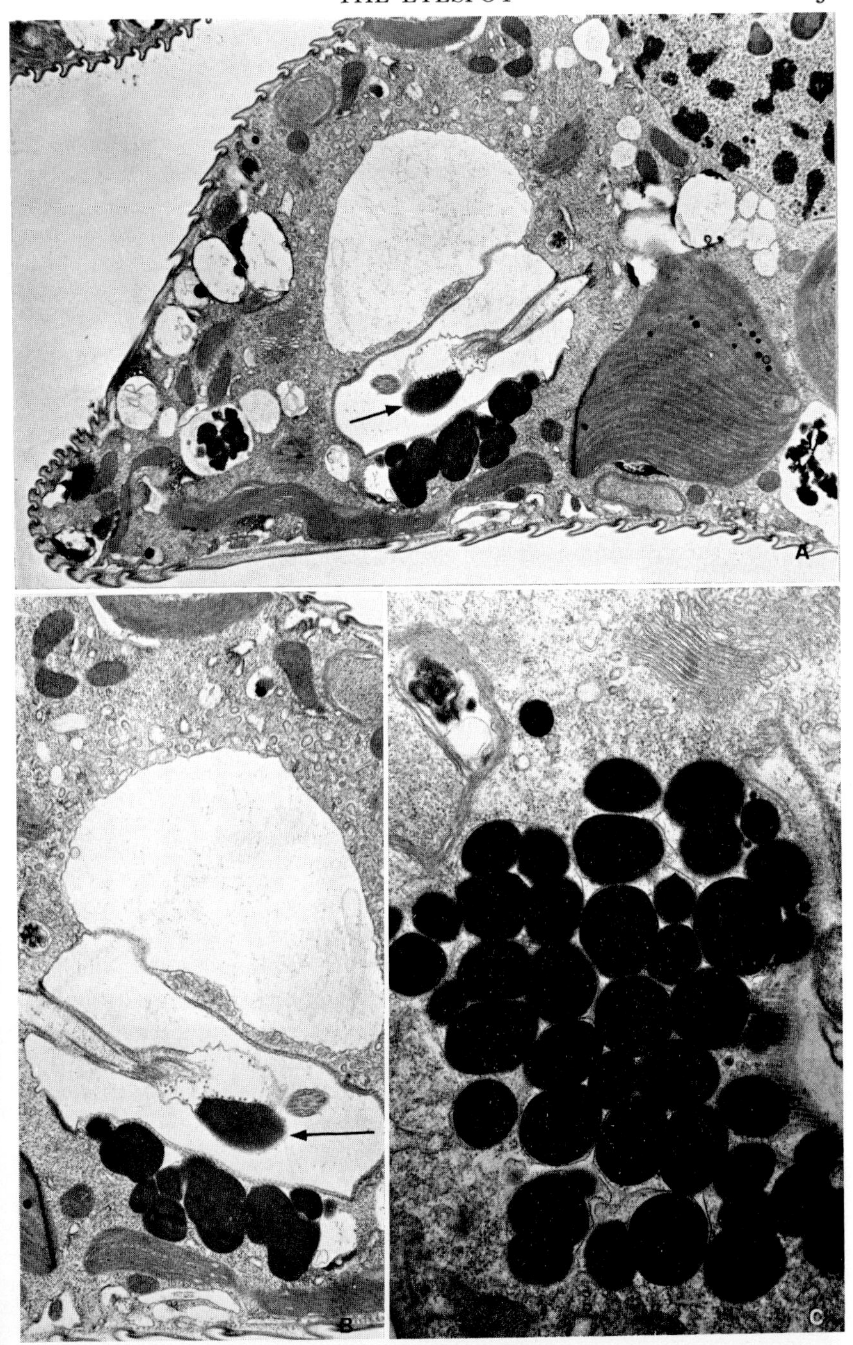

FIG. 6.6. The eyespot of *Euglena*. A, Anterior end of a cell showing non-linear arrangement of packets of granules to form the eyespot and the adjacent reservoir containing flagella and paraflagellar swelling (arrow); B, higher magnification of reservoir and contractile vacuole regions shown in Fig. 1; the relationship of the paraflagellar body (arrow) to the eyespot granules is seen more clearly, as is the non-linear, packeted arrangement of the granules themselves (×11,550); C, tangential section of eyespot of *Euglena granulata* showing 50–60 polygonal-hexagonal granules, many arranged in packets of 3–5. Microtubules are seen in association with the reservoir. (×18,000) Micrographs by courtesy of P. L. Walne (from Walne and Arnott, 1967.)

light the eyespot has returned to normal. Some studies have suggested that the globules are individually bounded by unit membranes (Leedale *et al.*, 1965) and in *Euglena granulata* groups of 2–5 are often enclosed within a common membrane (Walne and Arnott, 1967). It has also been suggested that there may be a single membrane enclosing all the granules (Kivic and Vesk, 1972). The number of granules ranges from one or two in species of *Eutreptia* and *Khawkinea* (Leedale, 1967) to between 20 and 60 in species of *Euglena*. Where there are few granules, as in *Trachelomonas* (Walne, 1971), these are normally arranged in a single curved layer adjacent to the wall of the reservoir but when a large number are present they generally form an irregular clump. The hexagonal close-packing reported from *Euglena gracilis* (Wolken, 1956) appears to have been a misinterpretation of a thick section.

Just as the size of the eyespot is very variable (from under 1 to 8 μm diameter) so too is the size of the individual granules, even within a single eyespot. Walne (1971) in a survey of euglenoid eyespots give the following data for the diameter of eyespot granules:

Euglena granulata	240–1200 nm
E. sanguinea	240–1000 nm
Lepocinclis buetschlii	240–700 nm
Trachelomonas cingulata	240–800 nm

It is obvious from these data that even the smallest granules here are much larger than the normal size found in types A and B.

Closely adjacent to the eyespot are the microtubules which run around the wall of the reservoir (Walne, 1971) and within the reservoir the long locomotory flagellum bears a large lateral swelling (or paraflagellar body) immediately opposite the eyespot. This swelling, which is about 1 μm in diameter, contains a para-crystalline body with a distinct lattice-like repeating pattern which in *E. granulata* is at about 7·5 nm intervals (Walne and Arnott, 1967). The swelling is bounded by the flagellar membrane and there is no direct connection between it and the eyespot. The crystalline body and the flagellar axoneme are joined together by what Walne (1971) calls a connecting body but it is likely that this is simply a part of the paraflagellar rod (cf. flagella chapter) as is certainly the case in *Euglena gracilis* (Kivic and Vesk, 1972).

In the organisms which have recently been placed in the new class Eustigmatophyceae (Hibberd and Leedale, 1970, 1972) the eyespot of the zoospores consists of an irregular group of globules situated at the anterior end of the cell, not surrounded by a membrane, and quite independent of the single chloroplast. Normally only one of the two flagella is emergent and this, in addition to the bilateral array of stiff hairs, bears a swelling at its proximal end. This swelling, which consists of an anchor shaped expansion of the flagellar sheath, is closely applied to the cell membrane over the site of the eyespot and has been seen to contain layers of electron-dense material in the curved side adjacent to the cell (Hibberd and Leedale, 1972).

TYPE D. VARIOUS TYPES OF EYESPOT AND OCELLUS FOUND IN THE DINOPHYCEAE

When algal eyespot fine structure was first reviewed (Dodge, 1969a) two distinct types of eyespot had been discovered in the Dinophyceae. Subsequent work has revealed the presence of two other types which do not clearly fit into the categories A–C described above. Because of this uncharacteristic diversity within a single algal class these types will all be grouped together for the purposes of description.

The most simple arrangement is that found in *Woloszynskia coronata* (Crawford and Dodge, 1973) where the large eyespot consists of an ovoid mass of osmiophilic globules situated behind the longitudinal groove or sulcus. There is no membrane around the eyespot and during fixation the globules often become merged together (Fig. 6.8). No swelling has been noted on the longitudinal flagellum which passes in front of the eyespot. This eyespot has a certain resemblance to type C as found in the Euglenophyceae.

In *Wolozynskia tenuissima* (Crawford *et al.*, 1971) and *Peridinium westii* (Messer and Ben-Shaul, 1969) the eyespot forms part of one of the many chloroplasts. In this, a single layer of lipid globules 30–100 nm diameter, lies at one side of the chloroplast just beneath its envelope (Fig. 6.7). As in *W. coronata* the whole structure is situated adjacent to the sulcus. This eyespot has some similarities with both types A and B as described above.

A more elaborate arrangement has been found in *Glenodinium foliaceum* (Dodge and Crawford, 1969b). Here, the eyespot has a somewhat trapezoid shape, but with an anterior finger-like process, and is situated behind the top of the sulcus. The longitudinal flagellum emerges through the gap in the eyespot and is directed over the main part of the structure but appears not to bear any swelling. The eyespot consists of two layers of osmiophilic globules which are separated by granular material (Fig. 6.9). It is surrounded by a triple-membrane envelope, like that characteristic of chloroplasts in this class, but there are no connections with true chloroplasts and thylakoids are not found in the eyespot. Each eyespot contains a vast number of globules which range in size from 80–200 nm in diameter. The flagellar bases lie near to the eyespot and it is possible that tubular flagellar roots connect with the eyespot before reaching the subthecal microtubular system. Also adjacent to the eyespot is a remarkable lamellar body. This may be 3 μm long and it consists of a stack of up to 50 flattened sacs each 16 nm thick and up to 750 nm wide. These lamellae may be connected together at their edges and also here, and at the ends of the stack, there may be connections with the rough endoplasmic reticulum system of the cell. Although the lamellar body is not directly connected with the eyespot its close proximity would suggest that it could function in association with it for the perception of unidirectional light stimuli.

The most complex and remarkable of all algal eyespots, and those most similar to animal light perception organs, are found in the family Warnowiaceae

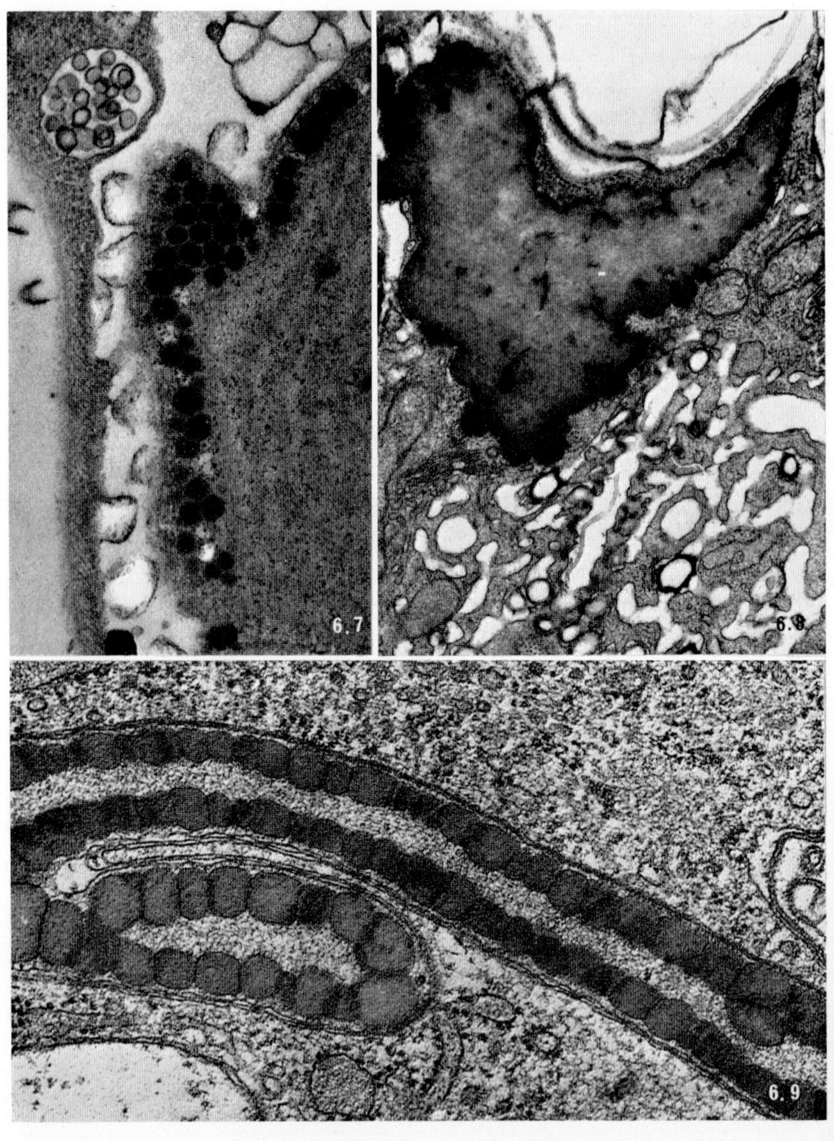

FIGS 6.7–6.9. Eyespots in the Dinophyceae. (6.7) The eyespot of *Woloszynskia tenuissima* which is found within the chloroplast (×38,000) (after Crawford, Dodge and Happey, 1971); (6.8) in the related organism *W. coronata* the eyespot is not part of a chloroplast and simply consists of a cluster of globules (here partly fused together) situated behind the groove termed the sulcus (×21,000); (6.9) in *Glenodinium foliaceum* the eyespot consists of a completely modified chloroplast which now contains two layers of eyespot globules. (×49,000)

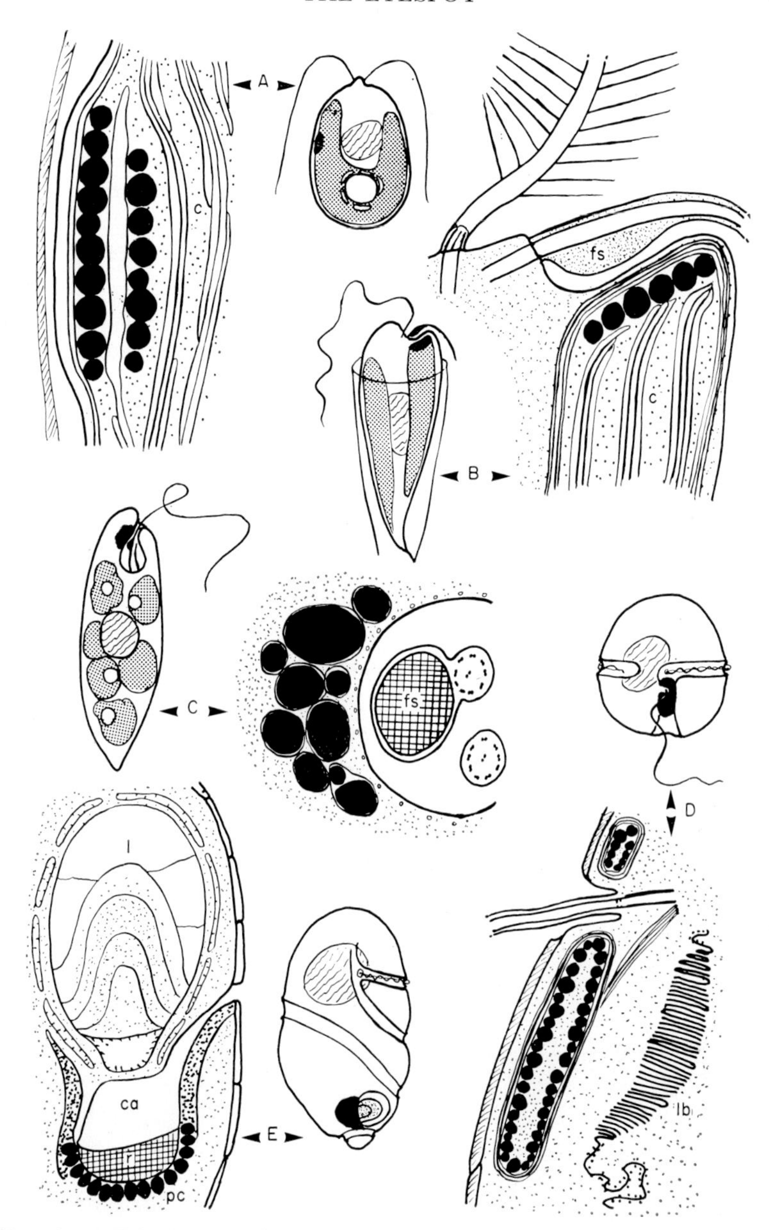

FIG. 6.10. Drawings to illustrate the main types of eyespot described in this chapter. In each case the small drawings indicate the location of the eyespots (as a black area) within the organism involved. A, Type A as in *Chlamydomonas*; B, type B as in *Dinobryon*; C, type C as in *Euglena*; D, the eyespot with lamellar body found in *Glenodinium foliaceum*; E, the complex ocellus found in *Nematodinium*. (c, chloroplast; ca, canal and chamber; fs, flagellar swelling; l, lens; lb, lamellar body; pc, pigment cup; r, retinoid) (after Dodge, 1969a).

of the Dinophyceae. These large bodies, up to 20 μm long and 6–15 μm in diameter, are generally termed *ocelli*. They are located towards the posterior end of the cell adjacent to the girdle furrow and with their long axis parallel to the short axis of the cell. With the light microscope they are seen to consist of two parts, a dark cup-shaped portion and a larger refractive structure or lens (Fig. 6.10E). In the several species which have been examined by electron microscopy (Greuet, 1965, 1968, 1970; Mornin and Francis, 1967) the lens is seen to make up nearly two-thirds of the eyespot and it consists of a number of closely packed swollen vesicles. Beneath this lies a chamber which is connected by a canal to the outside of the cell (i.e. it is lined by an invagination of the plasmalemma). Under the chamber is a hemispherical *retinoid* consisting of closely packed, probably para-crystalline, material which is surrounded on its lower side by a cup made up of a single layer of large ovoid pigment granules 200–250 nm across by 500–700 nm long. Smaller, more dense, irregularly arranged pigment globules surround the chamber and may continue around the pigment cup. The presence of the single layer of granules forming the cup clearly makes this structure analogous to other algal eyespots but it is a very much more highly developed organelle.

II. Development and Replication of eyespots

The only study in this area appears to be a recent one on *Euglena gracilis* by Kivic and Vesk (1972). These workers found that following nuclear division the eyespot replicated, and segregated to opposite sides of the reservoir, at about the same time as the flagella duplicated.

In experiments with inhibitors, synthesis of the red pigment was inhibited by chloramphenicol, which is known to affect 70S ribosomes. This suggests that the carotenoid pigment may be produced in the chloroplasts before transfer to the eyespot. In an experiment with cycloheximide, which affects protein synthesis by 80S ribosomes, the eyespot was reduced in size and contained several apparently empty membrane envelopes. Some eyespot granules may be scattered in the cell. This is rather similar to what may be found in senescent cells.

III. Eyespot Structure and Taxonomy

In general the types of eyespot fit in very well with the currently accepted ideas on classification and phylogeny in the algae. Thus the Chlorophyceae and Prasinophyceae which have eyespots of type A (Fig. 6.10A) both possess chlorophyll *b* and have starch as the major food reserve. They are often placed together in the phylum Chlorophyta. Similarly most of the organisms which have been placed in the phylum Chrysophyta (Xanthophyceae, Chrysophyceae and Phaeophyceae) have eyespots of type B (Fig. 6.10B). These classes share other features such as pigment composition and flagellar structure. The Euglenophyceae although possessing chlorophyll *b* has long been known to differ from the true 'green' algae for the nucleus has a unique structure and the paramylon food reserve material is only found in this class. These differences are reflected in the distinct type C eyespot (Fig. 6.10C) although this is now known to be also present in the Eustigmatophyceae a class which has certain characteristics of the Chrysophyta and certain unique features. Further work may perhaps reveal more similarities with the Euglenophyceae than have yet been discovered.

The Cryptophyceae is here said to have a type A eyespot which rather unnaturally associates this distinctive class with the green algae. Possibly the internal eyespot of the Cryptophyceae should be regarded as a unique type for this class has few features in common with other classes.

The Dinophyceae is also a unique class with many features such as nucleus, flagella and cell covering which indicate its homogeneity. It is therefore surprising to find such a diversity of eyespot structure in this class. Clearly further work is required here to see if eyespot type can be linked with revised genera or orders within the class.

IV. The Function of the Eyespot

This has generally been presumed to be the perception of light stimuli. However, it is well known that certain organisms without eyespots are phototactic and early shading experiments (Engelman, 1882) suggested that in *Euglena* the cell showed no response when the eyespot was shaded. In this case it appeared that the eyespot was itself acting as a shade for the flagellar swelling which was therefore thought to be the photoreceptor. This theory could also apply to the organisms of type B which all have flagellar swellings. However, as these do not appear to have the crystalline structure seen in the flagellar swelling of the euglenids it would perhaps seem more likely that the swelling acts as a shading device over the eyespot instead of vice-versa. The problem of function is more difficult to deduce for type A eyespots for here there is no obvious connection with the flagella. There have been suggestions that the membrane over the eyespot is important as the photoreceptor (Arnott and Brown, 1967) and it has been pointed out that the plasmalemma or the microtubules of the flagellar roots may be involved. This is an area which needs much more investigation before we can clearly state the function of the eyespot. The various theories have been reviewed by Haupt (1959), Bendix (1960) and Halldal (1964).

7. The Nucleus and Nuclear Division

I. The Interphase Nucleus

A. NUCLEAR ENVELOPE

In the majority of algae the interphase nuclei appear to be of the normal eucaryotic type found in most animals and plants. The nucleus (Fig. 7.1) is surrounded by a two-membrane envelope which is continuous with the cell ER system and each membrane is 7–8 nm thick. Between the two membranes there is a narrow perinuclear cavity about 20 nm wide, which at times may be expanded in connection with special activities such as flagellar hair construction (see Chapter 3). The nuclear envelope is perforated by numerous pores or annuli (Fig. 7.2) which appear to be similar in size, structure and distribution to those of higher plants (cf. Roberts and Northcote, 1970). In *Bumilleria* (Xanthophyceae) (Massalski and Leedale, 1969) the pores are apparently arranged in straight lines and the individual pores are 800–900 nm in diameter and often show an electron dense central spot and peripheral granules. In *Zonaria* (Phaeophyceae) there are bands of the envelope with numerous close-packed pores and other areas with none (Neushul and Dahl, 1972b).

In the Dinophyceae, which have unique chromosomes (see below), the nuclear envelope usually has the standard type of construction. Wecke and Giesbrecht (1970) have shown that in *Prorocentrum* the pores are arranged in closely packed hexagonal groups (Fig. 7.2) and the individual pores show hexagonal symmetry. In *Glenodinium* a random distribution of the pores was found. In two other dinoflagellates *Noctiluca* (Afzelius, 1963; Soyer, 1969; Zingmark, 1970) and *Gymnodinium fuscum* (Dodge and Crawford, 1969a) normal pores appear to be absent. In the former, annulated vesicles are found within the nuclear envelope and it is possible that these carry out the transport function of nuclear pores by fusing with the envelope and thereby discharging material to the exterior. By way of contrast Soyer believes they provide a reserve of nuclear envelope for use after division. In *Gymnodinium* a vesicular structure is built into the nuclear envelope and pores are only seen in the inner side of the vesicles (Fig. 7.3). In the large primary nucleus of *Bryopsis* (Chlorophyceae), Burr and West (1971b) have found a fibrillar reticulum, composed of 20 nm diameter fibrils, which lies just beneath the nuclear envelope in which there are normal nuclear pores.

B. NUCLEOLUS

All algal nuclei possess nucleoli (cf. Figs 7.1, 7.6, etc.) which are rather varied in shape, size and number per nucleus. As in higher organisms (Du Praw, 1970) these are composed of areas of ribosome-like granules about 20 nm diameter

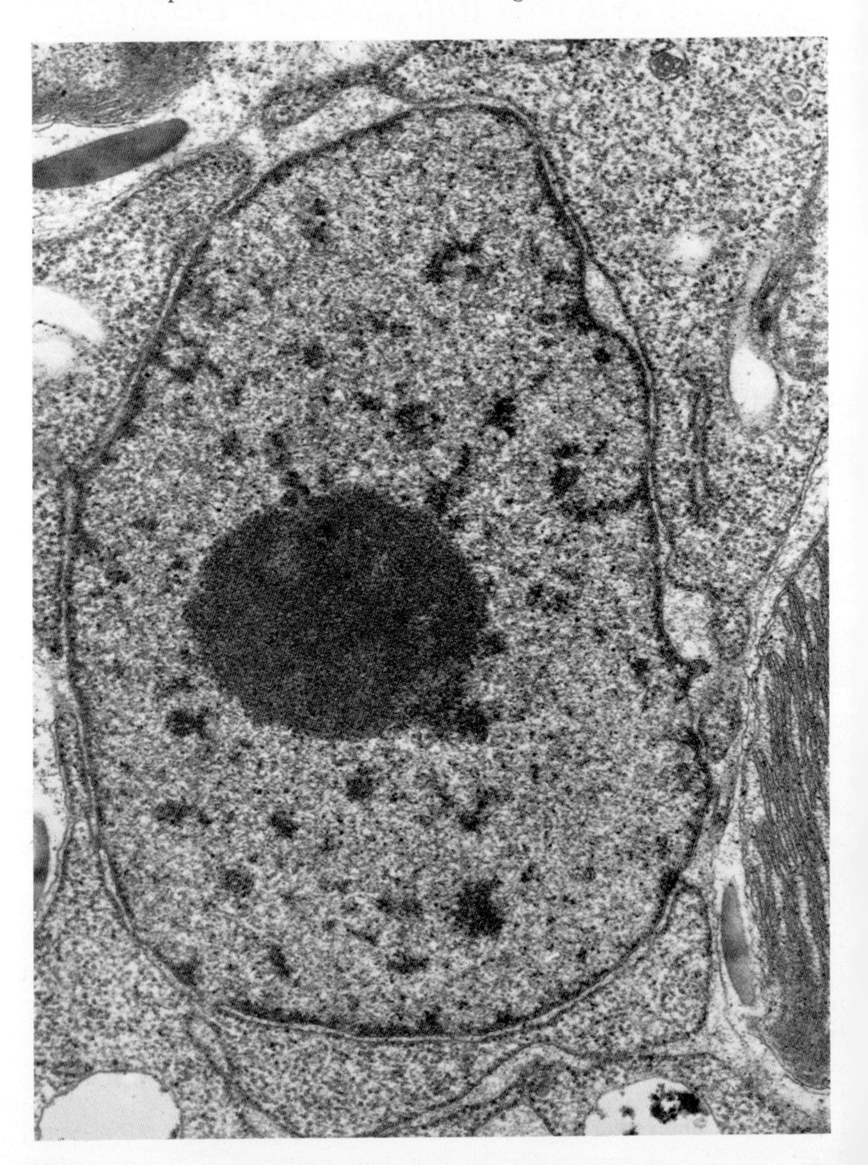

FIG. 7.1. A typical interphase nucleus, in *Cryptomonas* (Cryptophyceae). The nucleus contains a central nucleolus and scattered chromatin and is bounded by a double nuclear envelope connected to the ER system of the cell. (×30,000)

(Fig. 7.4); less dense regions containing 10 nm filaments; portions of chromosome (presumably nucleolar organizing chromosome) partially embedded in the nucleolar matrix (Fig. 7.5). Electron micrographs of members of most algal classes suggest that algal nucleoli are similar in construction. In *Crypthecodinium*

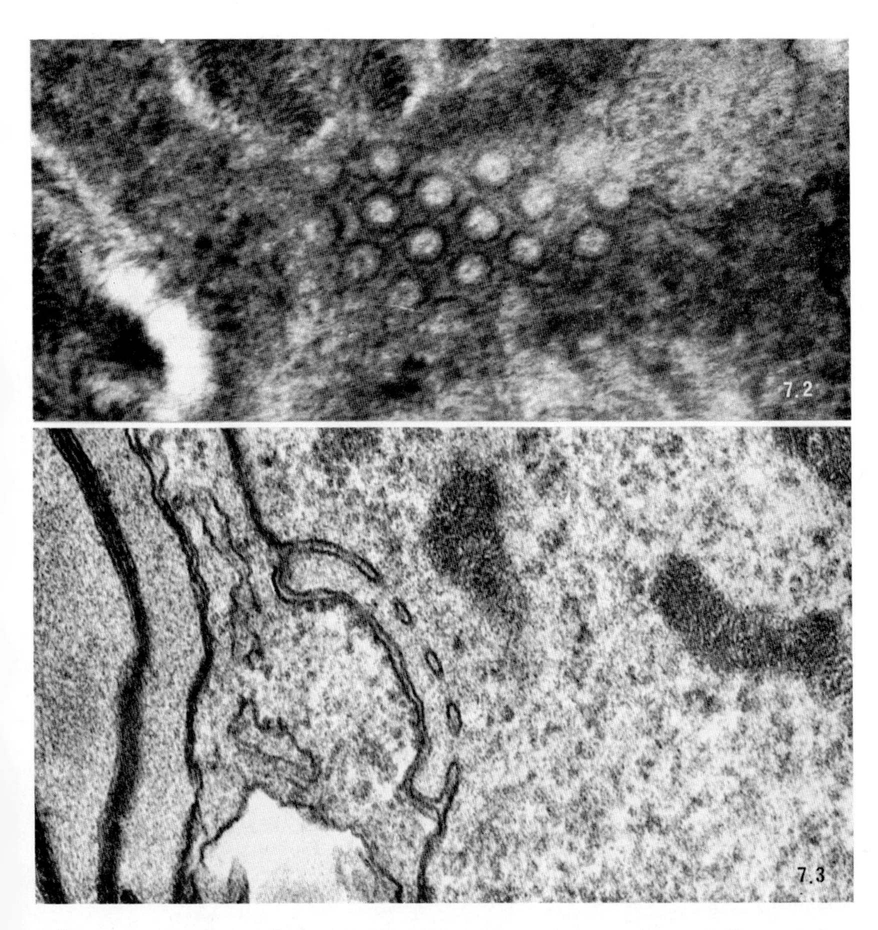

FIG. 7.2. A tangential section of part of the nuclear envelope of *Prorocentrum* (Dinophyceae) showing some characteristic nuclear pores in which a central spot and partially occluded periphery can be seen (×78,000) (after Dodge and Bibby, 1973); (7.3) a small part of the nuclear envelope in *Gymnodinium fuscum* (Dinophyceae) showing a vesicle built into the envelope and with pores only on its inner face. (×59,000)

(Dinophyceae) (Rae, 1970) nucleolar structure has been investigated in detail. It was found to consist of two types of ribosomal particles which sedimented as 16 and 25S, exactly the same as those from the eucaryote *Tetrahymena*. In the electron microscope the nucleolus of *Crypthecodinium* can be seen to have an extensive inner fibrillar component and a peripheral component consisting of *c.* 25 nm granules which can be dissolved by treatment with ribonuclease. Several chromosomes are intimately associated with the nucleolus. It has been suggested that the filaments and particles represent successive stages in the formation of ribosomes, the coding for which is provided by the genes of the

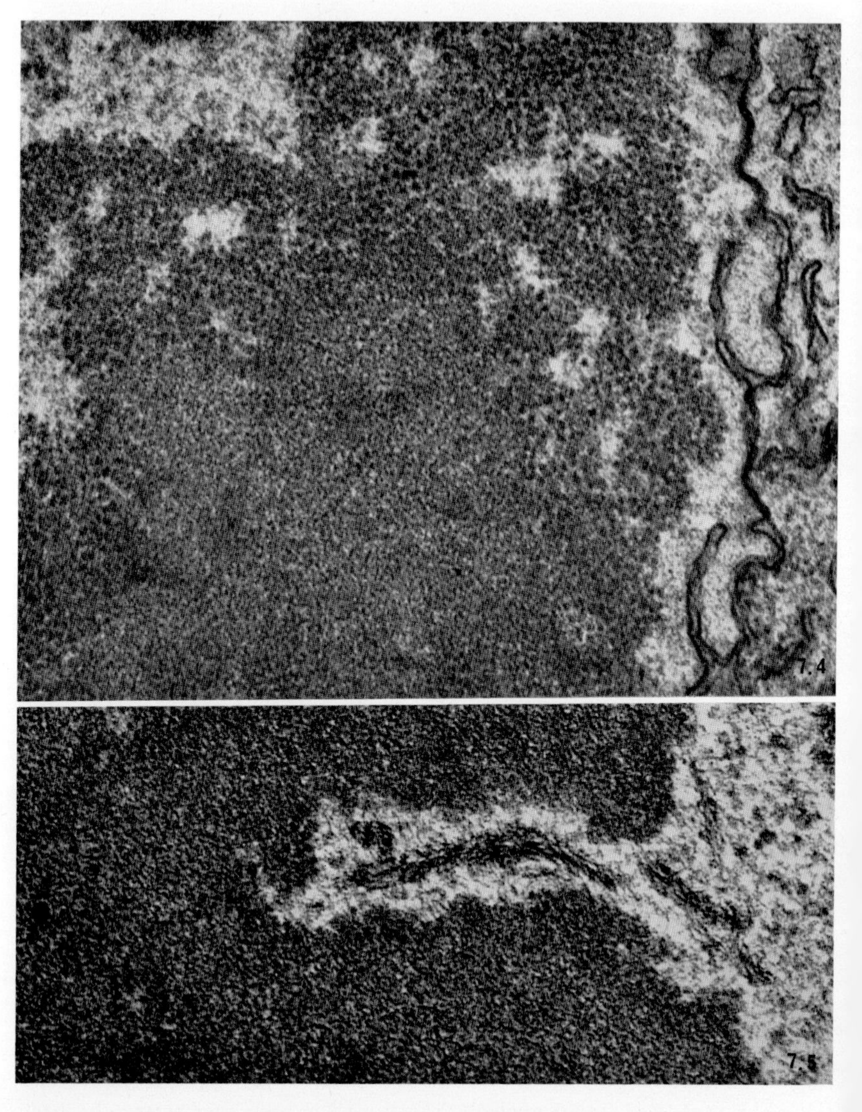

FIG. 7.4. The nucleolus in *Gymnodinium fuscum* with a dense region of fine granules and outer zone of larger granules (× 59,000) (after Dodge and Crawford, 1969a); (7.5) part of a chromosome (nucleolar organizer?) entering the nucleolus in *Glenodinium foliaceum* (× 59,000) (after Dodge, 1971a).

nucleolar organizing chromosomes. The green alga *Spirogyra* was also found to have granular and filamentous parts to its nucleolus and some associated chromosomes (Godward and Jordan, 1965; Jordan and Godward, 1969). This nucleolus persisted throughout nuclear division, wrapped around the chromosomes. In *Euglena* the nucleolus was found to contain coiled structures which were degraded by deoxyribose nuclease (O'Donnell, 1965). In the coccolithophorid *Hymenomonas* Riedmüller-Schölm (1972) has described a narrow channel which connects the nuclear envelope with the nucleolus. This is thought to allow for extrusion of material into the perinuclear space.

The behaviour of the nucleolus during mitosis has been reviewed by Pickett-Heaps (1970c) who finds that in most algae which have been investigated the nucleolus does not persist throughout division. Examples of algae where the nucleolus persists will be given in the later section dealing with mitosis.

C. CHROMATIN

The major part of the interphase nucleus is filled with chromatin which is sometimes seen as evenly dispersed, finely-granular material. This appearance may result from faulty technique for the more typical appearance (Figs 7.1 and 7.6) is of two distinct types of chromatin. The large volume is occupied by the finely-filamentous euchromatin and the more densely granular heterochromatin is found in scattered clumps throughout the nucleus and particularly around the periphery. No detailed fine-structural work appears to have been carried out on such nuclei.

In two algal classes a distinctly different nuclear structure is encountered. The Euglenophyceae (Ueda, 1960; Leedale, 1968) have an interphase nucleus in which distinct chromosomes (Fig. 7.7) are evenly distributed amongst nucleoplasm which is described as granular material amongst a less dense matrix. The chromosomes have a densely filamentous construction said to consist of a moderately coiled chromonemata of 10 nm diameter (O'Donnell, 1965). Thus, the chromosomes are much more highly condensed than is usual in a eucaryotic interphase nucleus.

In the Dinophyceae, which are mesocaryotic (Dodge, 1966), the chromosomes are quite distinct from those of any other organisms. They are very highly condensed and in longitudinal section reveal a series of transverse bands (Figs 7.8 and 7.9). The basic units are fibrils of 3–6 nm diameter but how these are arranged to give the chromosome is by no means clear. To date, some five theories have been propounded to explain the organization of the chromosomes (reviewed in Dodge, 1971b). What is quite clear is that unlike all eucaryotic chromosomes, those of dinoflagellates, like the nuclei of bacteria (Fig. 7.10), lack histone protein. This has been demonstrated by cytochemical staining but even more convincingly by dissolving the chromosomes with deoxyribose nuclease prior to embedding and sectioning (Leadbeater, 1967). In the resulting pictures (Fig. 7.11) the cell and nuclear matrix are seen to have their normal

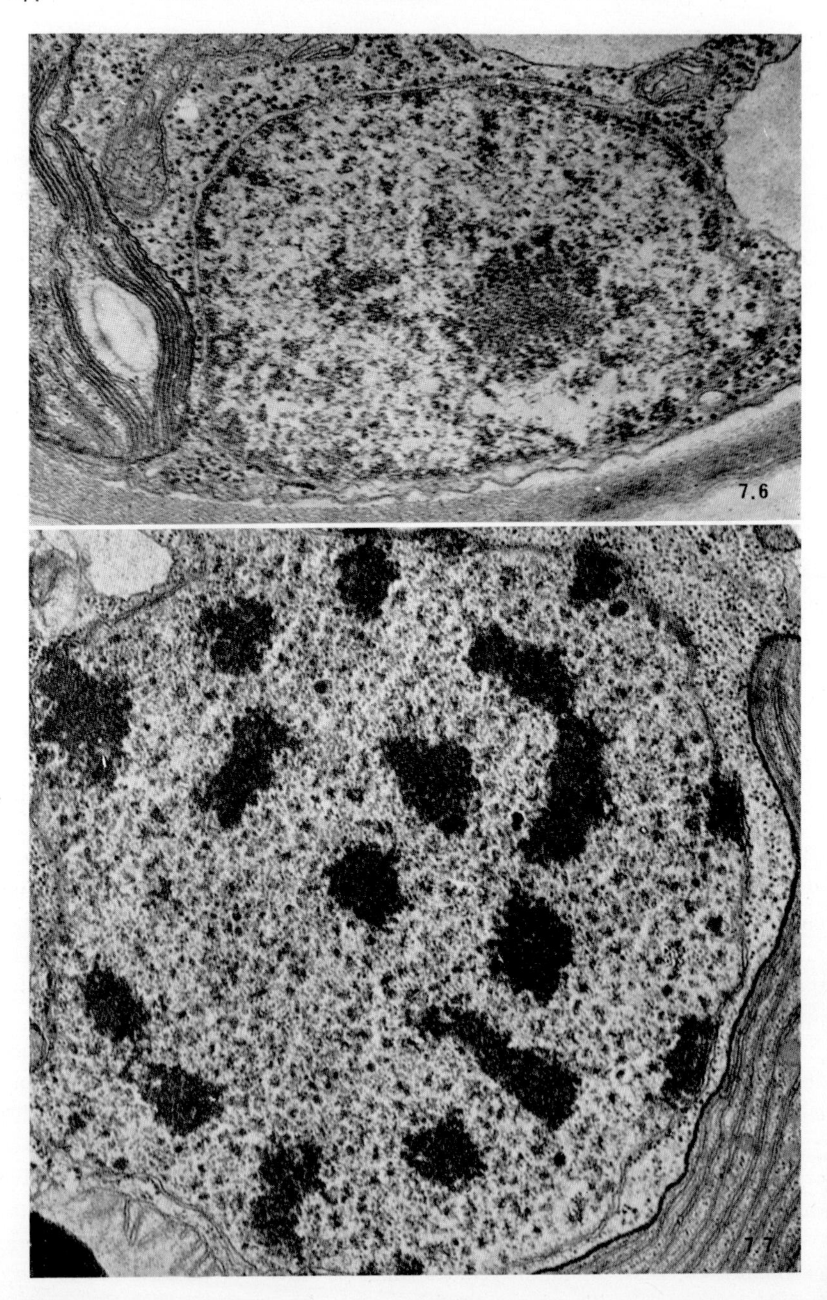

FIG. 7.6. A typical interphase nucleus in the green alga *Ankistrodesmus*. The nucleolus consists of two sizes of granules. Ribosomes are seen attached to the outer membrane of the nuclear envelope (×29,400); (7.7) the interphase nucleus of *Trachelomonas* (Euglenophyceae) with the chromosomes condensed, and quite distinct, and the nuclear matrix consisting of rather evenly dispersed granules. (×29,400)

structure but the chromosomes are completely removed. Several authors have noted the similarity in structure between dinophycean chromosomes and bacterial nuclei (e.g. Giesbrecht, 1962; Kellenberger, 1964).

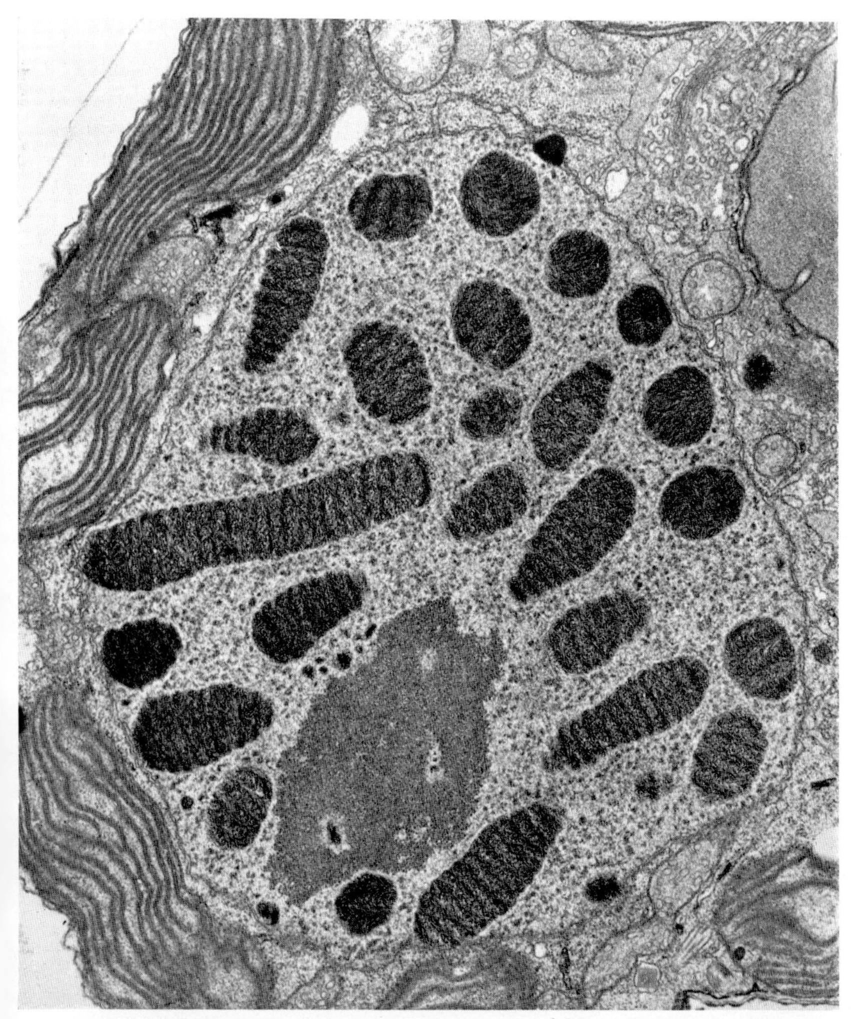

FIG. 7.8. Interphase nucleus in *Heterocapsa* showing the characteristic appearance of the mesocaryotic nucleus found in the Dinophyceae. Nuclear envelope and nucleolus have the typical construction but the chromosomes are condensed and have a banded appearance. (× 17,500) (After Dodge, 1971b).

In one dinoflagellate Dodge (1971a) has reported the presence of a normal dinophycean nucleus plus another polymorphic body which has an appearance like that of a typical eucaryotic nucleus. It has a nuclear envelope with pores, nucleolus composed of several types of material, and granular chromatin. The significance of the presence of these two nuclei within a single cell is not yet understood.

II. Nuclear Division—Mitosis

As yet the ultrastructural changes associated with mitosis have only been studied in very few algae, and most of these belong to the Chlorophyceae. It is likely that in most classes the process is not exactly similar to that found in higher plants. By way of illustration the mitosis of *Chlamydomonas* (Johnson and

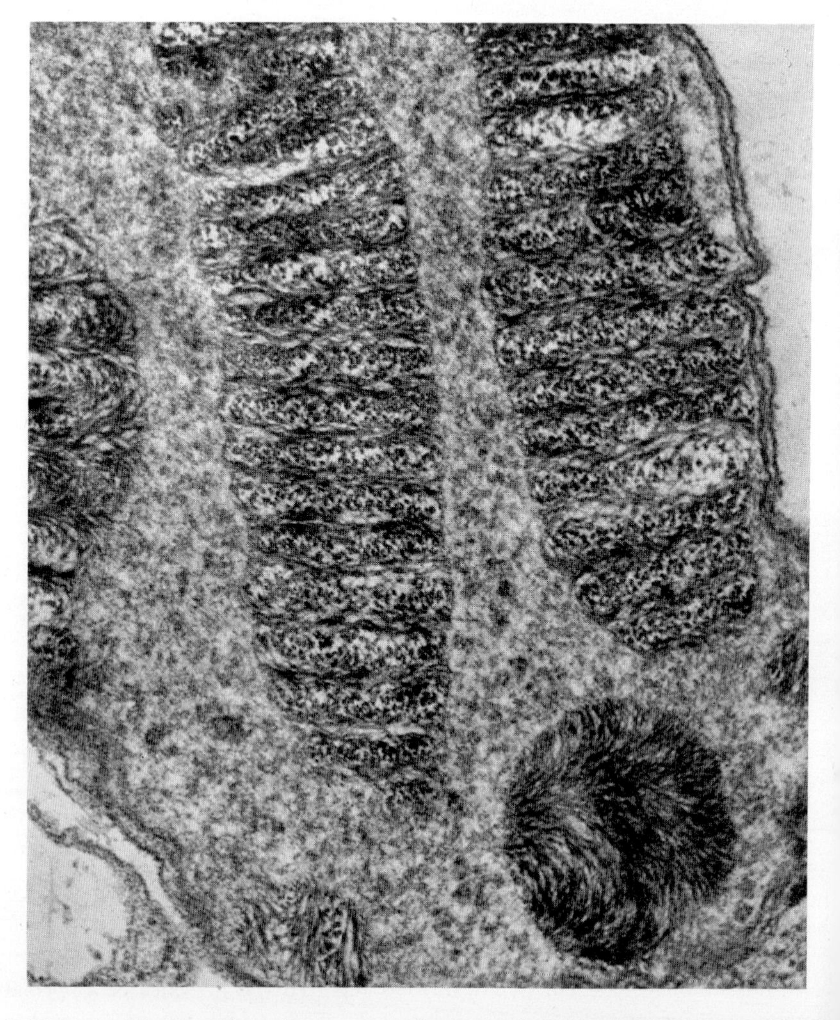

Fig. 7.9. Dinoflagellate chromosomes from *Amphidinium herdmani* more highly magnified and seen in both transverse and longitudinal sections. Note the arrangement of the DNA fibrils of which the chromosomes are composed. (×59,000)

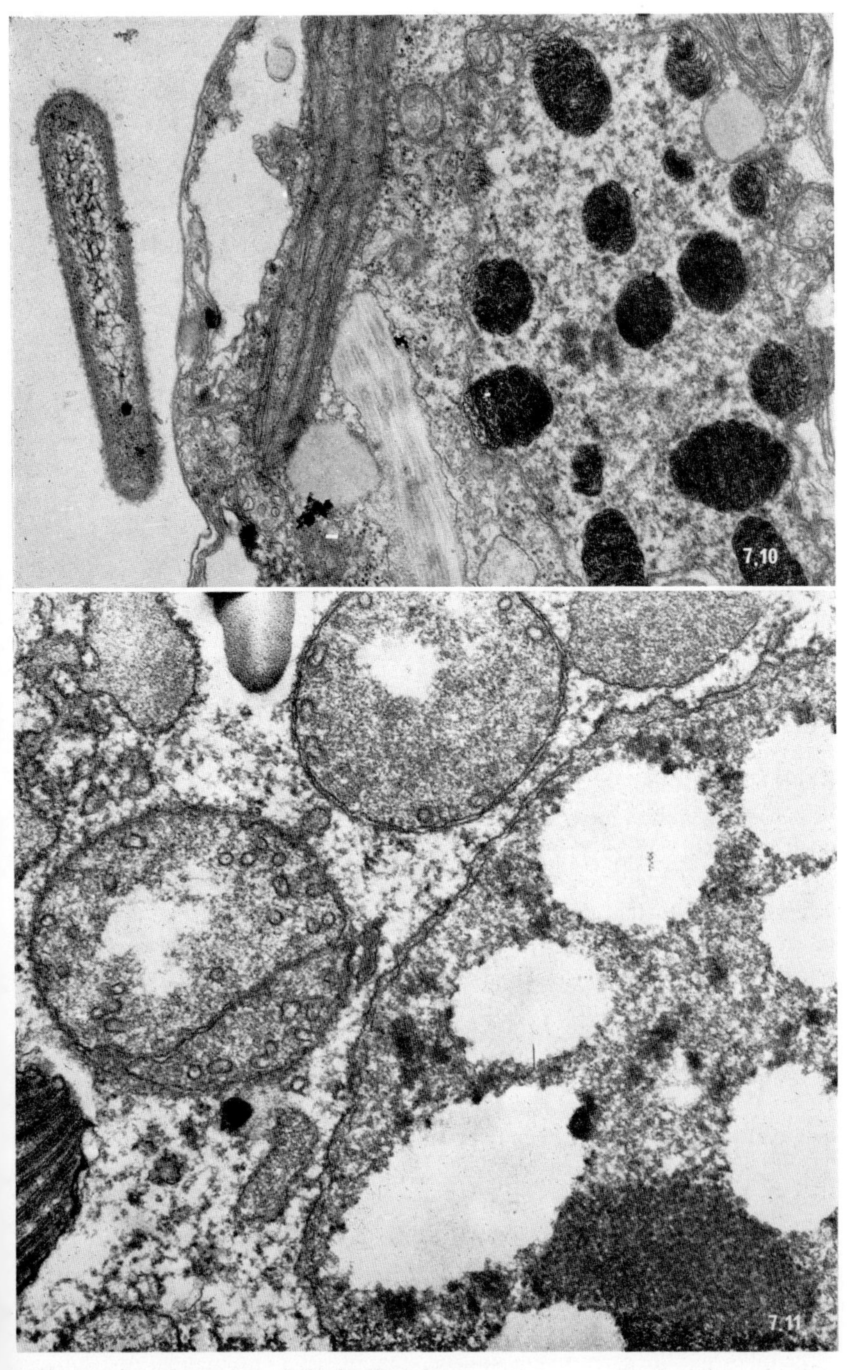

FIG. 7.10. A bacterium (left) fixed at the same time as a dinoflagellate, showing the rather similar appearance of the bacterial nucleus and the dinoflagellate chromosomes (×21,000); (7.11) part of a cell of *Gymnodinium* which was treated with deoxyribonuclease after glutaraldehyde fixation and before post-osmication. Note the complete absence of the chromosomes and of the DNA areas from the mitochondria. (×40,000)

Porter, 1968) will first be described and this will be followed by details of division in other algae where there are variations. Various aspects of this subject have been reviewed by Pickett-Heaps (1969, 1972e), Leedale (1970) and Du Praw (1970).

A. CHLOROPHYCEAE

In *Chlamydomonas* (Johnson and Porter, 1968) the onset of nuclear division is signalled by the replication of the flagellar basal bodies which have become separated from their flagella. The external parts of the flagella generally become detached from the cell. The four basal bodies remain in the anterior end of the cell and apparently do not act as centrioles. The nucleus moves near to one edge of the cell and the nucleolus disappears. As the chromosomes are condensed, and thus become visible, the nucleus assumes a spindle shape. At metaphase the chromosomes lie across a large median bulge in the nucleus. The nucleus is still surrounded by nuclear envelope except for two large gaps or fenestrations at the poles. Spindle microtubules are orientated towards the poles and may just protrude through the polar gaps in the nuclear envelope although the spindle has not been seen to connect with any centriolar-like structures. During anaphase the nucleus elongates as the daughter chromosomes separate and eventually the chromosomes come to lie against the nuclear envelope at opposite ends of the nucleus. The spindle tubules disappear and connections are formed across the nuclear envelope to separate the daughter nuclei. Cell cleavage then follows and this is apparently directed by the basal bodies with which are now associated bands of microtubules lying parallel to the cleavage furrow and along both sides of it. At the end of mitosis the chromosomes lose their condensation and the nucleolus reappears.

In the parenchymatous green alga *Ulva* (Løvlie and Bråten, 1970) the mitosis is very similar to that described above and the nuclear envelope also remains intact except for polar fenestrations. Here, a structure which is thought to be a centriole is located near to each pole of the spindle but not exactly on its longitudinal axis. Neither in *Ulva* or *Chlamydomonas* have distinct kinetochores (spindle attachment regions) been reported.

Mitosis in the coccoid green alga *Kirchneriella* has been studied by Pickett-Heaps (1970b). Here, in spite of the absence of flagellated stages in the life history, centrioles are present and these move to the polar regions during prophase. They definitely appear to be involved in spindle formation which at this stage is extranuclear. A perinuclear sheath of endoplasmic reticulum forms around the nucleus and small openings appear in the nuclear envelope adjacent to each centriolar complex. Microtubules and centriole enter the nucleus and an intranuclear spindle forms. Apart from small polar fenestrations the nuclear envelope remains intact during the considerable elongation of the nucleus which takes place following the early anaphase separation of sister chromatids. At telophase the re-forming nuclear envelope contracts tightly around each daughter nucleus and part of the envelope is laid down beneath the centriolar complex, thus excluding it from the nucleus. Pickett-Heaps (1972f) has recently described a similar mitosis in another coccoid green alga, *Tetraedon*, where there is also an extensive production of extranuclear ER prior to division. Centrioles have been discovered in *Chlorella* another member of the Chlorococcales (Atkinson *et al.*, 1971; Wilson *et al.*, 1973). In the interphase the centrioles occupy a depression in the nuclear envelope where it is nearest to the cell wall. They duplicate during prophase and, following migration to opposite ends of the nucleus, an intranuclear spindle forms. No flagella are ever found in this organism.

Another variant type of division is found in *Oedogonium* (Pickett-Heaps and Fowke, 1969, 1970a). In this filamentous green alga there are no centrioles and the spindle is entirely intranuclear. The nuclear envelope remains intact. The really unusual feature here is the formation of kinetochores on the chromatids during prophase. These are complex structures consisting of six or seven slightly curved layers. Rather simpler kinetochores have now been found in *Cladophora* and *Spirogyra* (Godward, pers. comm.), *Vacuolaria* (Heywood and Godward, 1972) and *Membranoptera* (McDonald, 1972). No similar structures have been found in any other plants although they are normally present in animals. During metaphase the kinetochores and their chromosomes become aligned on the metaphase plate and spindle tubules are attached to them. At anaphase the kinetochores split and migrate polewards with the rest of the chromatid trailing behind. Many microtubules connect the kinetochores with the polar regions of the nucleus. Interzonal microtubules form between the groups of chromatids, as they separate, and the nucleus becomes considerably elongated. Eventually the nuclear envelope contracts around the groups of daughter chromosomes and isolates them from the spindle, which then collapses.

Yet another type of mitotic behaviour has been found in the desmid *Closterium* (Pickett-Heaps and Fowke, 1970c). Here, at prophase the nuclear envelope breaks down and the nucleolus is dispersed. Small doubled chromosomes form a precisely aligned metaphase plate and an apparently normal spindle is present. During anaphase the plates of chromatids separate and the spindle becomes invaded by mucilaginous vesicles. Now a rather unusual thing happens. During telophase, as the spindle collapses, microtubules collect at each pole, being especially focused on a small region called the microtubule centre (MC). Following the completion of the septum dividing the cell into two, the new wall begins to expand to recreate the typical shape of the vegetative cell. At the same time the MC moves to the side of the cell and migrates towards the pointed end (see Fig. 8.2). Many microtubules connect the MC to the nucleus which begins to extend and follows the MC along the cell. Eventually the MC lodges in a cleft which has developed in the chloroplast nearly half-way down the cell. The nucleus joins the MC, the chloroplast splits into two, and the typical arrangement of the vegetative cell of *Closterium* is re-established. In the related alga *Spirogyra* (Fowke and Pickett-Heaps, 1969a) the mitosis follows a very similar pattern to that described for *Closterium* except that the anaphase chromosomes are often obscured by deposition of nucleolar material and the nuclear envelope does not completely disrupt until the spindle elongates. The envelope rapidly reforms and in telophase the daughter nuclei are again completely enclosed. Mitosis is followed by cytokinesis through the equatorial plane.

B. PHAEOPHYCEAE

In the Phaeophyceae little is known about the structure of dividing nuclei. However, a brief report suggests that in the antheridium of *Fucus* the nuclear envelope persists virtually intact and paired centrioles are present (Bouck, cited by Leedale, 1970). Recently Neushul and Dahl (1972b) have also shown the presence of centrioles in the nucleus of apical cells of *Zonaria*. Here, the nuclear envelope appears to remain intact except for polar fenestrations. During division microtubules radiate from the centrioles into the cytoplasm as well as through the nucleus. It is suggested by these authors that microtubule organizing substance surrounds the centriole when the nucleus is dormant.

C. CHLOROMONADOPHYCEAE

The first fine-structural study of a dividing nucleus in this class (Heywood and Godward, 1972) has indicated that in *Vacuolaria* the nuclear envelope is probably entire throughout mitosis. Each chromosome is attached to up to 6 tubules of the intranuclear spindle by a simple kinetochore 0·25–0·34 μm thick. The flagellar bases *may* act as centrioles. In telophase a new nuclear envelope forms around the daughter nuclei, within the parental nuclear envelope.

D. HAPTOPHYCEAE

For the Haptophyceae we have an account of mitosis in *Prymnesium parvum* (Manton, 1964e) which suggests that the division is very similar to that of higher organisms. Nuclear division is preceded by duplication of flagella, chloroplasts and Golgi body. During prophase the pairs of flagellar bases move apart and probably influence the orientation of the spindle, which forms parallel to a line drawn between them. The nuclear envelope breaks down completely and the chromosomes are aggregated into a compact metaphase plate. Anaphase appears to be very rapid and the separation of the groups of daughter chromatids is accompanied by considerable elongation of the spindle and extension of the cell in the same plane. During late anaphase the nuclear envelope begins to reform around the daughter nuclei, beginning at the polar side. Initially the nucleus has a rather irregular form because the envelope is laid down around the projecting chromosomes but, after decondensation of the chromosomes, the nucleus rounds off and this is followed by cell cleavage.

E. CHRYSOPHYCEAE

A recent study of mitosis in *Ochromonas*, the first fine structural investigation of division in the Chrysophyceae, has shown that this group exhibits several unusual features. Slankis and Gibbs (1972) found that during interphase the basal bodies of the two flagella replicate and the chloroplast divides by a constriction between its lobes. The rhizoplast, a striated root attached to the basal body of the long flagellum, runs under the Golgi body to the surface of the nucleus.

During pre-prophase the Golgi body replicates and also a second rhizoplast appears. Then the two pairs of flagella, with their associated rhizoplasts and Golgi bodies, separate across the anterior end of the cell. At this stage there is a considerable increase in the number of microtubules around the nucleus. The nuclear envelope begins to break down at the ends and this allows microtubules to enter and form the spindle. The possibly unique feature here is that one rhizoplast is associated with each pole and the spindle microtubules appear to be attached to the rhizoplasts. Some spindle tubules run from pole to pole whilst others are connected to the chromosomes. There do not appear to be any kinetochores. The remainder of the nuclear envelope now breaks down, except for the portions adjacent to chloroplasts. The chloroplast-ER remains intact throughout division. At anaphase the interpolar microtubules double in length and the chromosomes separate into two groups. In telophase the nucleolus reappears and the nuclear envelope is re-established, incorporating parts of the adjacent chloroplast-ER. Cytokinesis takes place by longitudinal fission which starts at the anterior end of the cell between the two pairs of flagellar bases.

F. BACILLARIOPHYCEAE

The few accounts we have of mitosis in the Bacillariophyceae suggest that a rather unique type of nuclear division is typical of this class. In the centric diatom *Lithodesmium* (Manton *et al.*, 1969a) there is a spindle precursor consisting of a rectangular body made up of a series of parallel plates, those at the ends being most dense. During prophase of mitosis in the spermatogonia the spindle proper is laid down between the precursor and the nuclear envelope. It consists of microtubules which rapidly increase in number and length as prophase proceeds. The poles of the spindle are marked by extensions of the terminal plates of the precursor and the remainder of the precursor breaks down, thus allowing the spindle to elongate. The nuclear envelope breaks down and the spindle sinks into the nucleoplasm and comes to lie through the centre of the mass of chromosomes. At this stage transverse sections of the spindle show approximately 100 microtubules near the poles and 200 near the equator, suggesting that there are in fact two half-spindles which overlap in the centre. A few microtubules may continue from pole to pole. During metaphase most of the microtubules aggregate into 16–20 bundles, each of which presumably serves a single chromosome. During anaphase the spindle elongates considerably and, even when the nuclear envelope has formed around the daughter nuclei, a tuft of 20–30 microtubules, representing the remains of the spindle, can be seen running between the two nuclei. This is finally broken and dispersed at cytokinesis.

In the pennate diatom *Surirella* Drum and Pankratz (1963) found a single spherical structure, 500 nm in diameter and consisting of coarse granules, located adjacent to the interphase nucleus. Filaments which are probably microtubules radiate from this structure which is thought to be the 'centrosome' described by earlier workers using light microscopy. No information was obtained about the behaviour of this centrosome during mitosis although it would seem likely that it does 'organize' the spindle.

G. XANTHOPHYCEAE

A recent study of mitosis in the coenocytic alga *Vaucheria* (Ott and Brown, 1972) has provided the first fine structural description of mitosis in the Xanthophyceae. A pair of centrioles are associated with each nucleus and, during prophase, one of each pair migrates to the opposite side of the nucleus which then becomes somewhat elongated. A centriolar plaque forms at each pole and cytoplasmic microtubules are seen to be associated with this. The nucleolus fragments. At metaphase the intranuclear spindle forms and this is enclosed by the intact nuclear envelope which persists throughout mitosis. In anaphase the chromosome separation is accomplished by the lengthening of continuous pole-pole microtubules which produce a very long interzonal spindle between the daughter nuclei. No kinetochores were observed on the chromosomes. The nuclear envelope contracts around the extended spindle and, at telophase, it invaginates to surround the daughter nuclei. The interzonal section of spindle is cut off and eventually degenerates. As the alga is coenocytic no cytokinesis follows and the daughter nuclei remain some distance apart.

This nuclear division is more similar to that of two fungi belonging to the Phycomycetes, *Catenaria* (Ichida and Fuller, 1968) and *Blastocladiella* (Lessie and Lovatt, 1968), than to that of any other algae yet described although the long interzonal spindle is reminiscent of the diatom *Lithodesmium* (Manton *et al.*, 1969a). It remains to be seen whether other members of the Xanthophyceae also have this type of mitosis for *Vaucheria* differs from them in many ways.

H. RHODOPHYCEAE

The first detailed study of mitotic fine structure in this class is that of *Membranoptera* by McDonald (1972). It was found that during prophase the nucleus became spindle-shaped and was more or less surrounded by a sheath of ER, as in many green algae. Microtubules appeared between the nuclear envelope and the ER. At each pole a structure termed a 'polar ring' becomes evident and now the nucleus becomes somewhat indented at its poles. The polar ring is a short hollow cylinder about 70 nm high and with a maximum diameter of 190 nm.

Its walls are about 25 nm thick. Often, microtubules appear to focus on the polar ring although this does not always seem to be the case. At metaphase the nuclear envelope remains intact except for large numbers of pores at the poles and, through these, spindle microtubules pass to connect with the chromosomes which are now arranged across the equator. The chromosomes each possess a pair of kinetochores which are relatively simple in construction. Other spindle microtubules pass directly from pole to pole whilst some remain outside the nucleus. The nucleolus becomes dispersed at this stage. During anaphase the nucleus elongates, the continuous microtubules apparently increasing in length whilst those attached to the chromosomes shorten. The nuclear envelope constricts in the middle and the daughter nuclei are separated. Cytokinesis then begins by centripetal furrowing of the parent wall.

In a study of the nucleus in the unicellular red alga *Porphyridium* Chapman *et al.* (1971) found, at a certain stage in the growth cycle, dense arrays of microtubules within the nucleus. Later, circular regions of the nucleus were found to be bounded by membrane and these were interpreted as cytoplasmic invaginations or channels. These so-called 'concentrosomes' contained four circular profiles of membrane. The function of the microtubules and concentrosomes was not determined although it seems likely that they form part of the mitotic apparatus.

I. EUGLENOPHYCEAE

In the Euglenophyceae we find a mitosis which differs even more from the typical pattern than the examples quoted above. In *Euglena* (Leedale, 1968) the premitotic nucleus migrates towards the anterior end of the cell. There is, however, no evidence that the flagellar bases act as centrioles although Sommer and Blum (1965b) do suggest that in the related organism *Astasia* the flagellar bases are 'division centers', for they lie in a depression of the nuclear envelope during prophase. This has also been observed in *Trachelomonas* (Dodge, unpub. obs.). In *Euglena*, prophase is marked by the appearance of microtubules in the nucleoplasm. The nucleolus (often called an endosome in this class) elongates along the plane of the division axis and the chromosomes become arranged in a rather loose equatorial band around it. This is the equivalent of metaphase. The nucleolus elongates and microtubules are seen to run near its surface. Bundles of microtubules also run from pole to pole but do not apparently ever contact the chromatids which have no centromere or kinetochore. The chromatids now slowly segregate to the poles and the endosome appears to become pulled into two portions, possibly assisted by the microtubules which run along its surface. The nuclear envelope, which is entire throughout mitosis, becomes constricted around the equatorial region and the two daughter nuclei are separated. The chromosomes remain condensed throughout the nuclear cycle but it is not clear when they replicate. According to Leedale (1968) the sister chromatids have separated from each other before metaphase.

J. DINOPHYCEAE

In the Dinophyceae there are some similarities with the mitosis of *Euglena*, in particular we find that the nuclear envelope is continuous throughout the division. The mitosis has, as yet, only been described in detail from two marine dinoflagellates (Leadbeater and Dodge, 1967b; Kubai and Ris, 1969) but it would appear that the process is fairly similar in all free-living members of this class. The division commences with the chromosomes splitting into pairs of chromatids. This process commences at one end and gradually moves along, giving Y- and V-configurations, until the sister chromatids lie more or less side by side. The nucleus enlarges whilst this is happening and becomes rectangular, in median section, and not at all spindle shaped. A number of rather irregular invaginations of the nuclear envelope penetrate into the nucleus (Figs 7.12 and 7.13) and some at least pass completely through. Microtubules are formed in these tunnels and ribosomes and occasional mitochondria may be found in them (Fig. 7.12). The microtubules run right through the nucleus and just into the cytoplasm at each side but they do not appear to connect with any other structure. This metaphase condition now leads to anaphase in which the nucleus extends and the chromatids begin to move towards opposite 'poles'. As the nucleus extends the envelope remains intact and the tunnels are drawn out. Eventually the nuclear envelope constricts around the 'equator' and separates the daughter nuclei. Residual tunnels may remain in the daughter nuclei for some time but they have normally disappeared by the time that cell cleavage is complete. The nucleolus persists throughout mitosis and is presumably drawn into two parts like that of the Euglenophyceae. It has been suggested (Kubai and Ris, 1969) that segregation is effected by the chromatids being attached to the nuclear envelope where it surrounds the tunnels (cf. Fig. 7.13). It is very interesting that in these organisms the equivalent of the spindle, i.e. the microtubules, is entirely cytoplasmic whereas in some algae, such as the Euglenophyceae, it is entirely intra-nuclear and in other organisms it is part cytoplasmic and partly nuclear.

K. CRYPTOPHYCEAE

Very recent work on *Chroomonas salina* (Oakley and Dodge, unpub. obs.) has given some information about the, clearly unique, form of mitosis in this class. Before division starts microtubules are present in the cytoplasm outside the nucleus. By metaphase the nuclear envelope is completely broken down but the area in which mitosis takes place is roughly delimited by sheets of endoplasmic reticulum. The chromosomes form a single dense metaphase plate on the 'equator'. Many spindle tubules pass right through the plate and others may be attached to it. No definite spindle-organizing structure or centriole has been observed. At anaphase the daughter chromosomes separate as two distinct clumps in which individual chromosomes cannot be distinguished. In telophase

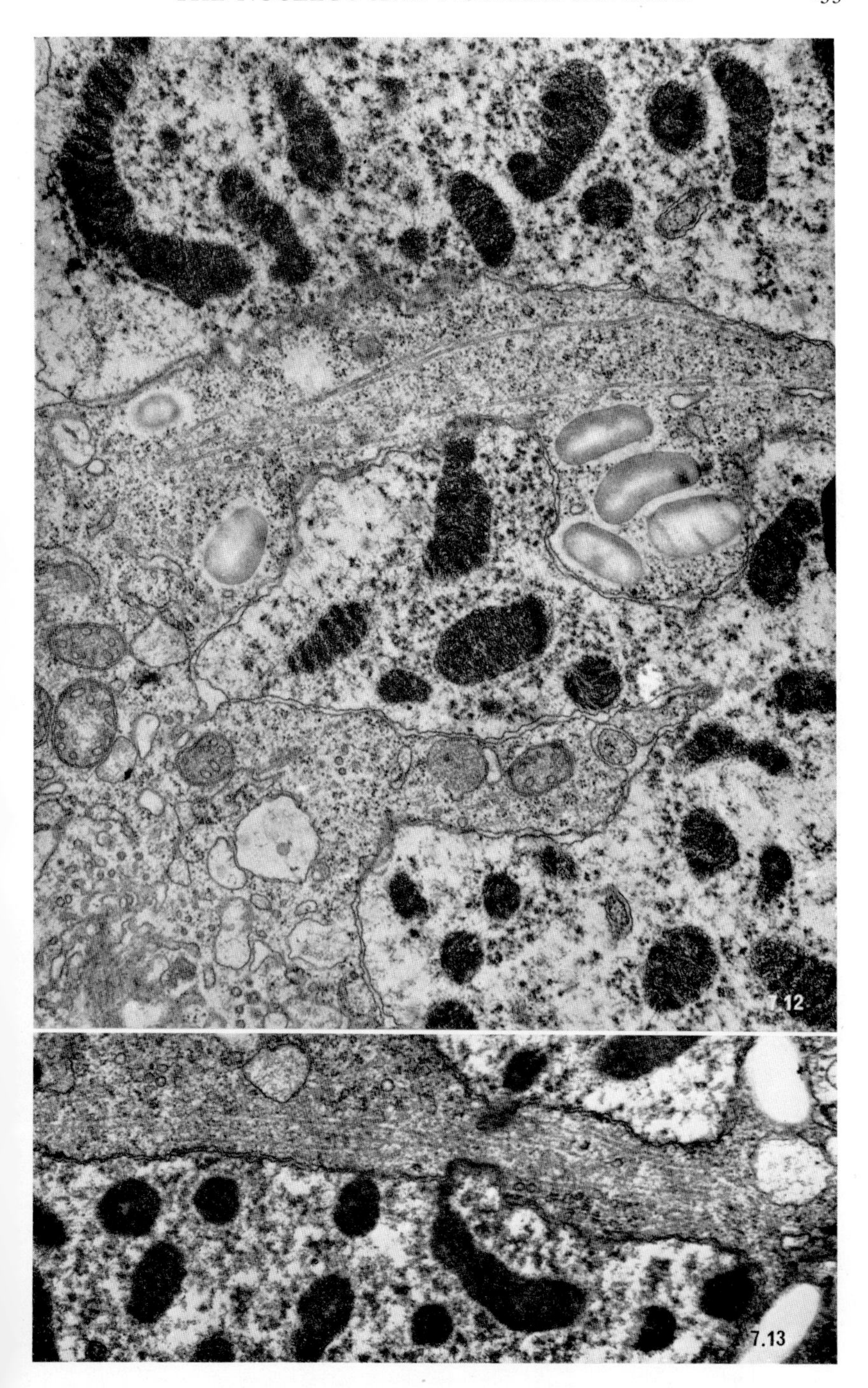

FIG. 7.12. A longitudinal section of part of a metaphase nucleus in *Heterocapsa* showing parts of two cytoplasmic tunnels through the nucleus, one of which contains several microtubules (×18,000) (after Dodge, 1971b); (7.13) a small portion of a cytoplasmic tunnel in a dividing nucleus of *Glenodinium*, showing two notches where chromosomes appear to be attached to the nuclear envelope bounding the tunnel. (×21,000) (After Dodge, 1971a.)

TABLE V

Summary of the present information on mitosis in algae

	Nuclear envelope			Mitotic ER sheath	Polar structure	Spindle type	Kinetochores
	Intact	Polar gaps	Breaks down				
Chlorophyceae:							
Volvocales		+			?	Nuc. + Cyt.	?
Chlorococcales		+		+	Centrioles	Nuc. + Cyt.	—
Ulvales		+			Centrioles?	Nuc. + Cyt.	—
Chaetophorales					?	?	
Oedogoniales	±				—	Nuc. + Cyt.	+
Zygnemales			+		?	Nuc. + Cyt.	+
Euglenophyceae	+				?	Nuclear	—
Chloromonadophyceae	+				Centrioles	Nuc. + Cyt.	+
Xanthophyceae	+				Centrioles	Nuclear	—
Chrysophyceae			+		Rhizoplast	Nuc. + Cyt.	—
Haptophyceae			+		?	Nuc. + Cyt.	—
Cryptophyceae			+	+?	—	Nuc. + Cyt.	—
Bacillariophyceae			+		Spindle precursor	Cytoplasmic?	
Phaeophyceae		+			Centrioles	Nuc. + Cyt.	
Dinophyceae	+				—	Cytoplasmic tunnels	—
Rhodophyceae		+		+	Polar rings	Nuc. + Cyt.	+

the clumps of chromosomes appear to be pushed back against part of the ER sheath surrounding the chloroplasts and this probably forms the first section of the nuclear envelope.

This type of mitosis appears to have some similarities with that of the Chrysophyceae (e.g. microtubules outside nucleus before prophase) and with that of the Haptophyceae (e.g. nuclear envelope broken down; no centrioles etc.). It is absolutely different to and distinct from the mitosis of the Dinophyceae and Euglenophyceae and fairly unlike that of the Rhodophyceae. Thus, the information presented here about mitosis in the Cryptophyceae will clearly necessitate some rethinking of the currently accepted ideas concerning the phylogenetic relationships of the algal classes.

As will be noted from the above account, we have as yet no details of nuclear division fine structure in the Eustigmatophyceae and Prasinophyceae. It will be very interesting to have these in order to complete the picture which is summarized in Table V. The evolutionary and phylogenetic implications of the various types of algal nuclear division, in relation to those found in the fungi, Protozoa and other groups, have already been thoroughly reviewed (Pickett-Heaps, 1969, 1972a; Leedale, 1970) and will not be discussed here.

III. Nuclear Division—Meiosis

To date this has been mainly studied in one organism, the marine centric diatom *Lithodesmium* (Manton *et al.*, 1969b, 1970a,b) for which we have a very detailed account. Meiotic division appears very similar to the unique form of mitosis found in this organism (see above). One difference is that the number of microtubules appears to be increased during first division, but in the second division it is reduced to approximately half and the spindle is smaller. Sections near the end of the mitotic spindle revealed 103 microtubules as compared with 166–190 at Meiosis I and 82–139 at Meiosis II. In all cases the microtubules were grouped into *c.* 16 bundles. During interkinesis between the first and second meiotic divisions of spermatogenesis, flagellar bases formed near the poles of the future spindle. These produced typical external portions of flagella during the early prophase, the spermatozoa being provided with a single flagellum in this organism.

A study of meiosis has recently been carried out on the green alga *Ulva* (Fig. 7.14) (Bråten and Nordby, 1973). Apart from prophase this did not differ significantly from mitosis in this alga (see IIA above and Løvlie and Bråten, 1970). In the prophase a distinct synaptonemal complex was observed. These structures have recently been observed in early prophase nuclei of tetraspore mother cells in a number of red algae (Kugrens and West, 1972a). They are flattened bodies consisting of two lateral, dark-staining, elements from which small fibrils extend inwards to form a central, less densely stained, element. This structure is almost identical to that found in meiotic nuclei of many animals and some plants (cf. Dupraw, 1970).

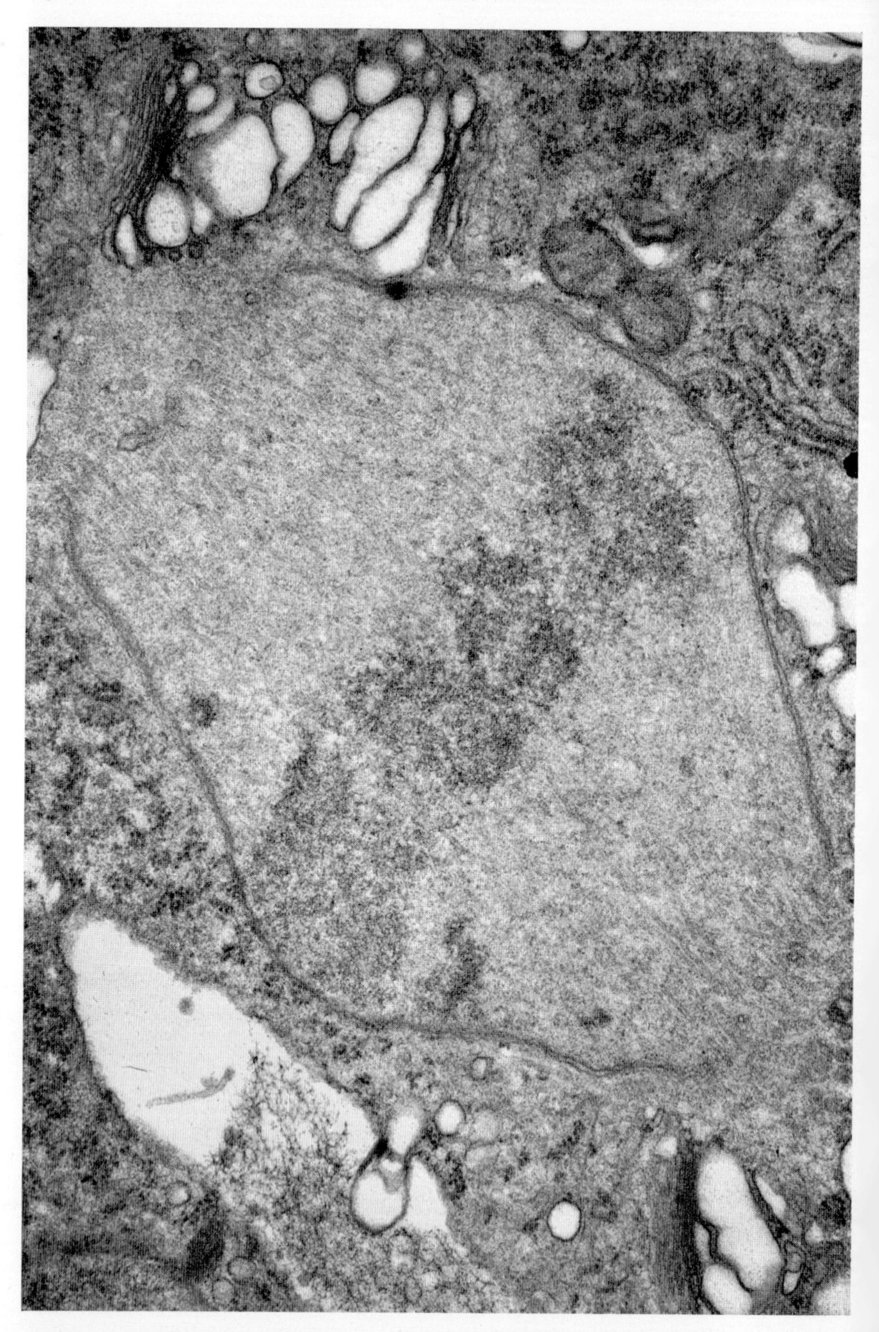

FIG. 7.14. Metaphase of meiosis in the 'slender' mutant of *Ulva mutabilis* (Chlorophyceae). Note that the nuclear envelope is complete except for the polar fenestrae. The chromosomes are all collected at the equator. (×38,500) (Micrograph provided by T. Braten.)

8. *Cell Division (Cytokinesis)*

I. Unicellular Algae

A. FLAGELLATES

In many simple unicellular algae nuclear division is immediately followed by cleavage of the parent cell into two daughter cells. The process is probably always directed, if not brought about, by microtubules. This clearly is the case in

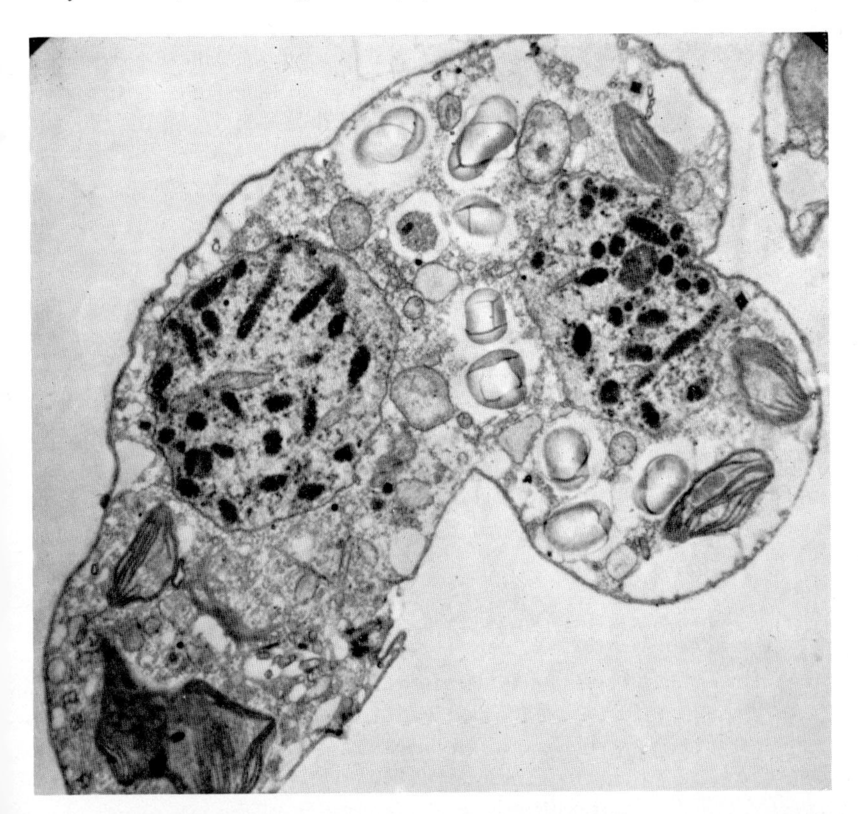

Fig. 8.1. Telophase in the division of a small dinoflagellate with the daughter nuclei separate but remnants of cytoplasmic tunnels still remaining. Cytokinesis is well under way, having commenced at the posterior end of the cell. (×7000) (After Leadbeater and Dodge, 1967.)

Chlamydomonas (Johnson and Porter, 1968) where the special set of microtubules which forms the cleavage apparatus appears in the cell at the end of telophase. These tubules radiate from the basal bodies and extend along both sides of the cleavage furrow just under the cell membrane. The furrow commences near the anterior end of the cell and soon extends all around it. A very similar process takes place in *Ochromonas* (Slankis and Gibbs, 1972) where the first constriction appears between the two sets of flagella at the anterior end of the cell. The constriction gradually deepens passing between the daughter nuclei and chloroplasts and eventually divides the leucosin vacuole. The two daughter cells may remain connected by their extended tails for a considerable period of time. Presumably a somewhat analogous process of cleavage takes place in *Woloszynskia* (Leadbeater and Dodge, 1967b) but here the involvement of microtubules has not yet been ascertained. In this flagellate, cleavage commences at the posterior end of the cell and works towards the anterior end (Fig. 8.1). In the examples mentioned above there is no appreciable increase in cell volume during division. However, two daughter cells having the same total volume as one parent cell will have a greater surface area. It is not clear where the increased cell membrane is synthesized although Johnson and Porter (1968) have noticed small vesicles, which they think have come from the Golgi, lying along the line of the future cleavage plane.

B. DESMIDS (CHLOROPHYCEAE)

In the desmid *Closterium* cell division is brought about by an ingrowing annular furrow into which vesicles appear to be discharged (Pickett-Heaps and Fowke, 1970c) (Fig. 8.2). No microtubules were seen associated with this wall although many were found orientated transversely, near to the wall, after it was completed. These tubules are probably involved in the expansion process by which the new portion of wall extends to re-establish the bilateral symmetry of the desmid. Numerous bundles of microtubules have been noted in *Micrasterias*, another desmid, immediately after cell division (Kiermayer, 1968).

C. COCCOID ALGAE (CHLOROPHYCEAE)

In *Chlorella* the process of cell plate formation has been described by Wanka (1968). Here, cytokinesis commenced with the appearance of small electron transparent droplets in a plane between the two daughter nuclei. These fused into a thin cell plate which extended to the cell periphery. Addition of colchicine prevented cell division by disturbing the orientation of the cell plates. Thus, although this work does not report the presence of microtubules in the developing cell plates the fact that colchicine has such a disruptive effect is strong evidence that they are in fact involved here, as in normal plant cell division. A later study

(Wilson *et al.*, 1973) has clearly revealed the presence of microtubules. Wanka also thought that the Golgi bodies, which he found situated at the opposite side of the cell to the developing cell plate, were not contributing to the deposition of wall material. If this is so it contradicts the work of Bisalputra *et al.* (1966) and Staehelin (1966), all with *Chlorella*, who reported the involvement of the Golgi bodies in the process of wall formation.

TABLE VI

The types of cell division in algae

	Cell cleavage in flagellates			Cell plate formation		
	Starts at anterior		Starts at posterior	Furrow type		Phycoplast type
Chlorophyceae:						
Volvocales	+	&	+			
Chlorococcales						+
Ulvales				+		
Chaetophorales						+
Oedogoniales				+	&	+
Zygnemales				+	&	+
Euglenophyceae	+					
Chloromonadophyceae	+					
Chrysophyceae	+					
Haptophyceae	+					
Bacillariophyceae				+		
Phaeophyceae				?		
Dinophyceae			+			
Rhodophyceae				+		

More recently, detailed accounts of cell division have been provided for two other coccoid green algae, *Kirchneriella* and *Tetraedron* (Pickett-Heaps, 1970b, 1972f), and these give some light on the early stages of cytokinesis. In these algae one or more nuclear divisions take place and then arrays of what are termed 'phycoplast' microtubules proliferate between the pairs of daughter nuclei (cf. Figs 8.3 and 8.4). Other microtubules encircle the cell, demarcating the future cleavage plane. The cytoplasm is now divided up by membrane furrows which run through the zone of phycoplast tubules (Fig. 8.4) and form the primary septum. The origin of these septa is obscure. They often appear first as an elongated two-membrane structure near the centre of the dividing cell but other examples suggest that they also might derive from invaginations of the plasmalemma. Any chloroplasts which lie in the way of the septa become constricted and then divided to allow the passage of the septum. The mother cell thus becomes cut up into a number of autospores and, as wall material is secreted into these cleavage septa, so the spores take on their characteristic morphology. A similar type of cell division appears to take place in *Ankistrodesmus* (Pickett-Heaps, 1972e) and *Scenedesmus* (Dodge, unpubl. obs.).

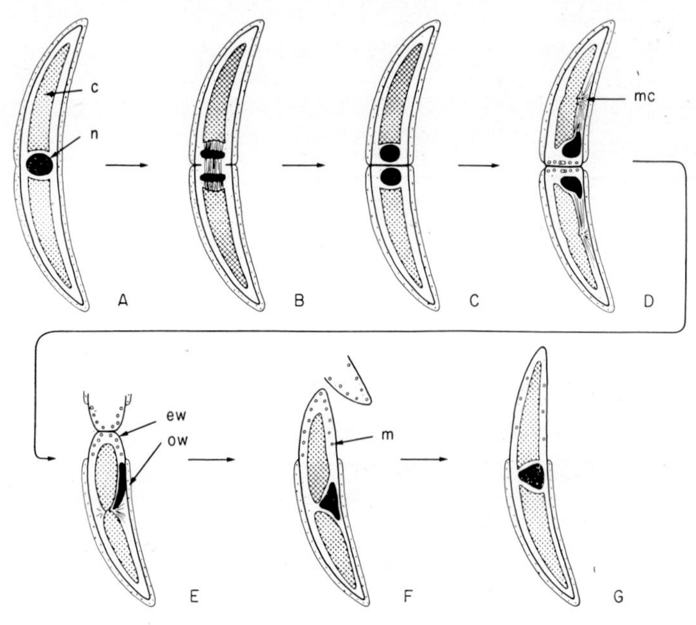

FIG. 8.2. Diagrams to illustrate cytokinesis in the desmid *Closterium*. A–B, Division of the nucleus (n); B–C, the septum forms between the daughter cells; D, transverse microtubules (m) appear near the septum and others half-way along each daughter cell, associated with a microtubule centre (mc); E–F, the nuclei migrate along the new cells and come to lie in a newly formed break in the chloroplast (c); F–G, the daughter cells expand (ew) at the septum and each forms a new semicell which extends out of the old wall (ow). Thus the typical form of the vegetative cell is re-established. Reproduced, with permission, from Pickett-Heaps and Fowke (1970c).

The striking difference between cell division in these coccoid algae and higher plants is that here the septum and wall form in the same plane as the microtubules whereas in higher plants they always form at right angles to the microtubules, which appear to remain in the position they had occupied during mitosis when they formed the spindle.

II. MULTICELLULAR ALGAE

A. FILAMENTS

In the filamentous alga *Spirogyra*, cell cleavage appears to be of a type intermediate between that described above and the phragmoplast type as found in

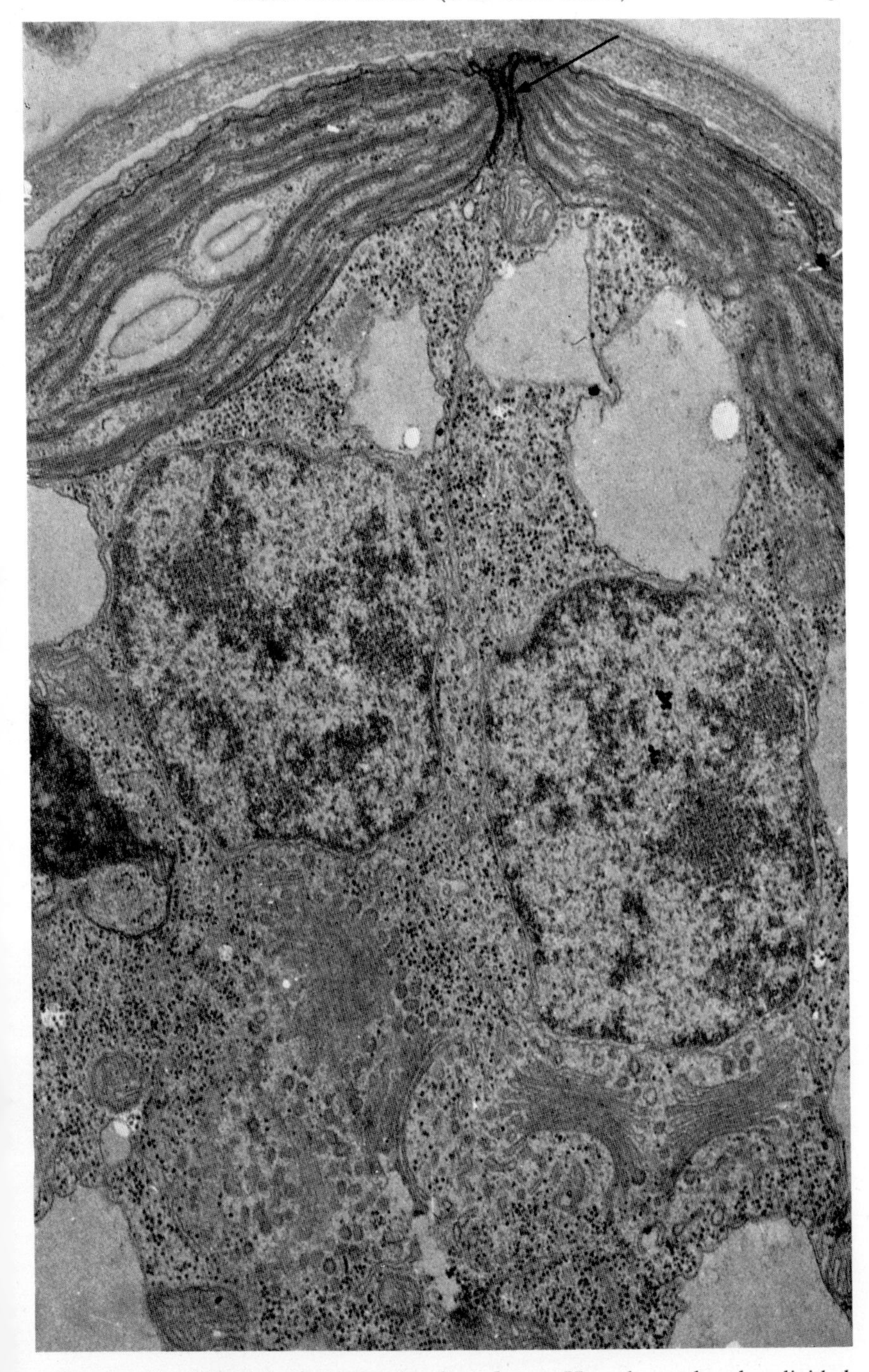

FIG. 8.3. Cytokinesis in the green alga *Scenedesmus*. Here the nucleus has divided and the Golgi duplicated. A break has recently formed through the peripheral chloroplast (arrow). ($\times$ 28,000)

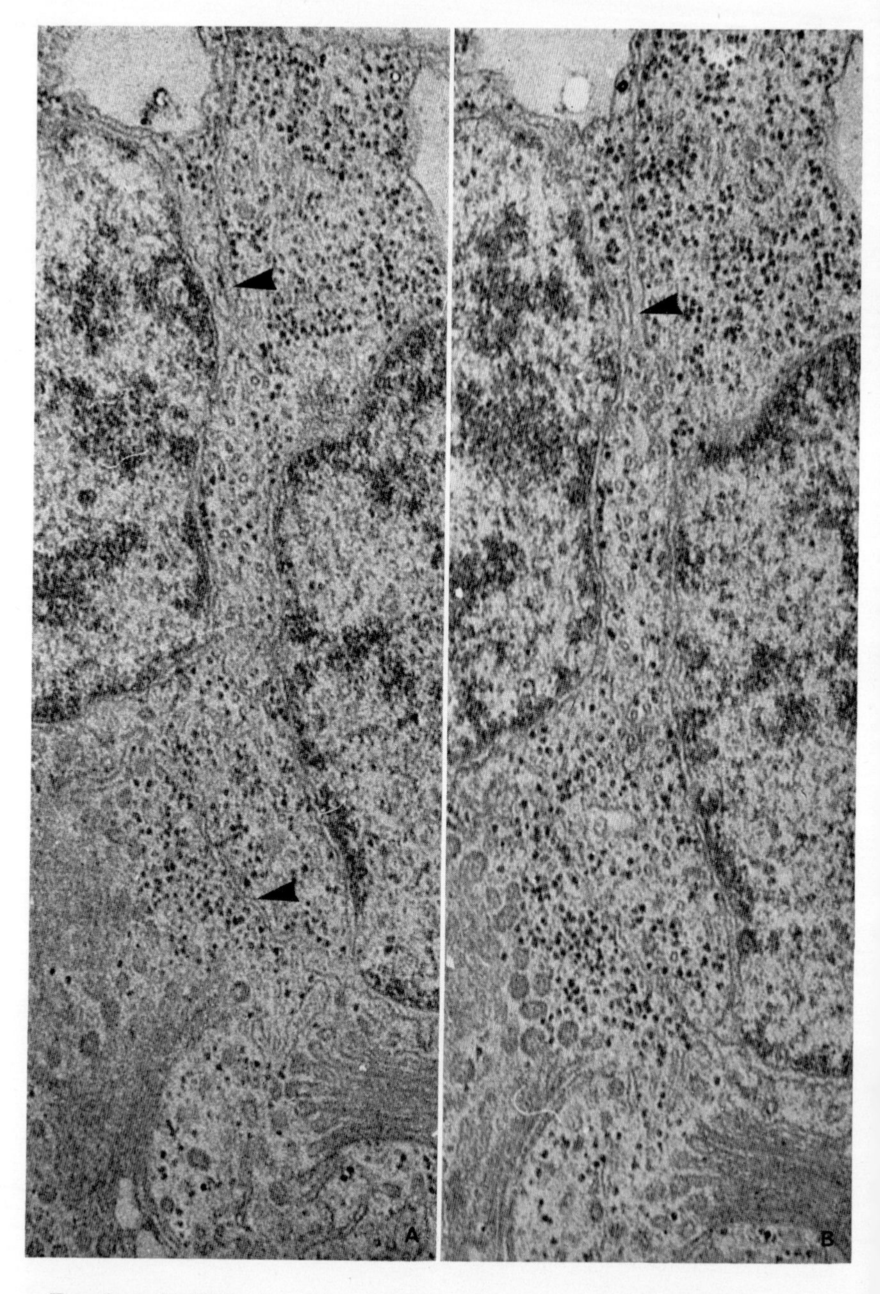

FIGS 8.4A–B. Two of a series of sections of cell plate formation in *Scenedesmus* showing the numerous transversely orientated phycoplast microtubules between the daughter nuclei. In places flattened vesicles appear to be lining up (small arrows) to form the cleavage fissure between the cells. (A, ×43,400); (B, ×43,700)

higher plants. Fowke and Pickett-Heaps (1969a,b) found that after nuclear division the new cross wall commenced as an annular ingrowth from all around the cell wall. This formed a septum which extended into the vacuole. When the septum was about half-completed the cytoplasmic strands which joined the two daughter nuclei bulged at their centre and joined with the cytoplasm which was being pushed forward by the advancing septum. The strands then fused to give a cylinder of cytoplasm between the nuclei. Longitudinal microtubules were present in the cytoplasmic strands. Vesicles collected in a line across the aperture in the septum and they were associated with electron dense material. The longitudinal microtubules passed through this area which forms a type of phragmoplast. The cell plate was then formed, presumably as a result of coalescence of the vesicles.

Another unusual type of crosswall formation is found in *Oedogonium* (Chlorophyceae) (Hill and Machlis, 1968; Pickett-Heaps and Fowke, 1969, 1970b). Before and during mitosis a ring of material is laid down near one end of the cell. When nuclear division is complete the parent wall splits around the ring and this becomes drawn out to form the outer layer of the new cell wall of one of the daughter cells. A septum is produced across the cell between the two nuclei at the end of mitosis and this contains many transverse microtubules interspersed with vesicles. Following the splitting of the wall the septum moves up the cell, like a diaphragm, until it reaches the bottom end of the split ring and here the vesicles fuse together to form a new crosswall. The interesting feature here is that the microtubules are all orientated transversely and not at all longitudinally in the way that they are in *Spirogyra* or in a normal phragmoplast.

In the green filamentous alga *Klebsormidium* (Floyd *et al.*, 1972b) the transverse septum begins to form, by ingrowth from the plasmalemma and cell wall, as early as metaphase of the nuclear division. During late telophase this grows across the cell at a rapid rate, cutting completely through the central vacuole. No microtubules were observed in the vicinity of the developing septum. A thick ring of wall material begins to be deposited in the septum even before mitosis is over and this is completed, in a centripetal direction, immediately following cytokinesis.

In *Stigeoclonium* Pickett-Heaps (1972e) found that the septum or cell plate was formed by coalescence of numerous vesicles which collected in the space between the daughter nuclei. The crosswall appeared to grow in a similar way to that of higher plants, however, a typical algal phycoplast of transverse microtubules was seen between the two daughter nuclei before the septum formed. It is interesting that in another member of the same order (Chaetophorales), *Fritschiella*, McBride (1967, 1970) has reported the presence of a 'typical cell plate'. This he says commences to form in the centre and develops towards the edges of the cell. A considerable amount of Golgi activity was noted, adjacent to the developing plate.

B. PARENCHYMATOUS ALGAE

The only parenchymatous alga that has been studied in any detail is *Ulva mutabilis*. Here, Løvlie and Bråten (1968, 1970) report that at the start of mitosis the nucleus and Golgi region sink to the centre of the bottom of the cell. Following nuclear division the Golgi body divides and the two portions become orientated facing each other between the two daughter nuclei. The division furrow appears to commence in the plasmalemma adjacent to the pair of Golgi bodies and then gradually extends up through the vacuole. There is no evidence of the furrow being formed by vesicle coalescence. When the septum reaches the chloroplast this appears to be bisected and the two halves slide down to lie against the side walls of the cell. The nuclei and Golgi also move, to lie adjacent to the chloroplasts. The division furrow is completed and then wall material is deposited in it. The rather unusual feature of this division is the way in which the furrow appears to develop from only one side of the cell and is therefore neither completely centripetal in development, like many other filamentous algae, or centrifugal, like the higher plants.

The phylogenetic and evolutionary significance of the various types of cytokinesis in the algae, fungi and other plants has been discussed by Pickett-Heaps (1972e) in a recent review.

9. *Ejectile Organelles*

In at least six of the algal classes some species possess organelles which on stimulation by contact, heat or chemicals discharge a structure such as a thread or tube from the surface of the cell. All of the organisms with such structures are unicellular and normally motile. In no case is the purpose of these organelles understood and very little is known about their mode of action.

I. Trichocysts in the Dinophyceae

Most dinoflagellates possess structures which are in many ways similar to the trichocysts of ciliates (Ehret and de Haller, 1963; Selman and Jurand, 1970) although they are less complex in construction. In the undischarged state they lie around the periphery of the cell and roughly at right angles to the cell surface. Each cell may contain just a few (as *Gymnodinium simplex*) or many hundreds (as *Oxyrrhis marina*; Dodge and Crawford, 1971b). The trichocysts are fusiform structures which clearly consist of two parts (Figs 9.1 and 9.5). The neck, which makes up about one-third of the length, is narrower than the rest and is attached to the thecal membranes which surround the cell. In organisms with thecal plates the trichocysts lie immediately beneath pores which are usually present in a characteristic arrangement on each plate. The neck (Fig. 9.5) contains a number of slightly twisted fibres which connect at their proximal end with the main body of the trichocyst. The main part of the organelle consists of a square or rhomboidal shaped shaft some 2–4 μm long which tapers away from the neck. Thin cross-sections reveal that this is a para-crystalline structure (Fig. 9.2) composed of proteinaceous granules arranged in a grid of *c.* 50×50. Each granule is about 6 nm in diameter. The whole trichocyst is surrounded by a single membrane which in some organisms shows transverse corrugations (Bouck and Sweeney, 1966). Occasionally, threads or fibrils are present between the membrane and the body of the trichocyst.

When stimulated the trichocysts shoot out threads, several microns long, which are always square or rhombic in section, but varying in width between 40 and 160 nm. When stained with phosphotungstate (Fig. 9.3) or when shadowed (Fig. 9.4) the thread can be seen to show a periodic transverse banding. The periodicity is extremely variable and presumably depends on both the environmental conditions during discharge, and on the method of preparation. In *Prorocentrum micans* (Bouck and Sweeney, 1966) major bands were spaced at

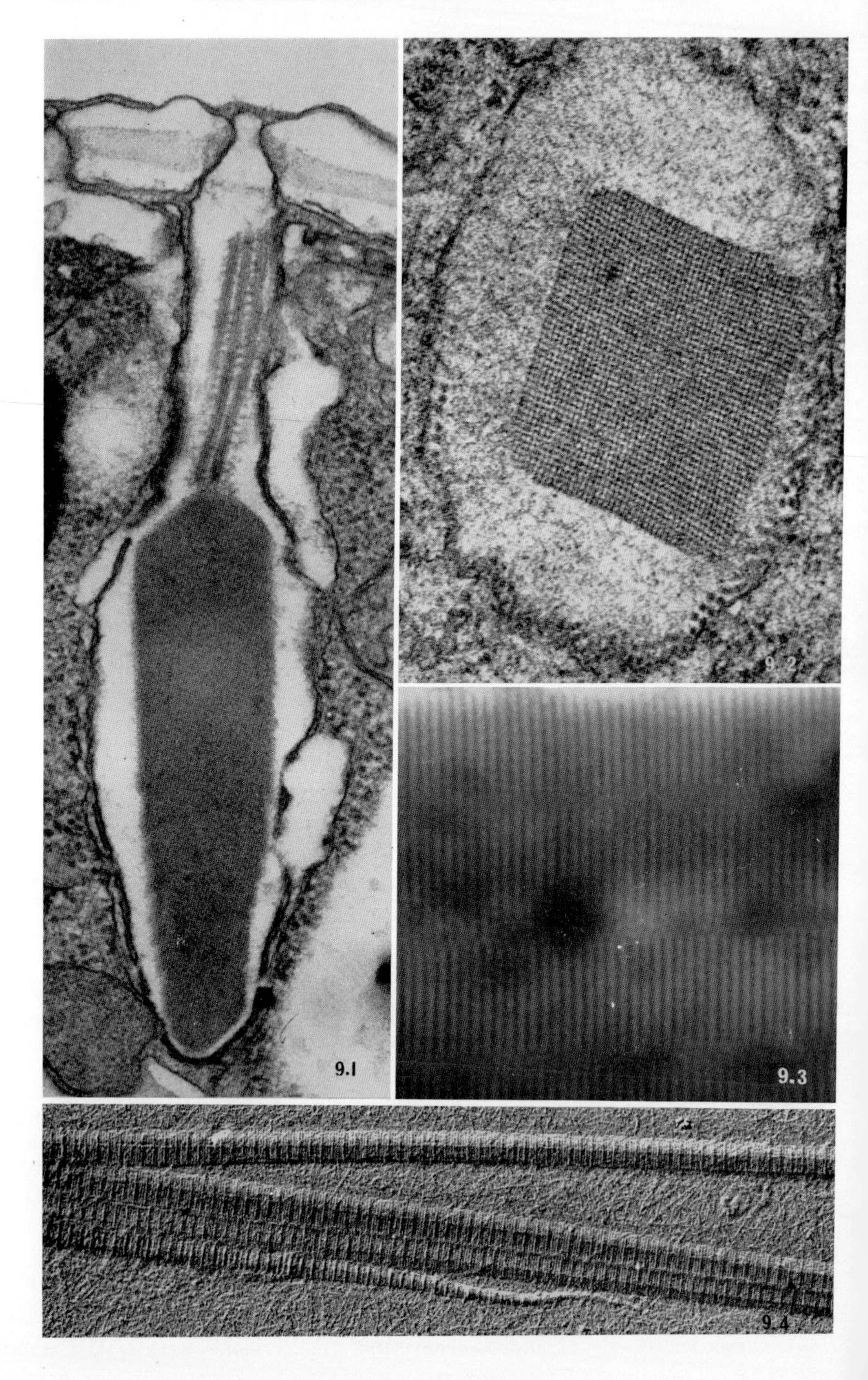
9.1
9.2
9.3
9.4

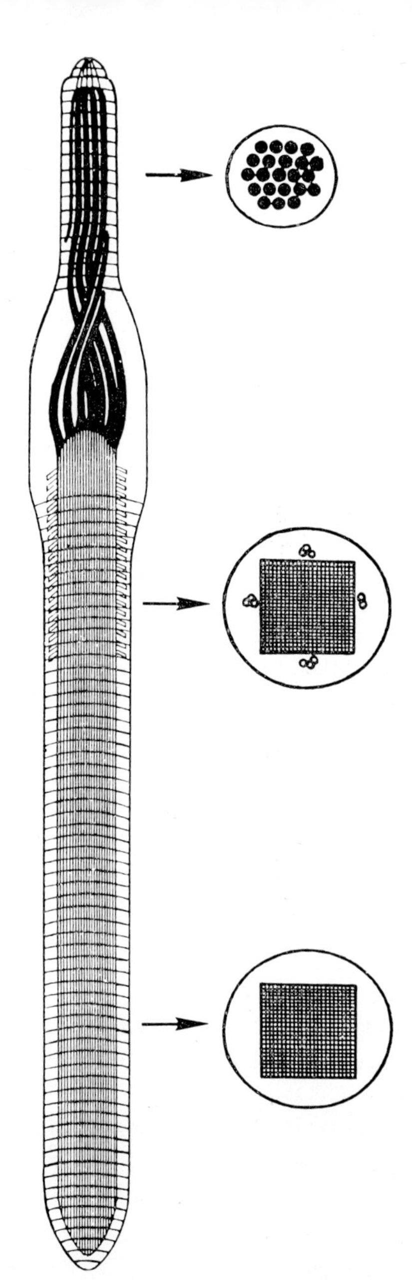

FIG. 9.5. A diagrammatic representation of an undischarged trichocyst of *Gonyaulax polyedra* as seen in longitudinal optical section and in transverse sections. Reproduced, with permission, from Bouck and Sweeney (1966).

FIGS 9.1–9.4. Trichocysts in the Dinophyceae. (9.1) A longitudinal section of a mature but undischarged trichocyst in *Woloszynskia coronata* showing the neck attached to the cell theca. (×59,000) (After Crawford and Dodge, 1973); (9.2) transverse section of an undischarged trichocyst in *Prorocentrum mariae-lebouriae* showing the regular packing of the protein subunits (×68,000); (9.3) discharged trichocyst threads in *P. micans* which have been stained with phosphotungstate. They show a regular 13 nm transverse banding (×126,000); (9.4) discharged trichocysts of *P. mariae-lebouriae* which have been shadowed. In this type of preparation every fifth band appears to be thicker than the intermediate ones. (×21,000)

82–85 nm with five minor bands of varying width between them. Yet the same species when differently prepared yielded regular bands of 13 nm (Fig. 9.3). In *Oxyrrhis marina* (Dodge and Crawford, 1971b) major bands varied from 37–70 nm and minor bands from 11–18 nm. In *Peridinium westii*, a freshwater dinoflagellate, major periods were of 65–70 nm with five irregularly spaced bands between (Messer and Ben-Shaul, 1971), and in this organism a longitudinal arrangement of fibrous subunits was also evident. The transverse banding of discharged trichocysts bears a strong resemblance to that of collagen but it has recently been shown by Messer and Ben-Shaul (1971) that the amino acid profile of the threads is not at all similar to that of collagen.

Trichocysts are initiated within swollen vesicles which arise in the vicinity of the Golgi bodies (Bouck and Sweeney, 1966; Leadbeater and Dodge, 1966). At first the vesicles contain dispersed fibro-granular material, but then a 'crystalline' structure begins to form at one side of the vesicle and as this extends so the dispersed material appears to be used up. Eventually the vesicle becomes extended by the trichocyst body which has formed in it. Now it migrates to the periphery of the cell and the neck region becomes delimited when the trichocyst is attached to the thecal membranes.

II. Nematocysts (Cnidocysts) of Dinoflagellates

More complex ejectile organelles are found in a few dinoflagellates and these are normally called nematocysts or cnidocysts because of their apparent similarity to the ejectile structures found in the Coelenterata (cf. Hovasse, 1965). Electron microscopy has shown that in *Nematodinium* (Mornin and Francis, 1967; Greuet, 1971) the undischarged nematocyst is 10–15 μm long and consists of a central core around which are arranged a group of about 14 ribs which are attached to it. The whole structure is enclosed by a membrane. The core appears to be pointed like a harpoon tip at one end and this lies towards the outside of the cell. Details of the fine structure of the slightly different, yet equally complex, nematocysts of *Polykrikos* have recently been provided (Greuet, 1972). No information is available concerning the development or the discharge of these organelles.

III. Ejectosomes of Cryptomonads

The ejectile organelles found in most members of the Cryptophyceae have been variously called *trichocysts* (Joyon, 1963b), *taeniobolocysts* (Mignot *et al.*, 1970) and *ejectosomes* (Anderson, 1962). Because they are completely different in structure from the trichocysts of dinoflagellates and ciliates it would seem appropriate to use a distinct word and *ejectosome* will be used here.

As first shown by Anderson (1962) for *Chilomonas*, the discharged ejectosome consists of a tube composed of membranous material. Shadowed preparations (Figs 9.6 and 9.8) reveal that although the tubes are mainly straight the tip is crooked (Dragesco, 1951; Schuster, 1968). More recently it has been shown that this region appears to consist of tube-membrane which is partly unrolled (Fig. 9.6b). When fully discharged the ejectosome is completely free from the cell and no connection remains.

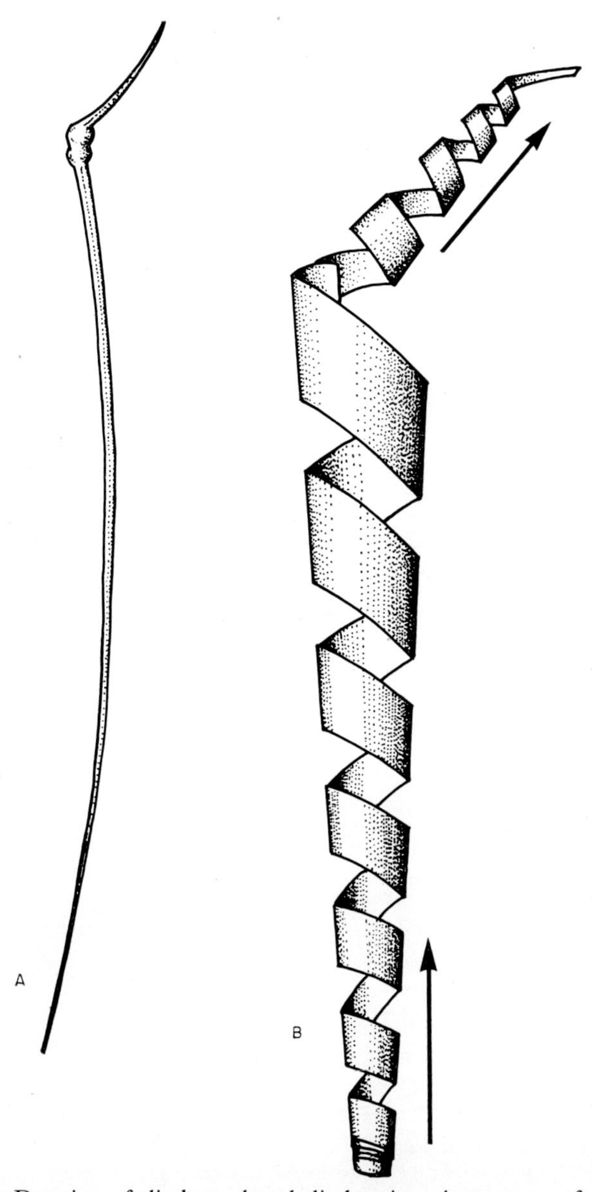

FIG. 9.6. Drawing of discharged and discharging ejectosomes of members of the Cryptophyceae. 9.6A shows a completely discharged structure (compare with Fig. 9.8) and in 9.6B this is stretched to show how it is constructed.

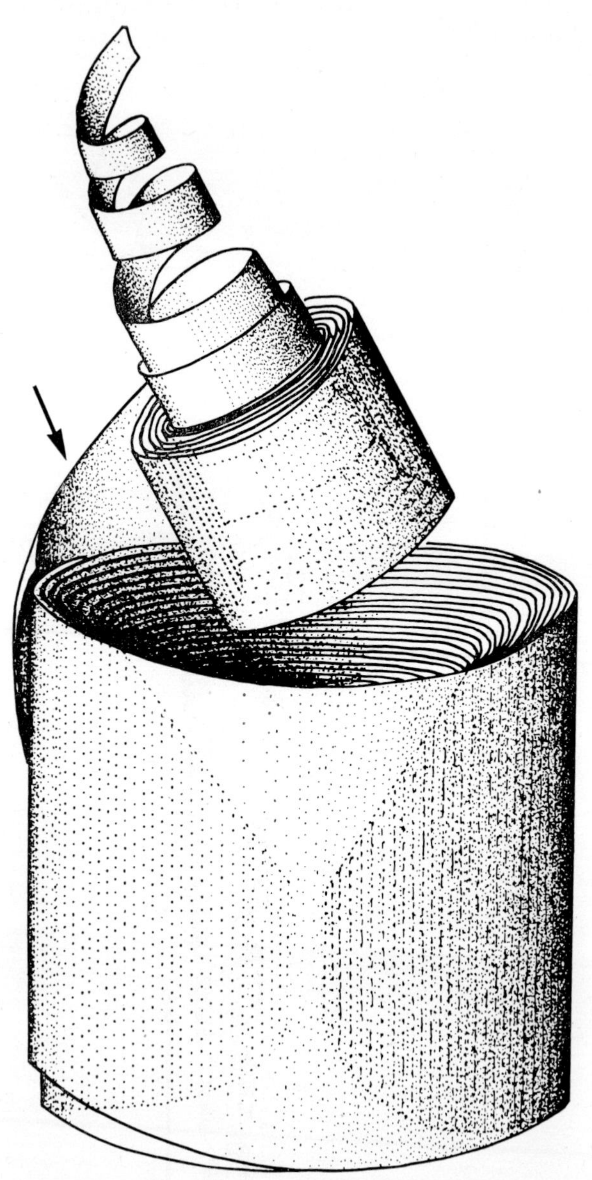

FIG. 9.7. This shows the reel of the undischarged ejectosome just beginning to be pulled out. (Figs. 9.6 and 9.7 reproduced, with permission, from Mignot, Joyon and Pringsheim 1968.)

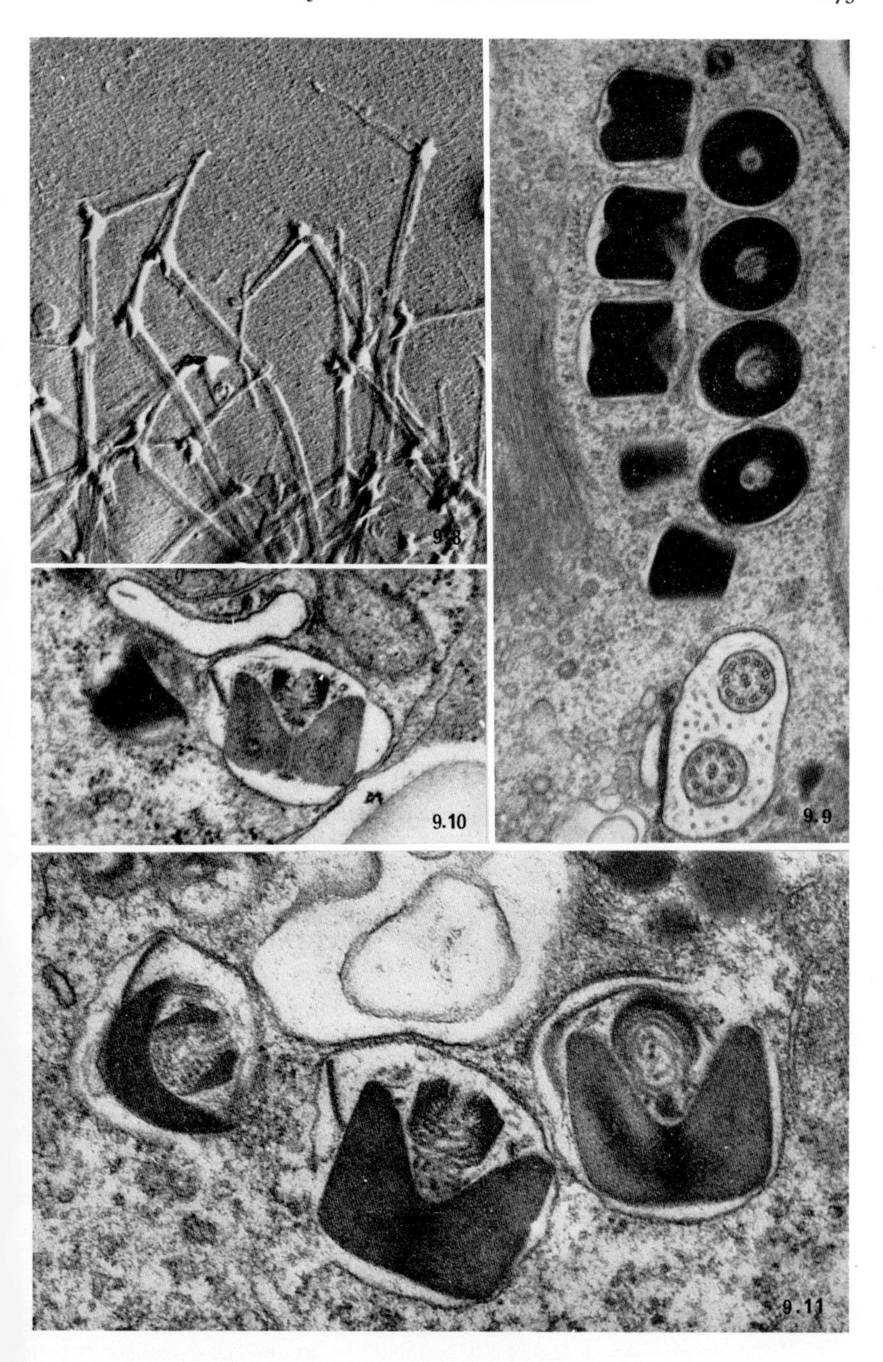

FIGS 9.8–9.11. Ejectosomes of Cryptomonads. (9.8) Discharged ejectosomes around a dried cell. (×21,000); (9.9) a group of mature ejectosomes ready for discharge in *Chroomonas*. These are seen in both longitudinal (left) and transverse (right) section (×28,000); (9.10) a rather unusual longitudinal section of an ejectosome of *Chroomonas* which appears to show a thin rod attaching the central body to the main reel (×43,000); (9.11) longitudinal and oblique sections of three ejectosomes in *Chroomonas mesostigmatica* which clearly show the central body. (×59,000) (Figs 9.9 and 9.11 after Dodge, 1969b.)

In all of the organisms which have been sectioned the undischarged ejectosomes consist of cylindrical structures surrounded by a single membrane. In *Chroomonas* (Fig. 9.9) they are about 500 nm diameter and 400 nm deep (Dodge, 1969b). They are mainly situated towards the anterior end of the cell adjacent to the depression or 'gullet'. In some organisms, such as *Chroomonas*, there are also small ejectosomes only 200 nm diameter which are located around the periphery of the cell, adjacent to grooves in the periplast (Fig. 2.1).

The internal structure of ejectosomes consists of two main parts. Most of the space is occupied by a cylinder of membranous material, coiled in a tight spiral, which surrounds a narrow channel and a V-shaped depression (Figs 9.7 and 9.11). In one case (Fig. 9.10) a short rod has been seen passing through the central channel. Partially sitting in the V-shaped depression and partly protruding is another small reel of membranous material which appears to have fibrous connections with the larger reel (Fig. 9.11). The possible ways in which the reels unwind or slide apart to give rise to the tubular discharged ejectosome have been thoroughly reviewed (Hovasse *et al.*, 1967). The generally accepted theory is that the central reel is shot out (by hydrostatic pressure?) and that as it leaves the cell it pulls out the ribbon of membrane from the larger coil starting from the outside.

The development of ejectosomes has been investigated by Wehrmeyer (1970a). He found early stages in their formation near to the single large Golgi body. These consisted of vesicles 100–160 nm in diameter containing a roll of only a few turns, plus a central body. During development the number of turns and the width of the membrane cylinder both increased. Wehrmeyer noted what appeared to be digestion of ejectosomes and suggested that those which are surplus to requirements might be digested by cytolosomes. Schuster (1970) found that when *Chilomonas* was 'starved' the ejectosomes appeared to be digested but when the organism was returned to a medium containing full nutrients then new ejectosomes were formed.

The basic similarity in construction between ejectosomes and the R bodies of killer *Paramecium* strains has been noted (Hovasse *et al.*, 1967). There is as yet no evidence that ejectosomes carry a toxin.

IV. Discobolocysts in the Chrysophyceae

In *Ochromonas tuberculatus* (Chrysophyceae) Hibberd (1970) has described an unusual form of ejectile organelle which he calls a *discobolocyst*. In the undischarged state these more or less ovoid organelles lie around the periphery of the cell partly protruding above the surface. Most of the discobolocyst is filled with fine fibrillar material, which is thought to be a mucopolysaccharide, except for the distal part which contains an electron-dense ring. On discharge the dense ring is shot some distance from the cell trailing behind it a delicate cylinder of fibrous material. It seems clear that these organelles arise from Golgi vesicles which swell and differentiate as they migrate towards the periphery of the cell.

V. Other Ejectile Organelles

Structures which are tubular when discharged have been reported from *Pyramimonas* (Prasinophyceae) (Manton, 1969) and *Entosiphon* (Euglenophyceae) (Mignot, 1963, 1966). In the case of *Pyramimonas grossi* the undischarged trichocysts are said to consist of a rolled ribbon, 5 nm thick and 500 nm wide, in some 15–25 turns. When discharged they form a hollow tube 100 nm wide and 35 μm long, tapered at both ends. The structure appears to be very similar to that of the ejectosomes of Cryptophyceae. The undischarged trichocysts of *Entosiphon* appear to be differently constructed. In transverse section they are seen to consist of a single thick circular membrane surrounded by what is probably a unit membrane. In longitudinal section they are tubular structures with rounded ends, about 150 nm in diameter and 4 μm long. They are quite distinct from any other ejectile organelles yet described.

More elaborate trichocysts, called *Alcontobolocysts* (Hovasse, 1965), have been noted in *Gonyostomum* (Chloromonadophyceae) (Mignot, 1967). These seem to have certain similarities with dinoflagellate trichocysts and in the undischarged state consist of a fusiform body with a narrow anterior region. As yet no details of the structure of the dense 'body' of these trichocysts have been reported, neither is it clear what form the discharged structure takes.

10. *Miscellaneous Organelles and Inclusions*

I. Mitochondria

All the groups of algae discussed in this book possess mitochondria and, as these appear to have the typical construction found in other plants and animals, they have proved of very little interest to electron microscopists. In shape they often have the common elongated ovoid form (Fig. 10.1) although in some groups bizarre and irregularly shaped mitochondria are common (Fig. 10.3). The inner mitochondrial membrane forms tubular cristae which have slightly constricted bases (Fig. 10.2). The cristae may be very numerous, so as to pack the lumen of the mitochondrion, or they may be very sparsely distributed as in the dinoflagellate *Oxyrrhis* (Dodge and Crawford, 1971b). In young zoospores of *Oedogonium* 'bristly' cristae have been observed (Pickett-Heaps, 1971b). There are only a few of these cristae in each mitochondrion and they are covered with pointed projections 22–31 nm long and 4 nm wide. It is possible that these bristles are related to the elementary particles or oxysomes which have been observed on the cristae in higher organisms (Fernandez-Moran *et al.*, 1964).

Electron transparent areas containing DNA fibrils (Fig. 10.4) have been found in the mitochondria of a number of algae including the dinoflagellates *Ceratium* (Afzelius, 1965) and *Wolozynskia micra* (Leadbeater and Dodge, 1966), the brown alga *Egregia* (Bisalputra and Bisalputra, 1967b) and various members of the Rhodophyceae, Phaeophyceae and Chlorophyceae (Yokomura, 1967b). DNA is probably present in all algal mitochondria but the amount per mitochondrion appears to vary considerably. The fibrils, which vary in thickness from 2–12 nm have been shown to be removed by DNAase treatment. An unusual tubular structure, composed of two or more helically arranged threads has recently been found in the mitochondria of the cells developing into carpospores in the red alga *Polysiphonia* (Tripodi *et al.*, 1972). This structure was removed by treatment with DNAase and presumably is the mitochondrial DNA in an unusually organized form.

Ribosomes have been noted in the mitochondria of *Ochromonas* (Smith-Johannsen and Gibbs, 1972). These workers were studying the effect of chloramphenicol on the structure of chloroplasts and mitochondria and they found that in the latter the drug caused a progressive loss of cristae which paralleled the decline in the growth rate of the cells. During treatment the mitochondrial matrix (in the lumen) became more dense after about 24 h treatment but the number of ribosomes per unit area decreased slightly.

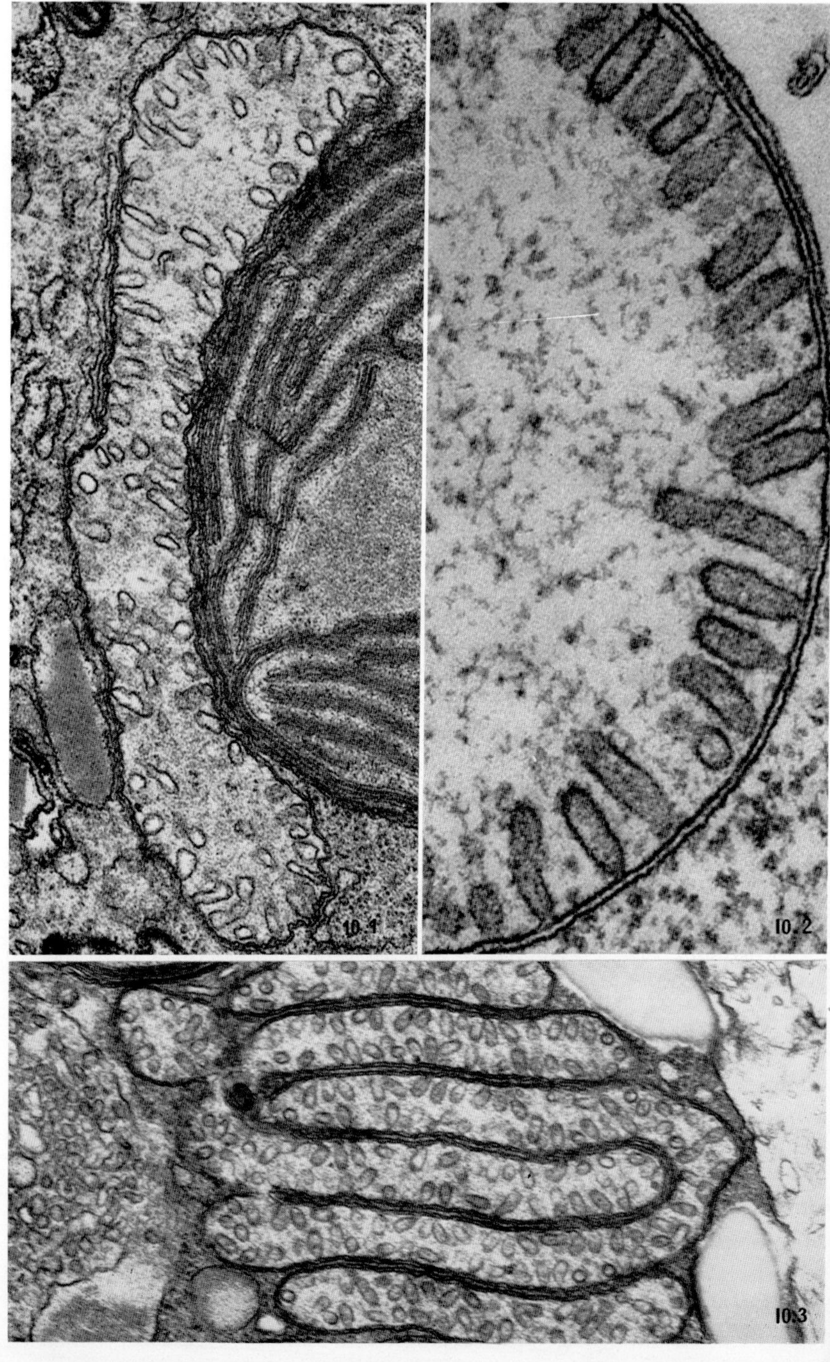

FIGS 10.1–10.3. Various mitochondria from dinoflagellates. (10.1) A 'typical' mitochondrion in *Glenodinium foliaceum* showing the two-membrane envelope and the tubular cristae (×43,000); (10.2) part of a mitochondrion from *Oxyrrhis marina* which shows the narrowed connection between the cristae and inner membrane and also granular contents of the cristae (×98,000) (after Dodge and Crawford, 1971b); (10.3) a rather unusual shaped mitochondrion or group of mitochondria in *Heterocapsa*. (×36,000)

II. Golgi or Dictyosomes

Typical Golgi bodies or dictyosomes are present in virtually all algal cells. In small unicellular organisms such as *Micromonas* (Manton, 1959a) each cell possesses just a single dictyosome (e.g. Fig. 10.5) but in large unicells, such as the diatom *Pinnularia* (Drum, 1966) and large cells of multicellular algae, the Golgi apparatus may consist of numerous dictyosomes (Fig. 10.6). Werz and

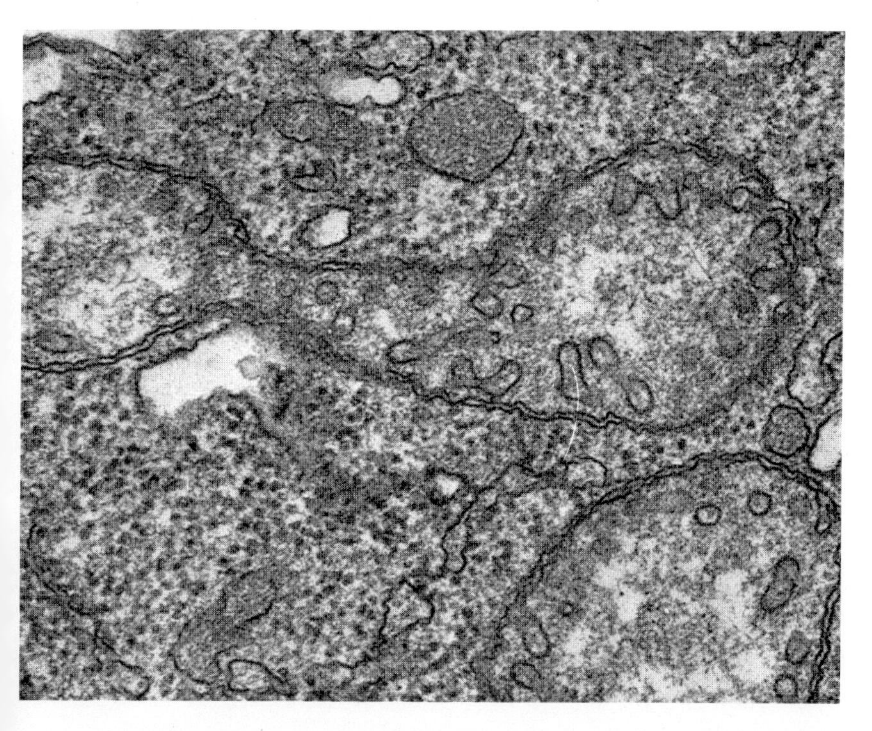

Fig. 10.4. Mitochondria of *Katodinium glandula* (Dinophyceae) with electron lucent areas containing delicate DNA fibrils (compare with Fig. 7.11). (×68,600)

Kellner (1970) used freeze-etching to examine the Golgi of *Dunaliella*. They found that the general form of the organelle was essentially the same as shown by the normal methods of chemical fixation and embedding but, in addition, it was possible to see particles on the surfaces of the cisternal membranes. It may be that they were looking at 'coated' swollen edges of the cisternae like those of *Chroomonas* shown in Fig. 10.5. Mollenhauer *et al.* (1968) have shown that in *Euglena* each of the numerous dictyosomes consists of a stack of 15–30 cisternae. At the centre of each dictyosome there is dense material within the lumen of the cisternae and in this region the intracisternal space is increased and the texture of the membranes seems to be altered. The significance of these observations has not been established, neither has those of Drum (1966) who showed that in *Pinnularia* the vesicles at the edges of the cisternae contain only lightly stained material during interphase of the nuclear cycle but that in prophase dense or darkly staining particles are present. Stahelin and Kiermayer (1970), using freeze-etching, found that there was a gradual increase in the density of particles on the membranes of successive cisternae from the forming to the maturing face

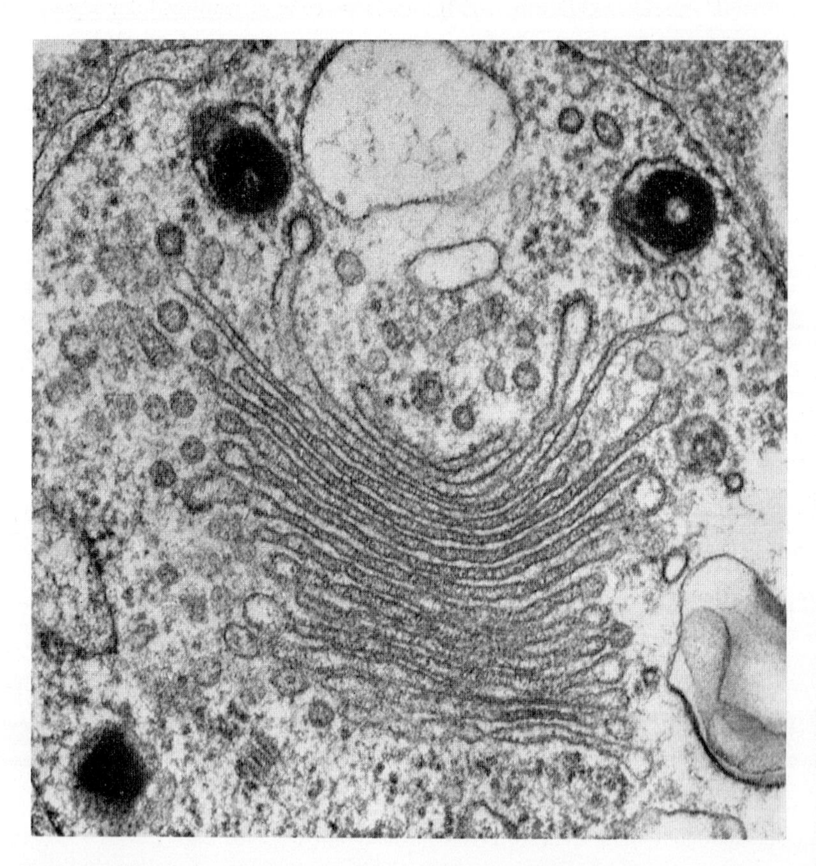

FIG. 10.5. Vertical section through the Golgi body of *Chroomonas mesostigmatica*. Note the swollen ends of the cisternae and the presence of small particles within and around some of these swellings and the adjacent Golgi vesicles. (× 59,000)

of the Golgi bodies in *Micrasterias*. When vesicles became separated from the cisternae at the edges of the Golgi there was a drop in the number of particles on the membrane which, it was thought, could be correlated with secretory products becoming packaged within the vesicles. An unusual microtubular 'crystal' has been found associated with the Golgi apparatus in *Pleurochrysis* (Brown and Franke, 1971). It is assumed that the microtubules are either being produced by the Golgi or alternatively they are involved in the production of scales which takes place in the Golgi of this organism. The role of Golgi bodies in thecal formation in the Prasinophyceae has been studied autoradiographically by Gooday (1971).

The role of Golgi bodies in wall formation, scale and coccolith morphogenesis, and trichocyst development, has been mentioned in other chapters. These clearly are very versatile organelles so far as the chemical composition and morphology

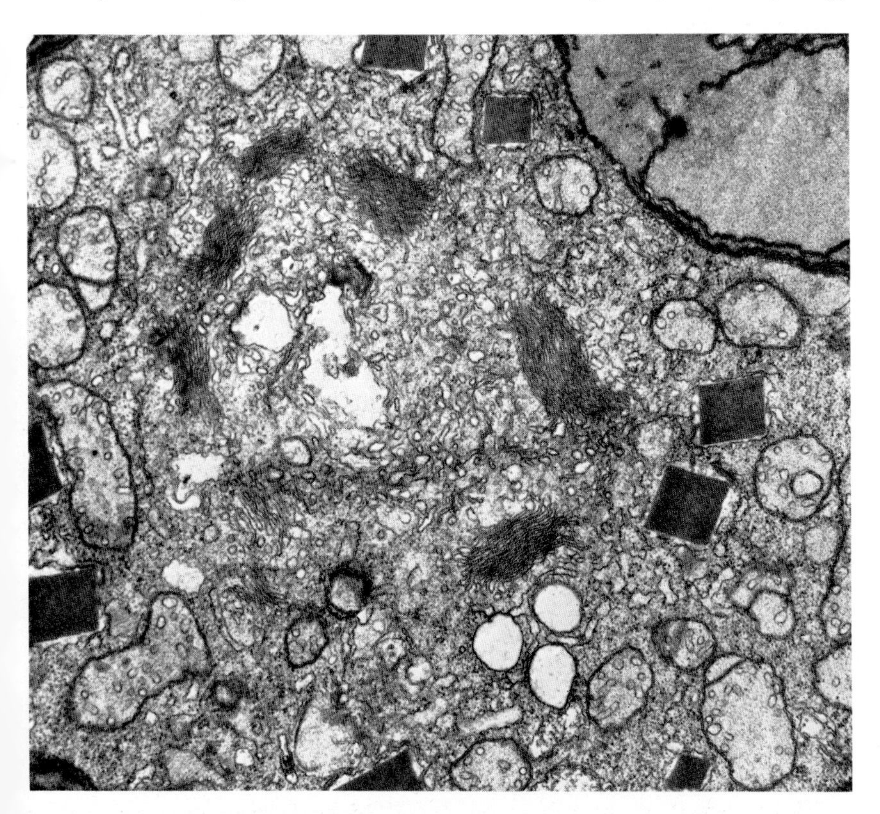

FIG. 10.6. The Golgi complex (or apparatus) situated above the nucleus in *Prorocentrum mariae-lebouriae*. In this transverse section numerous Golgi vesicles are seen both within and outside the ring of dictyosomes. A number of trichocysts are also present. (× 18, 400)

(if any) of the products are concerned. What has not yet been established is whether the dictyosome is merely a collecting point for materials synthesized in other parts of the cell, or whether the protein, silica, calcite, polysaccharide, etc. are actually synthesized in the cisternae.

Because electron microscopy deals with static images very little is known about the formation and continual rebuilding of the Golgi apart from the fact that they appear to increase in number by vertical division. However, in the Xanthophyceae, Falk and Kleinig (1968) and Massalski and Leedale (1969) have observed a relationship with the adjacent nuclear envelope which suggests a way in which dictyosomes may be continually built up. The nuclear envelope appears to be budding off numerous small vesicles which pass to the Golgi body where they are presumed to fuse together to form new cisternae. Thus, the side adjacent to the nucleus would be the forming face of the dictyosome and the distal side the producing face. It is a fact that vesicles with obvious contents (i.e. products) always seem to be most abundant near the 'top' or wider face of the Golgi body which is usually the side away from the nucleus. In single-celled algae the Golgi apparatus is invariably adjacent to the nucleus.

III. The Contractile Vacuole

Organelles concerned with osmo-regulation and probably also excretion—known as contractile vacuoles—are found in most freshwater unicellular algae which lack a rigid cell wall. They are also found in a few marine algae. Generally, contractile vacuoles are situated at the anterior end of the cell, between the Golgi apparatus and the plasma-membrane. In the silica scale bearing members of the Chrysophyceae, such as *Mallomonas* (Belcher, 1969b), they are found at the posterior end of the cell.

Fine-structural observations have been made on contractile vacuoles from members of the Chlorophyceae (e.g. *Stigeoclonium* zoospores: Manton, 1964b); Prasinophyceae (e.g. *Mesostigma:* Manton and Ettl, 1965; *Pyramimonas*: Maiwald, 1971); Euglenophyceae (e.g. *Euglena*: Leedale *et al.*, 1965); Chloromonadophyceae (e.g. *Gonyostomum*: Mignot, 1967); Xanthophyceae (e.g. *Bumilleria* zoospores: Massalski and Leedale, 1969); Haptophyceae (e.g. *Chrysochromulina*: Parke *et al.*, 1962); Chrysophyceae (e.g. *Chrysococcus*: Belcher and Swale, 1972a; *Chrysamoeba*: Hibberd, 1971); Cryptophyceae (e.g. *Chilomonas*: Anderson, 1962).

In most cases the main vacuole consists, when expanded (the stage termed 'diastole'), of a rounded vesicle bounded by a single membrane. Numerous small vesicles fuse with the membrane around the vacuole, thus discharging their contents into it. These small vesicles and others which lie around the contractile vacuole often have a layer of small particles on their outer surface—they are termed 'coated', 'hairy' or 'alveolate' vesicles. In *Poteriochromonas* the small vesicles are elongated and covered on the outside by a layer of 8–10 nm particles which are attached to the membrane by fine stalks (Tsekos and Schnepf, 1972).

In *Euglena* (Leedale, 1967) irregularly shaped accessory vacuoles lie around the contractile vacuole and fluids are thought to collect in these before being discharged into the main vacuole which, when full, fuses with the plasmalemma and thus discharges its contents to the exterior (the stage termed 'systole'). This is said to be a rapid process which may happen every 20 s.

In the chrysophyte *Ochromonas tuberculatus* the accessory vesicles are elongated and have rather thick walls (Hibberd, 1970) and somewhat similar tubular structures have been observed around the contractile vacuole of *Gonyostomum* (Mignot, 1967). In another member of the Chloromonadophyceae—*Vacuolaria*—Schnepf and Koch (1966) have described how numerous small vesicles arise from the extensive Golgi region (consisting of about 50 dictyosomes) which lies in the form of a shell around the anterior end of the nucleus. The vesicles fuse together in the usual way to form the contractile vacuole.

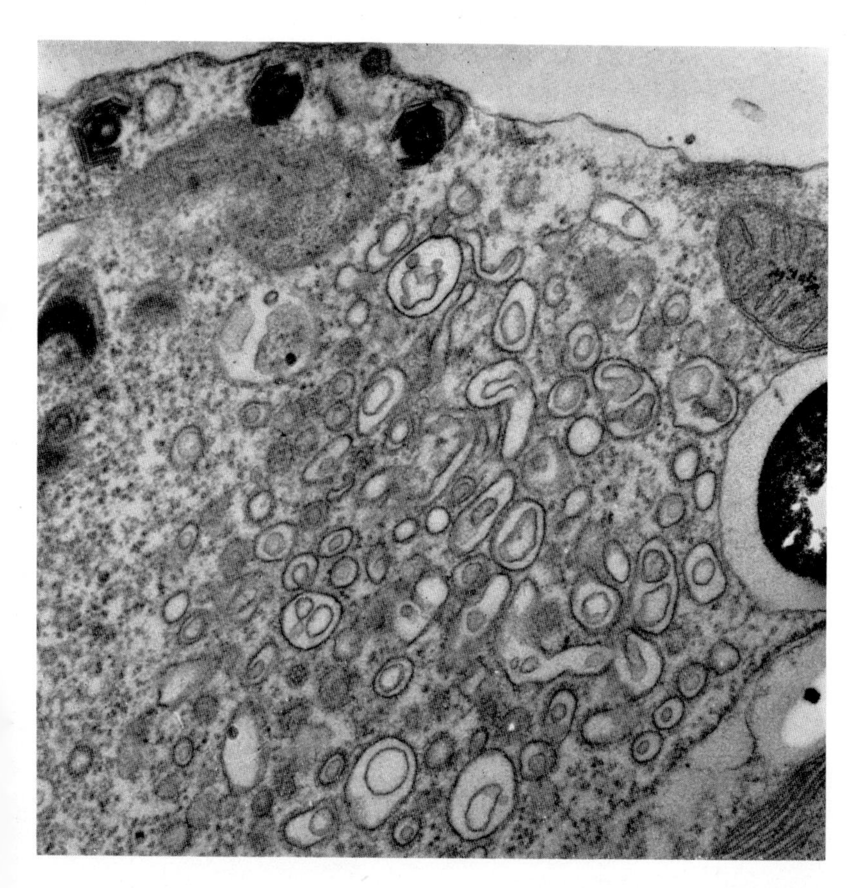

FIG. 10.7. The contractile vacuole region at the anterior end of the cell of *Cryptomonas acuta* containing what appear to be invaginated tubules. (×43,000)

In marine members of the Cryptophyceae the contractile vacuole appears to take various forms. In *Cryptomonas rostrella* and *C. calceiformis* (Lucas, 1970a) they have the typical structure as found in freshwater algae. In *C. reticulata* (Lucas, 1970a) and *C. acuta* (Dodge, unpublished observations) there is no large collecting vacuole but the anterior end of the cell is extensively vacuolated by irregular and sometimes interconnected sacs (Fig. 10.7). It appears that these structures move from the dictyosome and probably eventually fuse with the

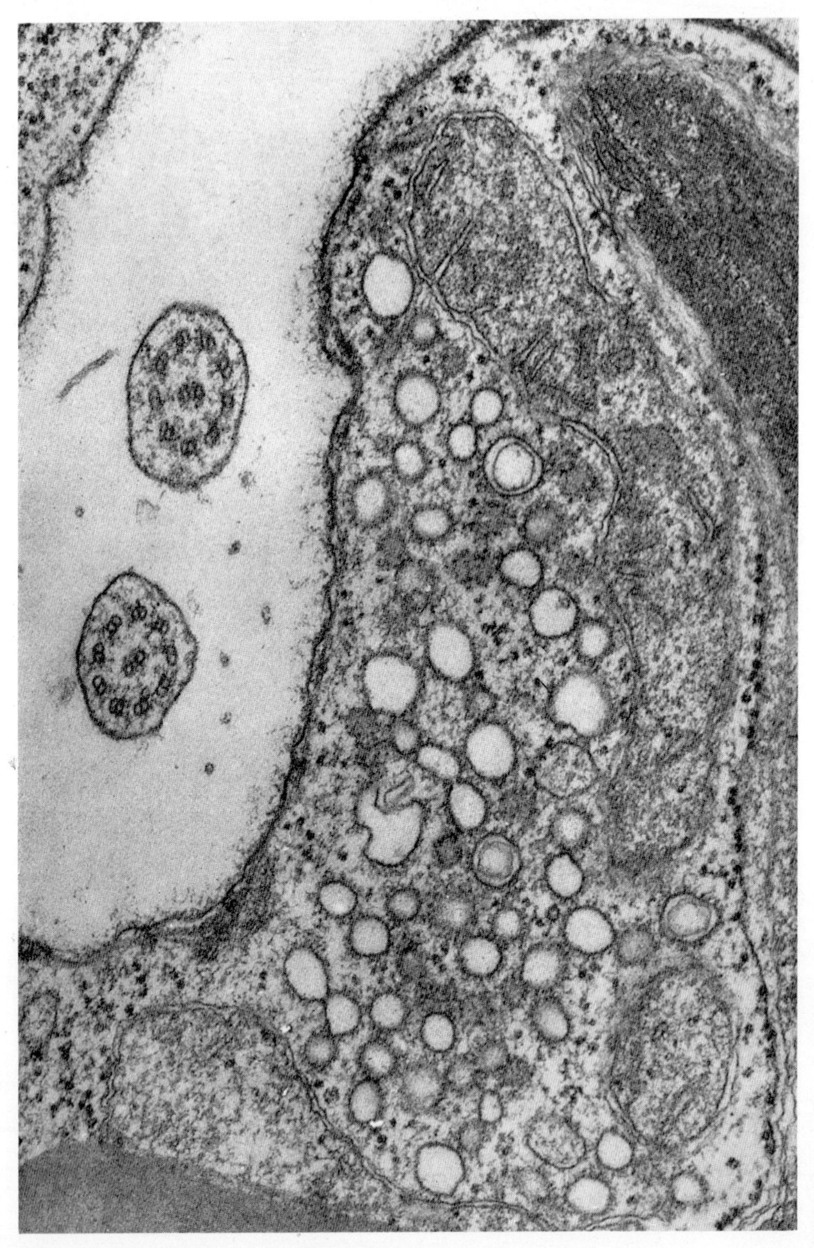

FIG. 10.8. Contractile vacuole region of *Chroomonas mesostigmatica* containing plain and alveolate vesicles which are sandwiched between mitochondria and the cell membrane. (× 59,000) (After Dodge, 1969b,)

plasmalemma. In *Chroomonas mesostigmatica* (Dodge, 1969b) the vesicles are less numerous and mostly spherical (Fig. 10.8) and they appear to be of the coated variety. It is perhaps noteworthy that here the vesicular area is bounded by large mitochondria.

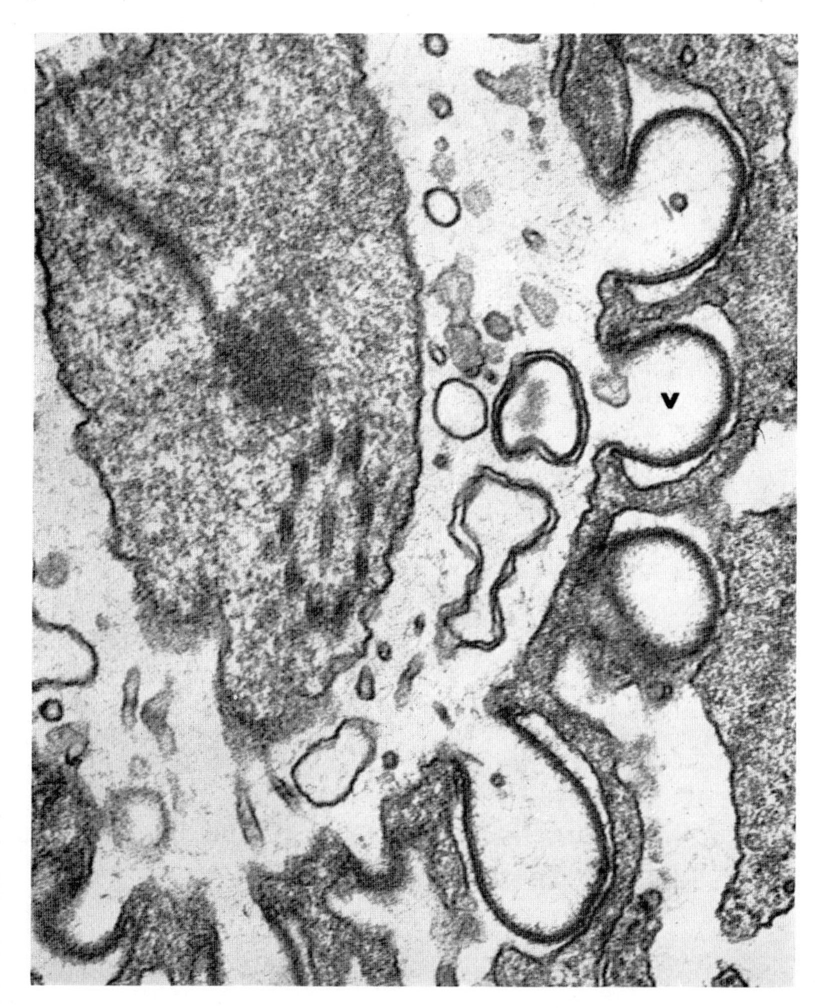

FIG. 10.9. The simple pusule of *Gymnodinium nelsonii* consisting of vesicles (v) around the flagellar canal which are ensheathed by cell vacuolar reticulum. (×57,000) (After Dodge, 1972.)

IV. THE PUSULE

In the Dinophyceae there is no simple contractile vacuole but many members of the group, both marine and freshwater, have permanent organelles called pusules. Although there is much variation in the form of the pusule the basic construction of this organelle is uniform. It consists of an invagination of the plasma-membrane from the base of the flagellar canal which is, in part at least, closely adpressed to a membrane derived from the cell vacuolar reticulum. Thus, the characteristic appearance of parts of the pusule is of a thick wall (composed of two membranes) which bounds a vesicle open to the cell exterior, and around the cell side of which there is a vacuolar sheath. It appears to be an ideal system for osmoregulation.

A full account of the various types of pusule has recently been published (Dodge, 1972) and these will only be briefly described here:

(a) Simple type, pusule vesicles open directly into the flagellar canal. (Example: *Gymnodinium nelsonii*) (Figs 10.9, 10.11A).

(b) Pusule collecting chamber, into which the vesicles open, branches off from the flagellar canal. (Example: *Amphidinium herdmani*).

(c) Pusule consisting of a collecting chamber situated near the centre of the cell and communicating with the exterior by way of a coiled tube. (Example: *Gymnodinium fuscum*, Dodge and Crawford, 1969a).

(d) Complex pusule which consists of a sinuous tube, into part of which the pusule vesicles open, then it continues for some distance with the inner membrane lined by a tomentum of small elongated projections. (Example: *Woloszynskia coronata*, Crawford and Dodge, 1973) (Fig. 10.10).

(e) Simple tubular pusule (no pusule vesicles) which is entirely constructed of the two closely appressed membranes. In some organisms the tubes are also much invaginated. (Example: *Glenodinium foliaceum*) (Fig. 10.11B,C).

(f) Sack pusule which consists of a convoluted structure with numerous invaginations. (Example: *Prorocentrum* (= *Exuviaella*) *mariae-lebouriae*) (Fig. 10.11D).

Pusules have been described from other dinoflagellates by Cachon *et al.* (1970), Mignot (1970) and Schnepf and Deichgräber (1972), and in all cases have the same basic structure although the form of the organelle shows much variation.

V. Microbodies (Peroxisomes)

It is likely that microbodies or peroxisomes are present in most algae but their recognition in these organisms has lagged behind the work on higher plants. To date they have been mainly found in green algae where they are known to be present in three strains of *Chlorella* (Gergis, 1971), in *Polytoma* and *Chlorogonium* (Gerhardt and Berger, 1971), and in the filamentous alga *Klebsormidium* (Stewart *et al.*, 1972). In the latter each cell contains a single microbody located adjacent to the nucleus. Outside the Chlorophyceae microbodies have only been found in *Euglena* (Graves *et al.*, 1971); in a number of dinoflagellates such as

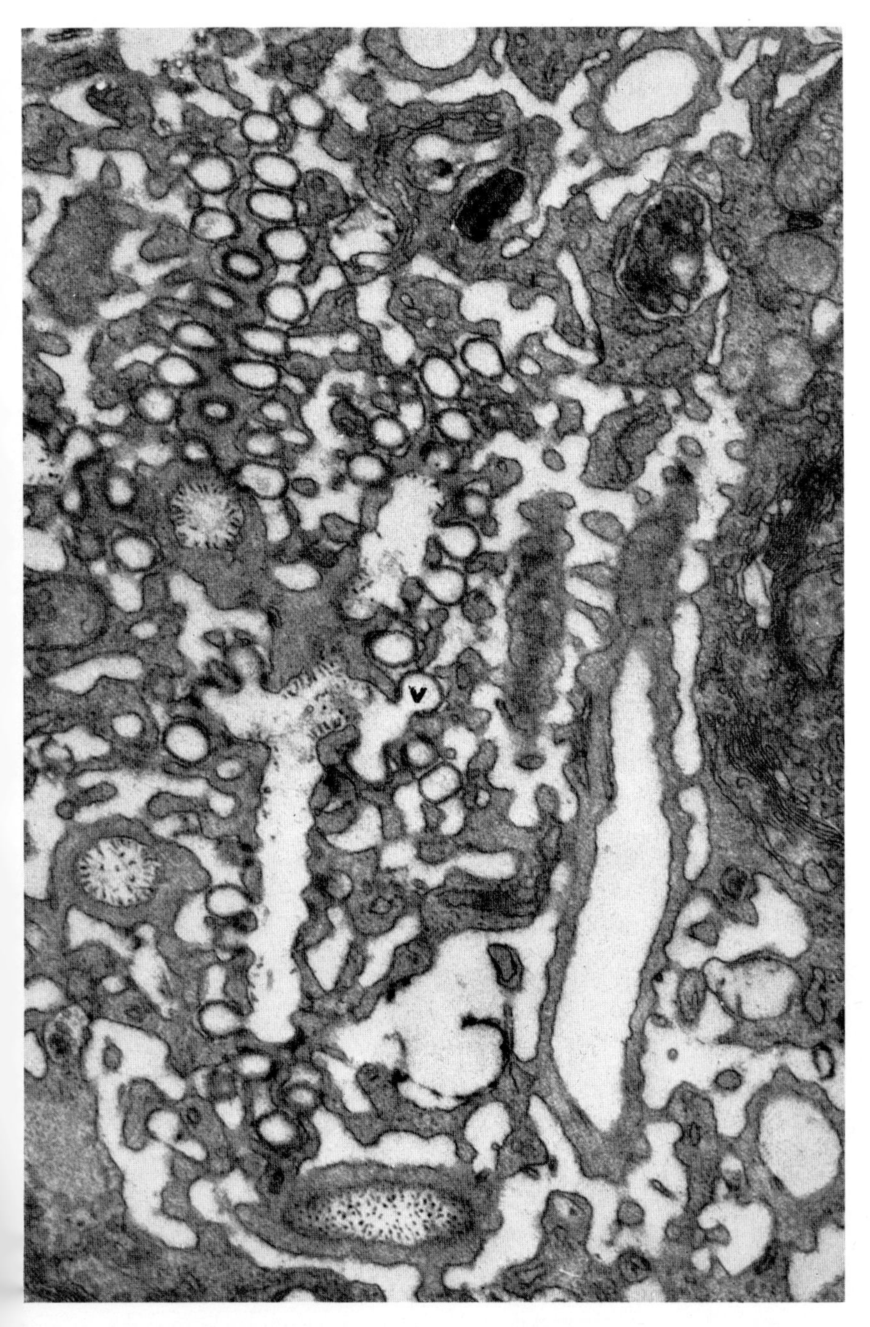

FIG. 10.10. The very complex pusule of *Wolosyznskia coronata* showing both smooth and rough tubules, numerous pusule vesicles (v) and the ramifying cell vacuolar system which surrounds both tubes and vesicles. ($\times$ 36,400)

Gymnodinium micra (Leadbeater and Dodge, 1966) where they were observed but not identified and in *Scrippsiella* (Fig. 10.13) (Bibby and Dodge, 1973); and in the unicellular red alga *Porphyridium* (Fig. 10.12) (Dodge and Oakley, unpubl. obs.).

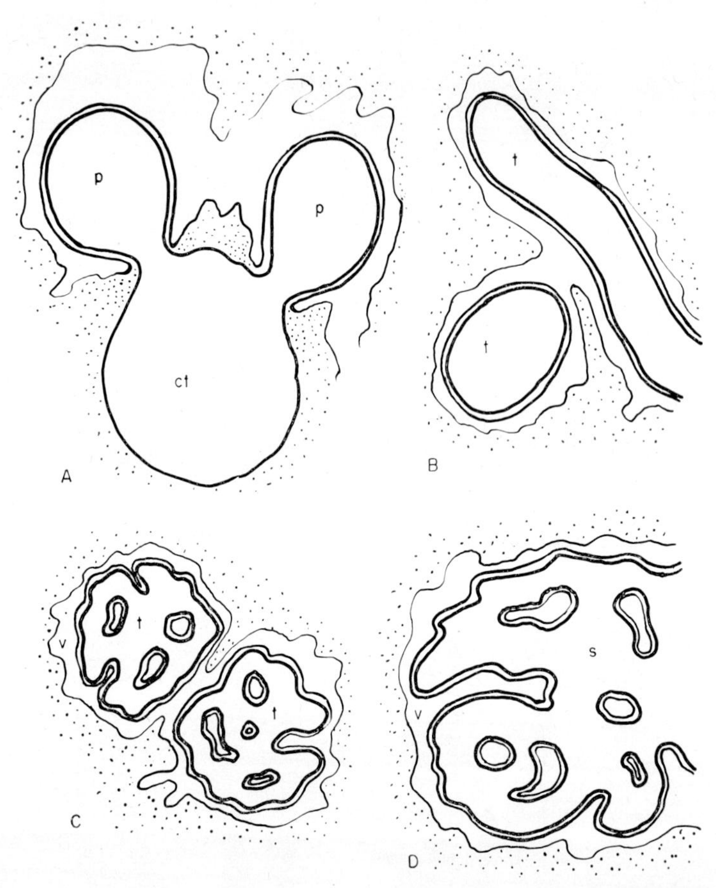

FIG. 10.11. Drawings to illustrate some of the basic types of pusule as seen in section. A, Pusule vesicles (v) opening into a collecting tubule (or flagellar canal); B, simple tubular pusule; C, pusule consisting of invaginated tubules; D, sack pusule. (After Dodge, 1972.)

The algal microbodies consist of a finely granular matrix, 0·25–0·8 μm in diameter, which is bounded by a single membrane (Figs 10.12 and 10.13). They are often situated adjacent to mitochondria and in *W. micra* they are usually packed into the centre of a cluster of several mitochondria. Apart from the peroxisome of *Klebsormidium* (Stewart *et al.*, 1972) those that have been tested have given a negative reaction to the diamino benzene/peroxide test for catalase and thus do not appear to be true peroxisomes. Graves *et al.* (1971) have found the microbodies to be rare in *Euglena* which has been grown on a glucose medium but abundant when the alga is grown on acetate or ethanol. They suggest that the bodies may be involved in the metabolism of 2-carbon substrates (the enzymes for which are known to be membrane-bound) and thus to be more analogous with the glyoxisomes of higher plants.

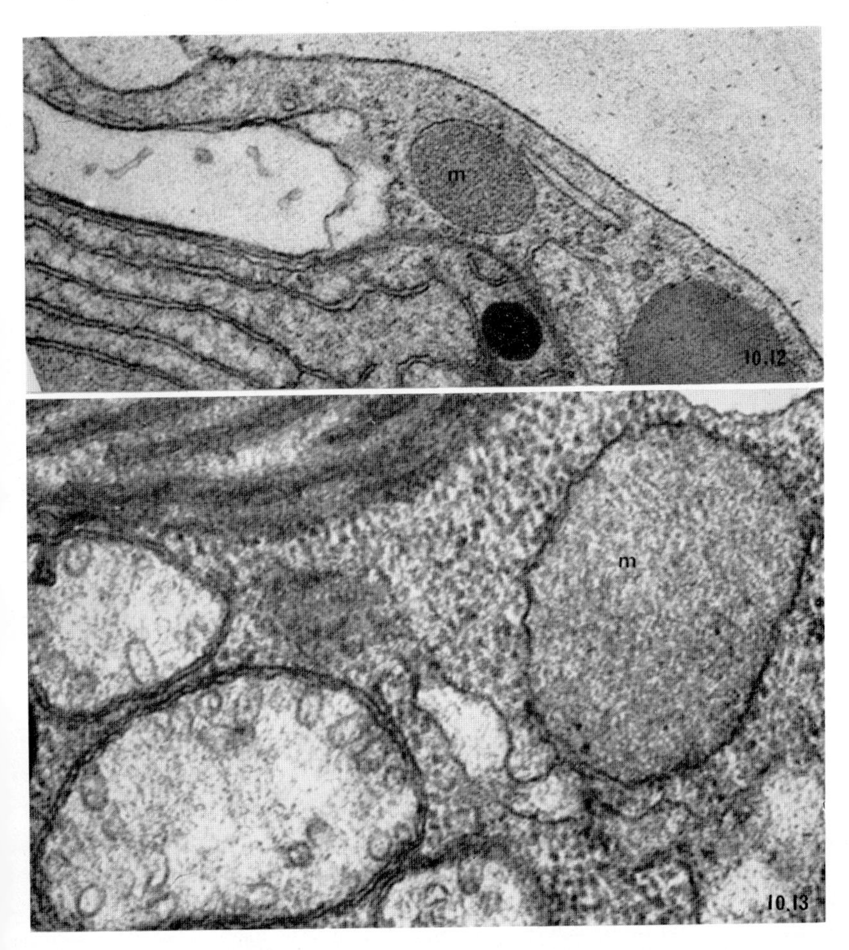

FIGS 10.12–10.13. Microbodies in algae. (10.12) A microbody (m) at the edge of a *Porphyridium* cell (×32,400); (10.13) a microbody (m) near to a similarly sized mitochondrion in *Scrippsiella* (Dinophyceae). (×58,800)

VI. FOOD STORAGE MATERIALS

A. POLYSACCHARIDE

Very many algae store polysaccharide food reserves in the form of starch grains and in the Chlorophyceae and Prasinophyceae these are always situated within the chloroplasts, just as starch grains are in higher plants. In the Cryptophyceae starch is found between the chloroplast envelope and the E.R. sheath which encloses the chloroplast (Dodge, 1969b). In other groups such as the Dinophyceae (Dodge and Crawford, 1971a) (Fig. 10.14) and Rhodophyceae (Fig. 1.10) (Peyrière, 1963) the starch grains lie free in the cytoplasm. Starch grains are rather variable in size and the shape may be spherical, ovoid, flattened, or dome-shaped when they form over the surface of a pyrenoid. In some species such as *Woloszynskia coronata* (Fig. 10.15) a large number of starch grains are often concentrated at the posterior end of the cell and fat droplets are found mainly at the anterior end (Fig. 10.14).

In the Euglenophyceae we find a chemically distinct polysaccharide called paramylon. This takes the form of variously shaped grains which lie free in the cytoplasm or form around a pyrenoid. Surface replicas of grains have revealed a helical organization of subunits (Leedale, 1967) which probably indicates the manner in which the grains are built up.

The liquid polysaccharide reserve material leucosin (chrysolaminarin) is found in members of the Chrysophyceae (e.g. *Ochromonas*: Hibberd, 1970) and the Haptophyceae (e.g. *Crystallolithus*: Manton and Leedale, 1963b) and *Pavlova* (Kreger and van der Veer, 1970). It occurs in the form of large droplets which are usually situated near the posterior end of the cell. Owing to its form and moderate electron density leucosin can easily be confused, in electron micrographs, with lipid droplets. Chardard (1972) has described the production of mucilage in the desmid *Cosmarium*. This material is shown to be a form of polysaccharide.

B. LIPIDS

Many dinoflagellates store fat or oil in the form of droplets (Dodge and Crawford, 1971a) and it is particularly abundant in organisms such as *Ceratium hirundinella* when they have been fixed from a natural habitat in which the conditions were very favourable for photosynthesis (Dodge and Crawford, 1970b). Fat is the main food reserve in diatoms where it is also thought to be important in providing a low average specific gravity to enable the planktonic species to float. Lipids always appear in the form of droplets of varied sizes (Fig. 10.15) but sometimes the droplets seem to consist of small globules which have coalesced during preparation.

C. PROTEIN

Proteinaceous food storage material does not appear to have been described from any algae and this may be because of the possible use of the pyrenoid for that function in many algae (see Chapter 5). Burr and West (1971a) have provided a detailed account of the formation of the protein bodies in *Bryopsis* (Chlorophyceae) which are known to be involved in wound healing and basal septum

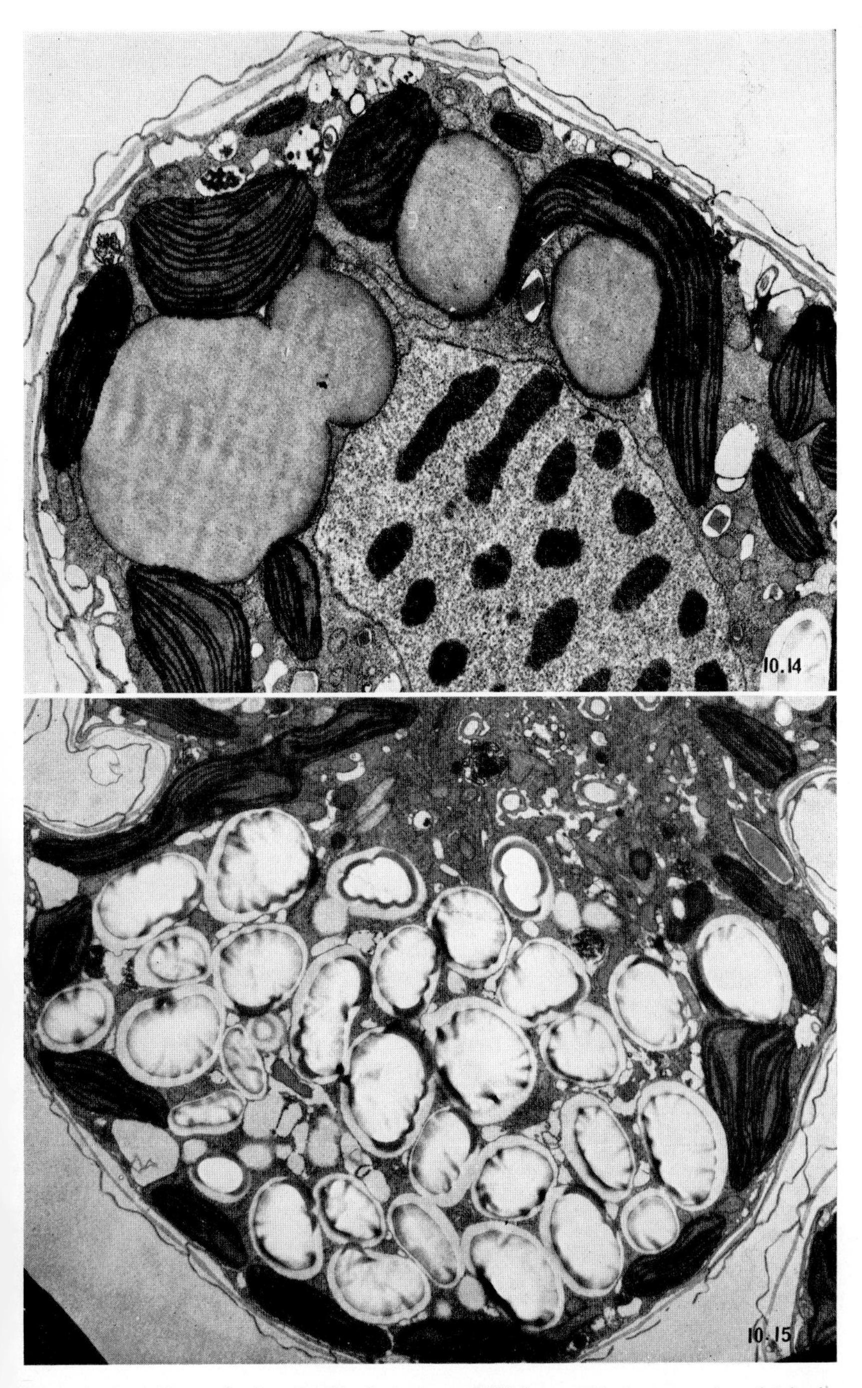

FIG. 10.14. Part of a longitudinal section of *Woloszynskia coronata* in which the epicone (anterior portion of cell) is seen to contain large lipid globules. (×8400); (Fig. 10.15) another section of *W. coronata* showing a hypocone packed with starch grains. (×6000) (Figs 10.14 and 10.15 after Dodge and Crawford, 1971a.)

formation. This proteinaceous material forms within swollen cisternae of the endoplasmic reticulum and later is released into the cell vacuole where it is broken down, first into rod-like structures and later into a granular-fibrillar substance.

VII. Polyphosphate Inclusions and Phosphatases

In a number of green algae more or less spherical electron-dense structures have been observed in the cytoplasm (Fig. 13.3). A recent study (Fisher, 1971) has shown that these granular inclusions consist of polyphosphate. It was found that when *Trebouxia* was grown in normal medium there were numerous granules but when the phosphate concentration of the medium was lowered, few were found. At the light microscopical level the inclusions were found to stain metachromatically with toluidine blue and to precipitate lead sulphide, both indications that they contain linear polyphosphate. Also by use of reactions which deposit lead, it has been shown by Micalef (1972) that acid phosphatase enzymes are present in the Golgi vesicles of *Ulva*, and Sommer and Blum (1965a) have located them in the Golgi, paramylon grains and in vesicles around the reservoir of *Euglena*. They have also been found in the 'residual' body, probably a digestive vesicle, of *Hymenomonas* (Pienaar, 1971b). These enzymes would appear to be of fairly general occurrence in algal cells and are not just associated with the polyphosphate inclusions.

VIII. Virus-Like Inclusions

It has been known for some time that blue-green algae may be infected with phage-type viruses but, until recently, no viruses were known from other groups of algae. Then, in 1966 Tikhonenko and Zavarzina reported a virus-like particle which was able to cause lysis in *Chlorella*. Similar particles are now being found in algae of various classes and it would seem that viral infections of algae may be much more common than was realized.

In the freshwater red alga *Sirodotia* Lee (1971b) found a spherical mass of 'viroplasm' in the cytoplasm. This granular material, which often had a central transparent zone, was surrounded by 50–60 nm diameter polygonal particles which, when numerous, formed a crystalline array. In apical cells of the alga the virus-like particles were also found in the nucleus, associated with the nucleolar material.

In the marine brown alga, *Chorda* (Toth and Wilce, 1972) and the freshwater green, *Oedogonium* (Pickett-Heaps, 1972a) a similar situation has been described. In both cases virus-like particles with distinct wall and core components have been found in young sporelings and these plants were showing signs of the breakdown of organelles and non-formation of the normal cell wall. In the cell of *Chorda* the particles were hexagonal in cross-section and 170 nm in diameter whilst those of *Oedogonium* were of similar shape but some 240 nm across. In both cases the authors suggest that the 'virus' probably infected the naked motile zoospores for they have only been found at the germling stage, before the alga has developed a protective wall.

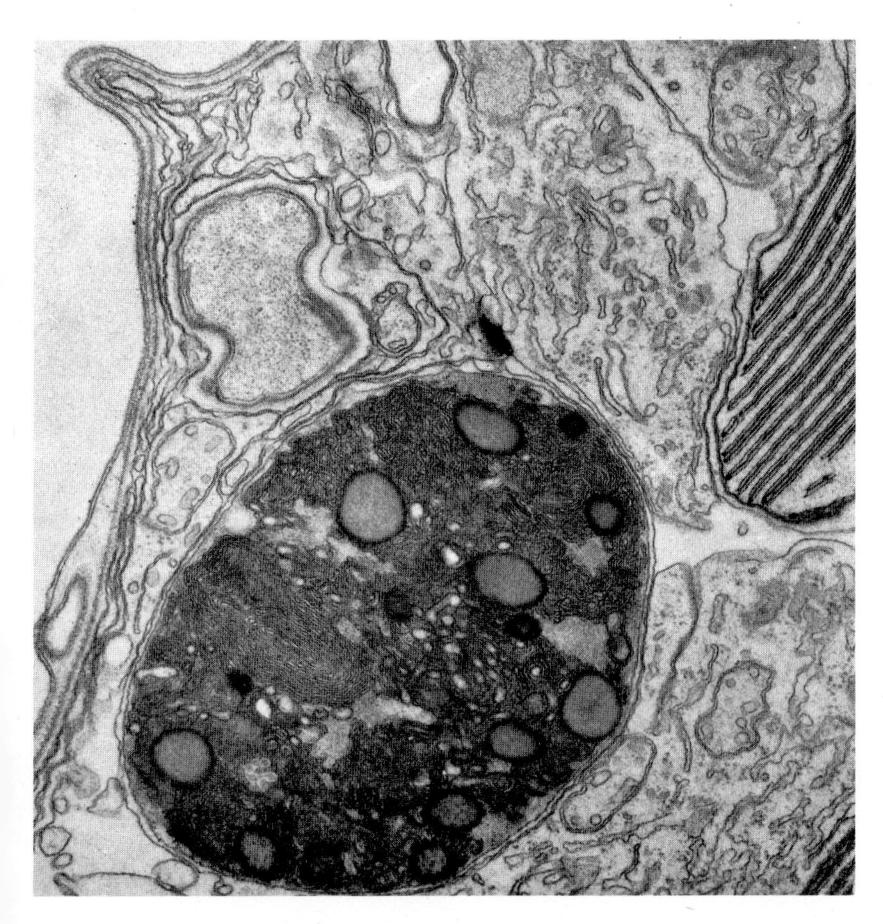

FIG. 10.16. A food digestion vacuole in *Ceratium hirundinella* (Dinophyceae) containing membranous material and lipid droplets, etc. At the left is seen the membranous lining of the ventral chamber, probably the site of ingestion of food particles as the rest of the cell is covered by thick plates. (×28,000) (After Dodge and Crawford, 1970b.)

In all cases the vital tests involving extraction of a cell free extract and infection of new plants remain to be carried out in order to prove that the structures described are in fact viruses.

IX. Membranous Extrusions and Inclusions, Digestion Vesicles

It has long been known that algae secrete material such as carbohydrates and proteins into the environment surrounding them. In the basically naked flagellate *Ochromonas* (i.e. bounded only by a membrane) Aaronson (1971) has suggested that the numerous membranous extrusions seen around the cells and flagella, in electron micrographs, might in fact be involved in the excretion of such materials. Aaronson *et al.* (1971) have also shown the presence of a variety of intracellular membrane-bounded structures and myelin-like figures within the cell of *Ochromonas*. How many of these result from the fixation procedure and how many were present when the cell was alive has not been determined but freeze-etching should clarify this point.

Internal membrane structures and vesicles with granular contents have been reported from *Hymenomonas* (Riedmüller-Scholm and Allen, 1972). Some of these would appear to be autolytic or digestive vesicles. Similar structures recently described from *Brachiomonas* (Gouhier, 1970) have been termed ‘cytolosomes’. These are thought to be portions of cytoplasm isolated within a Golgi vesicle. They either contain material ingested by the cell or cell material being broken down. Definite food digestion vacuoles have been found in the dinoflagellates *Ceratium* (Fig. 10.16) (Dodge and Crawford, 1970b) and *Oxyrrhis marina* (Dodge and Crawford, 1971b). These vacuoles are bounded by a single membrane but often have a complex membranous reticulum outside this. The vacuole may contain recognizable remains of small diatoms, blue-green algae and bacteria. Similar phagotrophic vesicles have also been described from *Chrysochromulina* (Haptophyceae) (Manton, 1972b) and in these various stages of ‘digestion’ were observed.

11. *Reproductive Structures*

I. Gamete Formation and Structure

A. PHAEOPHYCEAE

The first algal gamete to be examined by electron microscopy was the spermatozoid of *Fucus* (Manton and Clarke, 1950) and this gamete was later embedded and sectioned by the same authors (Manton and Clarke, 1956). The spermatozoid was found to have a hairy anterior flagellum, a smooth posterior flagellum bearing a swelling opposite the eyespot and a flattened anterior structure, the proboscis, consisting of concentric fibrillar strands enclosed in a membrane. Internally the structure is very simple and most of the cell is occupied by the large nucleus, in addition to which there are several mitochondria and a chloroplast which is virtually reduced to an eyespot (Fig. 11.1). The male gamete of *Ascophyllum* has been found to have a similar structure (Cheignon, 1964). In *Dictyota* (Manton, 1959b) the spermatozoid is more like a vegetative zoospore in structure and it lacks the eyespot and proboscis. There is a single anterior flagellum which has a unique structure, for it bears a row of short spines which appear to be attached to the median doublet on the upper side of the flagellum.

In *Zonaria* Liddle and Neushul (1969) have described the development of the oogonium, a process which takes a lunar month. During development the cell increases in size some 30–40 times and the number of organelles, including plastids and pyrenoids, increases with age. There is also an accumulation of metabolites such as oil and tannin-containing bodies (physodes). In mature oogonia of *Zonaria* and *Dictyota* (Neushul and Liddle, 1968) the main organelles are arranged in a very distinctive manner, with the nucleus in the centre surrounded by large numbers of plastids and the anterior and posterior ends of the cell occupied by vesicular structures. Before the mature egg is released an amorphous 'mucilage layer' develops inside the wall and, following fertilization, the plastids become uniformly distributed throughout the zygote. In the zygote of *Fucus* (Quatrano, 1972) there is also found to be random orientation of the organelles up to 14 h following fertilization but then the perinuclear region becomes highly polarized with dense accumulations of inclusions radiating towards the site of primary rhizoid formation. Numerous extensions of the nuclear envelope also appear on the rhizoid side of the nucleus.

B. BACILLARIOPHYCEAE

In the centric diatom *Biddulphia* (Heath and Darley, 1972) the male gametangium contains an irregular central mass of cytoplasm which includes all the plastids together with some mitochondria, Golgi, and ER. Around this mass the

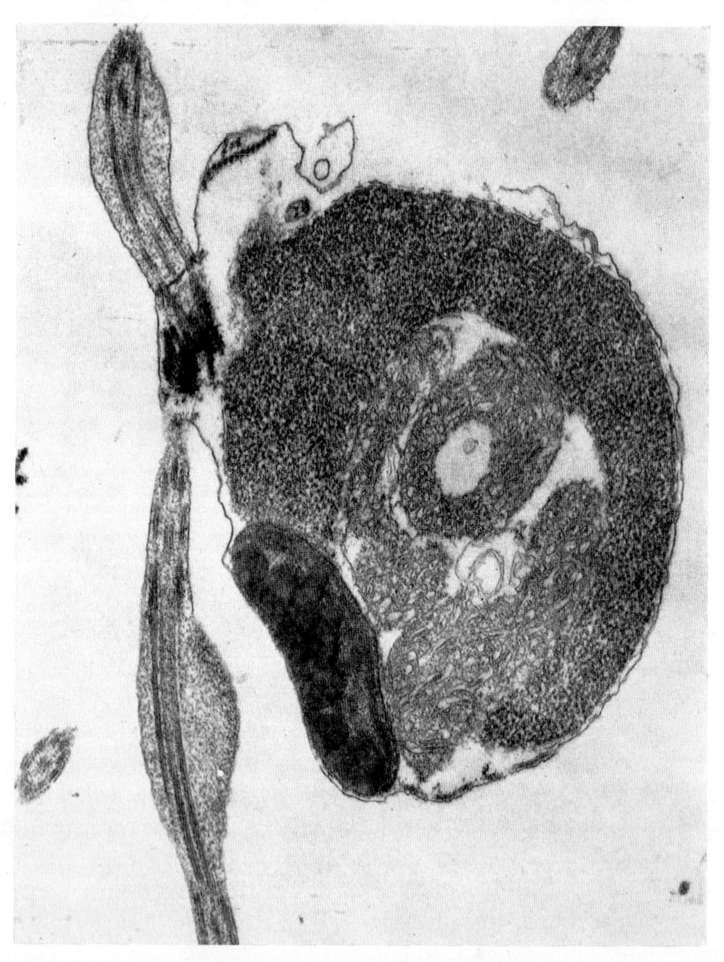

Fig. 11.1. A longitudinal section of the spermatozoid of *Fucus serratus* (Phaeophyceae) showing the two flagella (the posterior one with a swollen portion), the eyespot, large granular nucleus and several mitochondria. (×24,500)

spermatozoa develop and it remains behind after the gametes are released through a special pore in the girdle band, but then it degenerates. The gametes are uniflagellate fusiform cells and the axoneme of the flagellum lacks the central two tubules. A cone of microtubules radiates from the base of the flagellum to the nuclear envelope and the gamete contains no plastids or Golgi bodies.

In another centric diatom, *Lithodesmium* (Manton and von Stosch, 1966) the spermatozoid has much the same structure except that it possesses several chloroplasts. The nuclear divisions during spermatogenesis have been described in detail (Manton *et al.*, 1969a,b, 1970a,b) and are summarized in Chapter 7.

C. XANTHOPHYCEAE

In *Vaucheria* Moestrup (1970b) has shown that the spermatozoids are formed by abstriction from a central mass of cytoplasm. As in *Biddulphia* (see above) all chloroplasts and many mitochondria remain behind in the sporangium after the release of gametes. The spermatozoa consist of an elongated body with a pair of laterally inserted flagella, one of which bears two rows of stiff hairs. The anterior part of the body is extended in the form of a proboscis and the body itself contains only an elongated nucleus, several mitochondria, a single Golgi body and a band of microtubules. It is remarkably similar in structure to the spermatozoid of *Fucus*, simply differing in the lack of an eyespot.

D. CHLOROPHYCEAE

In the large coenobial alga, *Volvox* (Deason *et al.*, 1969) sperm packets, which are flattened plates of spermatozoids, develop in much the same way as vegetative colonies. The male initials cleave incompletely leaving cytoplasmic bridges between the developing spermatids. The whole structure is bounded by a two-layered membrane which encloses a thick layer of wall material, apparently derived from the wall of the initial cell. When released from the packet each sperm has two anterior flagella, a small anterior eyespot, two contractile vacuoles, mitochondria, one or more dictyosomes, a nucleus and a posterior chloroplast enclosing a pyrenoid. They are in fact virtually small scale replicas of the vegetative cells although in shape they are somewhat thinner.

In another oogamous green alga, *Golenkinia* (Chlorococcales) Moestrup (1972) has shown that the biflagellate spermatozoids are elongated structures in which the nucleus is attached to the flagellar bases. They contain a reduced chloroplast but no pyrenoid or eyespot. The flagella are unusual in that they contain only one central tubule, in spite of which they appear to move in the normal manner. During development of the sperm each is surrounded by a wall, but this breaks up when the antheridium dehisces. Released sperm, as is probably the case in all other algae, are bounded only by a single membrane.

In the thalloid green alga *Prasiola* Friedmann and Manton (1960) found the main difference between the two types of gametes was that the males possessed two flagella plus a tiny chloroplast whilst the much larger females had no flagella and a very large chloroplast. In the spermatozoid the flagellar bases are actually attached to the nucleus. During syngamy the first contact between the gametes seems to result in one flagellum of the male being drawn into the cytoplasm of the female. This is followed by complete fusion but the zygote remains motile using only the one remaining flagellum. Later the zygote rounds up and secretes a wall.

In the coenocytic green alga *Bryopsis* the process of formation of the anisogamous gametes has been described in detail (Burr and West, 1970). The first indication that a branch may be about to develop into a gametangium is the formation of a thick plug which separates that branch from the rest of the thallus. Within the branch cytoplasmic streaming ceases whilst the chloroplasts increase in number, accumulate much starch, and become compactly arranged in the peripheral cytoplasm. The plastids later undergo numerous divisions without any interim enlargement and at this time multiplication of the nuclei takes place. Centrioles are found at polar positions outside the nuclear envelope and spindle microtubules pass across the nucleus, probably emerging through nuclear pores. Both male and female gametes develop within the same gametangium, usually male in the proximal sector and female in the distal sector. The plasmalemma appears to become invaginated to enable cleavage to take place and the gametes delimited.

Whilst this is happening the flagella grow out and eyespots arise *de novo* within the chloroplasts of the female gametes. All the nuclei and chloroplasts become incorporated into gametes but a certain amount of residual cytoplasm, vesicles, ER and membranous structures, remain after the completion of cleavage. The two types of gametes differ in structure. The male gamete (Fig. 11.2) contains a large inverted-cup shaped mitochondrion which occupies much of the cytoplasm, a single nucleus, and may contain a much reduced chloroplast. The larger female gamete (Fig. 11.3) contains all the normal cell organelles and its large posterior chloroplast contains a very extensive uniseriate eyespot at one side. Often endogenous bacteria are found within the gametes.

II. Gamete Fusion and Zygote Development

The process of gamete fusion has been most extensively studied in *Chlamydomonas* which has more or less isogamous gametes (Gibbs *et al.*, 1958; Friedmann *et al.*, 1968; Brown *et al.*, 1968). Information was obtained about the initial fusion of the flagellar tips and the formation of a papillate fertilization tubule which grows out from one side of the anterior knob of one of the gametes. This tube connects the pair of gametes and later they merge together, round off, and the nuclei fuse. Cavalier-Smith (1970) has described the later process of chloroplast fusion which in *C. reinhardii* takes place some 5 h after the first stage of copulation. The two chloroplasts (Fig. 11.4) put out slender bridges which connect their basal regions together. The pyrenoid sections now move closer together and the bridges become traversed by thylakoids. By $5\frac{1}{2}$ h, most zygotes have a single chloroplast but two pyrenoids, one of which may be showing signs of degeneration. By $6\frac{1}{2}$ h, chloroplasts with two pyrenoids are rare. The zygote now secretes a thick ornamented wall inside the remains of the gamete wall.

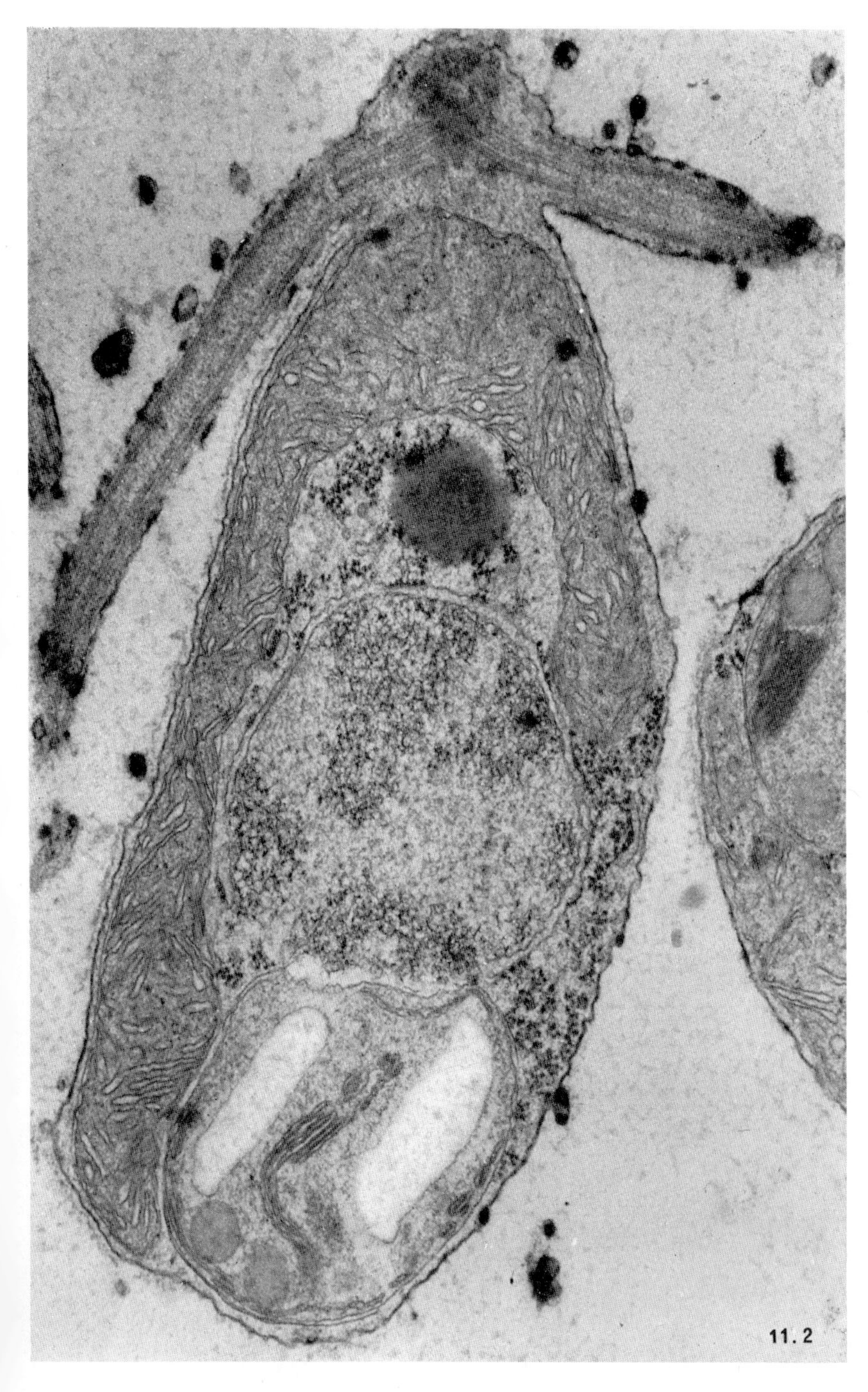

Fig. 11.2. Longitudinal section of a male gamete of *Bryopsis hypnoides* (Chlorophyceae) showing the two flagella inserted into an apical papilla, the giant mitochondrion, the nucleus and a much reduced chloroplast at the posterior end of the cell. ($\times$ 30,000) (Reproduced with permission, from Burr and West, 1970.)

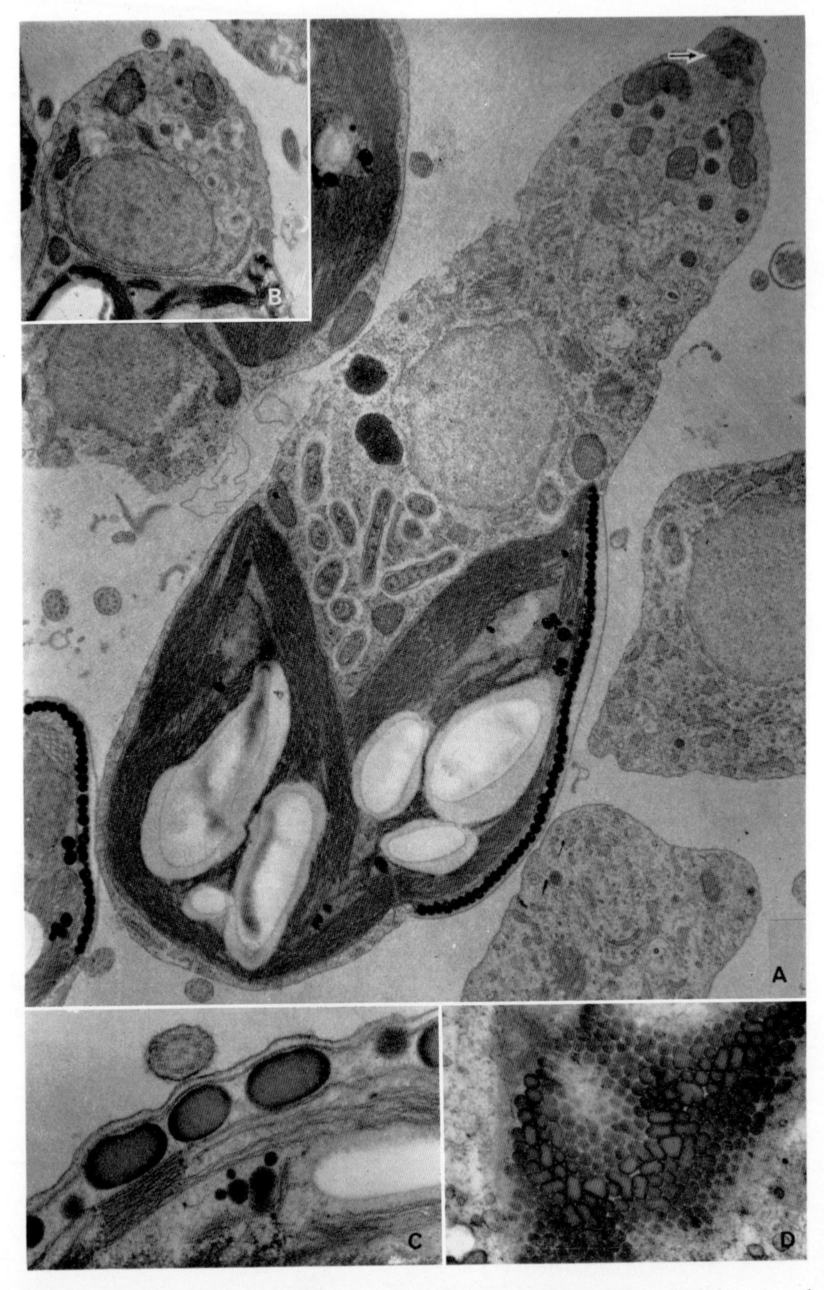

FIG. 11.3. A, Longitudinal section of a female gamete of *Bryopsis* showing the point of insertion of the flagella (arrow), the large multilobed chloroplast with eyespot at the right, numerous bacteria between nucleus and chloroplast, small mitochondria, Golgi, etc. in the anterior half of the cell (×17,000); B, anterior portion of gamete showing association between nucleus and rough ER (×8500); C, transverse section of small part of the eyespot showing a single layer of eyespot globules beneath the cell membrane (×25,500); D, tangential section of part of the extensive eyespot, showing that the globules have various shapes. (×23,000) (Reproduced, with permission, from Burr and West, 1970.)

In *Ulva*, where both gametes are also motile, the first contact and fusion appears to be similar to that of *Chlamydomonas* except that there is no special papilla (Bråten, 1971). After the initial cytoplasmic contact cell fusion takes only

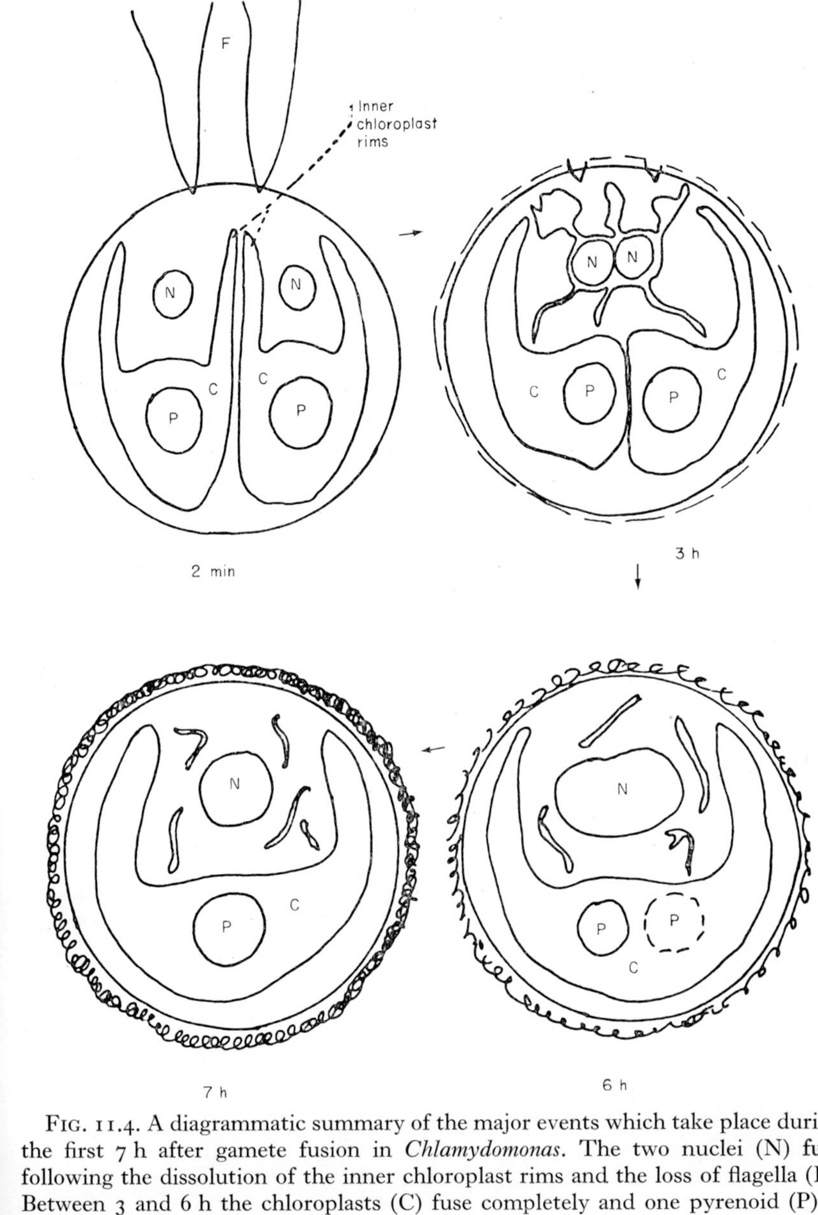

FIG. 11.4. A diagrammatic summary of the major events which take place during the first 7 h after gamete fusion in *Chlamydomonas*. The two nuclei (N) fuse following the dissolution of the inner chloroplast rims and the loss of flagella (F). Between 3 and 6 h the chloroplasts (C) fuse completely and one pyrenoid (P) is lost. The thick resistant zygote wall is gradually formed whilst these processes are taking place. The time scale indicated is only approximate. (Reproduced, with permission, from Cavalier-Smith, 1970.)

5 min and the nuclei fuse within 30 min of the start of copulation. The zygote now settles on a substrate and the four flagella are gradually drawn into the cell, as in *Prasiola* (Friedman and Manton, 1960). A cell wall is then produced and one of the two chloroplasts in the zygote disintegrates. This differs from the situation in the brown alga *Laminaria* (Bisalputra *et al.*, 1971) where there is said to be continuity of both paternal and maternal chloroplasts in the zygote.

In *Bryopsis* the fusion of nuclei following fertilization has been studied by Urban (1969). After the gametes have fused much endoplasmic reticulum appears around the larger female nucleus. As the two nuclei draw closer together they become connected by a strand of ER which then widens out to make a passage between the perinuclear cavities of the two nuclei. The ER connection shortens and the nuclei gradually coalesce together. The chromatin is then evenly dispersed through the zygotic nucleus.

A rather different type of gamete fusion, conjugation, has been described from *Spirogyra* (Pickett-Heaps and Fowke, 1971) and *Closterium* (Fowke and Pickett-Heaps, 1971). In *Spirogyra*, after a pair of filaments become attached by mucilaginous material, papillae grow out from opposite cells and eventually make contact. The end walls dissolve to open the conjugation tube and the whole contents of the 'male' cell, which has already shrunken away from the cell wall and rounded up, passes into the 'female' cell. The two gametes fuse and, after further shrinkage in size, a resistant wall is formed. During this process considerable Golgi activity is evident.

In the desmid *Closterium*, prior to conjugation the pair of cells become more dense and accumulate lipid droplets. The first stage of fusion consists of the formation of a circumferential strip of papilla-wall material around each cell. This then balloons out to form the papilla within which there is an expanding vesicle. As the papilla grows, so the contents of the cells shrink and various other changes occur in the cytoplasm. Fusion then takes place in the wide papilla and the zygote protoplast rounds up and shrinks. The thick wall is slowly secreted and lipid droplets appear to form an almost complete layer within this.

The formation of the resistant zygote wall of the desmid *Micrasterias* has been studied in detail by Kies (1970a). He found that only 30 min after the formation of the globular zygote a patterned arrangement, corresponding with the future pattern of spines, was visible in the peripheral cytoplasm. This was followed by secretion of the outer layer of the zygote wall—the exospore—which had less microfibrils and more matrix in the areas of spine initials. The spines push out with the assistance of turgor pressure and within them a zonal arrangement of organelles and vesicles can be seen. When the spines have reached their maximum length the secondary exospore, a layer 1·4–1·6 μm thick, is laid down. This consists mainly of crossed bands of parallel microfibrils. A 60-nm thick amorphous layer also forms outside the secondary exospore and within two to three days of conjugation the primary exospore disintegrates and is shed. Kies (1970b) has also described the development of the inner layers of the zygote wall, termed the mesospore and endospore.

III. Spore Formation and Structure

As yet very few examples of sporogenesis have been studied with the electron microscope and there is very likely much still to be discovered in this area.

In the xanthophycean alga *Pseudobumilleriopsis* Deason (1971a) has described the formation of the numerous zoospores within a sporangium. The process simply appears to involve the segregation of the cell components into a number of groups (up to 16) followed by cleavage and the production of a pair of flagella on each unit. The zoospores are released by dissolution and separation of the wall at the point where the two wall sections overlap.

A very different type of sporogenesis is found in the Oedogoniales (Chlorophyceae). Here, the production of the single zoospore per sporangium, and its structure, has been described in *Bulbochaete* (Retallack and Butler, 1970b, 1972) and *Oedogonium* (Pickett-Heaps, 1971a, 1972b). The process involves the rounding off of the sporangial contents and synthesis of material, between the plasmalemma and the cell wall, which later forms the temporary zoospore vesicle when the sporangium opens. The zoospore of these organisms is an ellipsoidal or spherical body surmounted by a clear dome-shaped head which is surrounded by a ring of flagella joined together by a complex root system. In *Bulbochaete* (Retallack and Butler, 1972) the zoospore contains a reticulate chloroplast containing stacked thylakoids and several pyrenoids. The usual organelles are present together with various dense bodies. The head region is packed with mitochondria and also contains ER, Golgi and vesicles thought to contain mucopolysaccharides. The flagellar ring is similar to that of *Oedogonium* which has been described in detail by Hoffman (1970) and Hoffman and Manton (1962) (see Chapter 3). Pickett-Heaps (1972c,d) has also given details of the differentiation of the germling, following settlement upon a substrate, and its first cell division.

A very detailed account is available of the production of zoospores and their aggregation into a daughter net in *Hydrodictyon* (Marchant and Pickett-Heaps, 1970, 1971, 1972a,b). Following cleavage of the parental cell the zoospores link together, still within the original cell wall, and a conspicuous feature at this stage is the presence of bundles of microtubules beneath the plasmalemma. Also involved in the aggregation is material, presumed to be adhesive in nature, in the intercellular spaces between the zooids. After the zoospores join together the flagella are retracted and both flagellar microtubules and basal bodies disintegrate.

The somewhat similar process which takes place in *Pediastrum*, to form a new plate-like coenobium, has been described by Hawkins and Leedale (1971). Spherical zooids are released into a special zoospore vesicle which extends from the vegetative cell that became a sporangium. The zoospores change shape to first become oval and then almost rectangular just before they join together. Internally there is an unusual arrangement of organelles with the flattened nucleus lying immediately under the plasmalemma, at one side of the flagella, and the Golgi situated beneath the nucleus. Millington and Gawlik (1970) have found parallel arrays of microtubules under the plasmalemma at this stage. The cells become arranged into a symmetrical, one cell thick, flat plate and lose their flagella. They change shape again to that characteristic of the vegetative cell and, when this is mostly accomplished, wall formation begins. As the two-layered wall is laid down, the cells lose their contractile vacuoles and develop a large pyrenoid. The wall material appears to stick the members of the coenobium firmly together. Gawlik and Millington (1969) and Millington and Gawlik (1967, 1970) have described the development of a complex reticulate pattern of resistant material as part of the wall. This first appears as plaques arranged on the plasmalemma in a hexagonal array. The sites of these plaques appear to correspond with clusters of ribosomes on endoplasmic reticulum under the plasma-membrane. On completion of the outer wall a thicker inner wall is deposited and in this the reticulate pattern is seen as strips of greater electron density. It is suggested that the pattern of the wall is templated by the plasmalemma.

In another green alga, *Stigeoclonium*, Manton (1964b) has described the structure of the zoospore and the changes which take place when it settles to develop into a sporeling. The zoospore contains an eyespot, contractile vacuole and much endoplasmic reticulum, in addition to the normal organelles. When the spore begins to settle on a substrate the flagella are withdrawn and these lie for a while beneath the plasmalemma before they are broken down. Then the flagellar bases sink into the cell and lie adjacent to the nucleus where they presumably act as centrioles. The cell surface of the settling zoospore becomes sticky with a layer of flocculent material and the shape and arrangement of the organelles alters. As the wall begins to form, the Golgi appear very active and at the same time the contractile vacuole, together with its attendant 'hairy vesicles', disappears. Nuclear division now takes place and the filamentous condition is re-established.

The process of settlement has been studied in the alga *Enteromorpha* (Evans and Christie, 1970), an organism which is a great nuisance because of the way in which it grows so well on the bottoms of ships. In the motile zoospore three types of vesicle have been noted: small vesicles situated peripherally at the anterior end of the cell; larger vesicles with electron opaque contents which fill much of the anterior half of the cell; small vesicles lying adjacent to the Golgi bodies. Newly settled zoospores lose the larger vesicles and at the same time rapidly secrete a fibrillar substance which is probably adhesive in nature. Enzyme tests have shown that the material in the larger vesicles contains protein and other tests suggest that the sticky substance contains polysaccharide. It is therefore suggested that it is muco-polysaccharide. Within 30 min of contact with a firm surface a thin cell wall is visible between the fibrillar material and the plasmalemma and this wall continues to increase in thickness for some time, being up to 25 nm thick after 1 h and 200 nm after 4 h. By this time the sporeling is both firmly attached to the substrate and also well protected by its wall.

The structure of zoospores of the green algae *Draparnaldia* and *Chaetomorpha* has been studied in shadowed whole-mounts (Manton *et al.*, 1955). Information

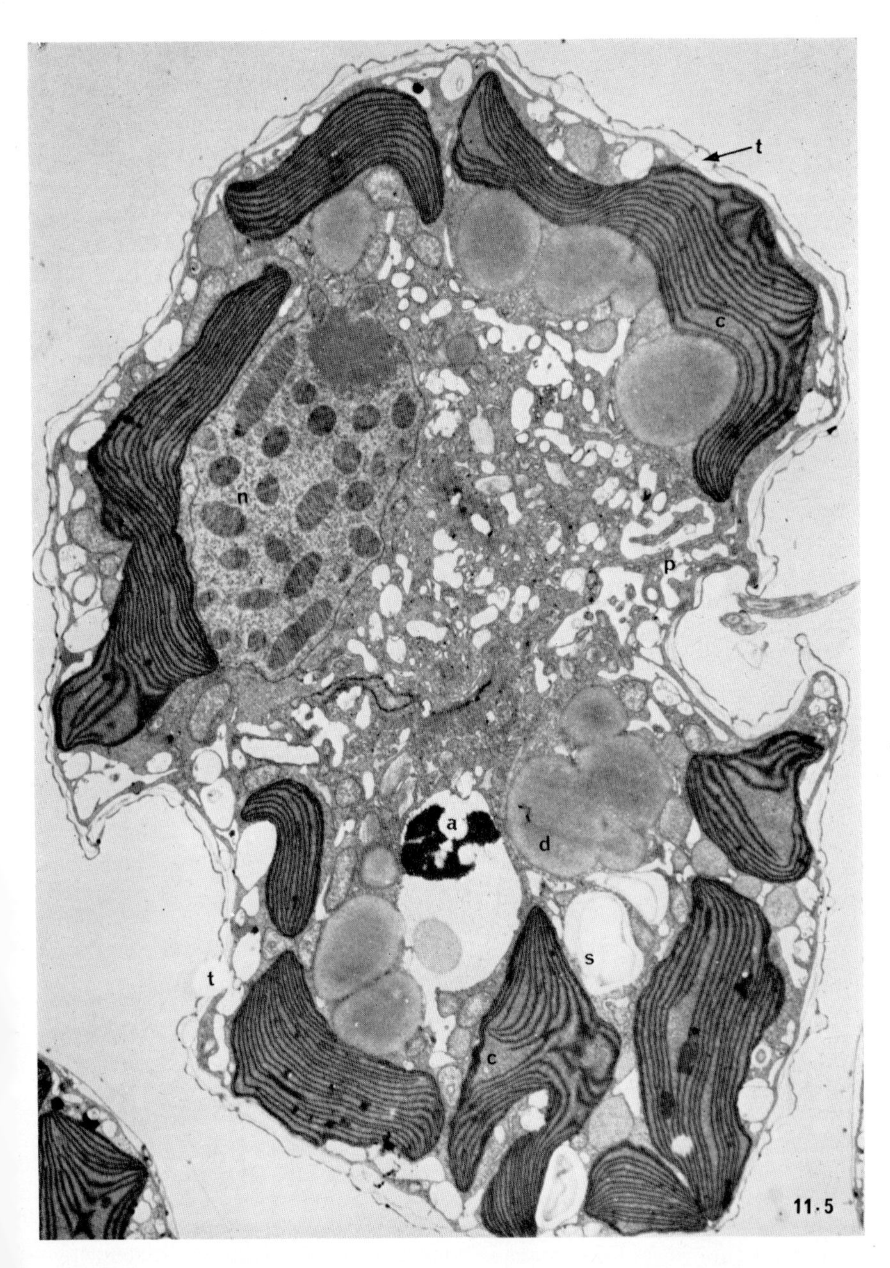

FIG. 11.5. Longitudinal section of a motile cell of *Woloszynskia tylota* (probably a planozygote) showing the nucleus (n), lipid droplets (d), starch grains (s), pusule (p) and small accumulation body (a). The cell is bounded by a typical theca (t) containing fairly thin plates. (×6000)

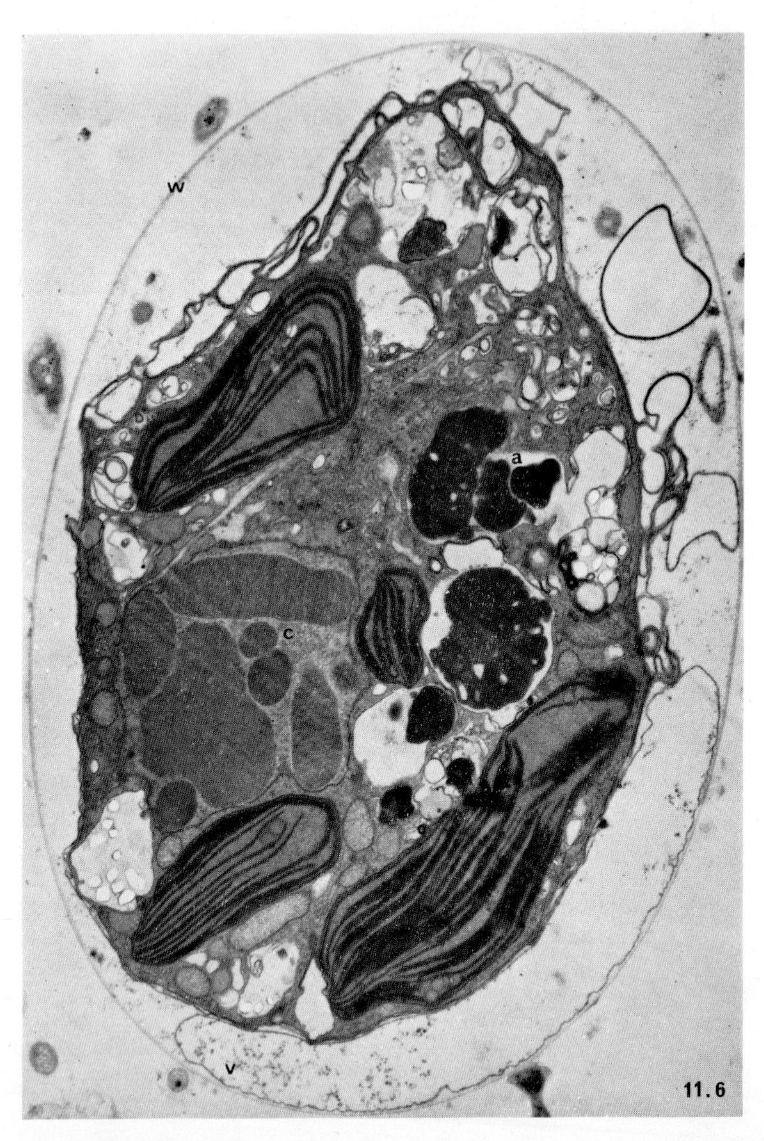

FIG. 11.6. *W. tylota*. Transverse section of a maturing cyst showing the protoplast to have shrunken away from the entire cyst wall. In the nucleus the chromosomes (c) are still visible, the chloroplasts are reduced in number, the accumulation body (a) is much increased in size and the remainder of the cytoplasm is beginning to look rather degenerate. (× 11,500)

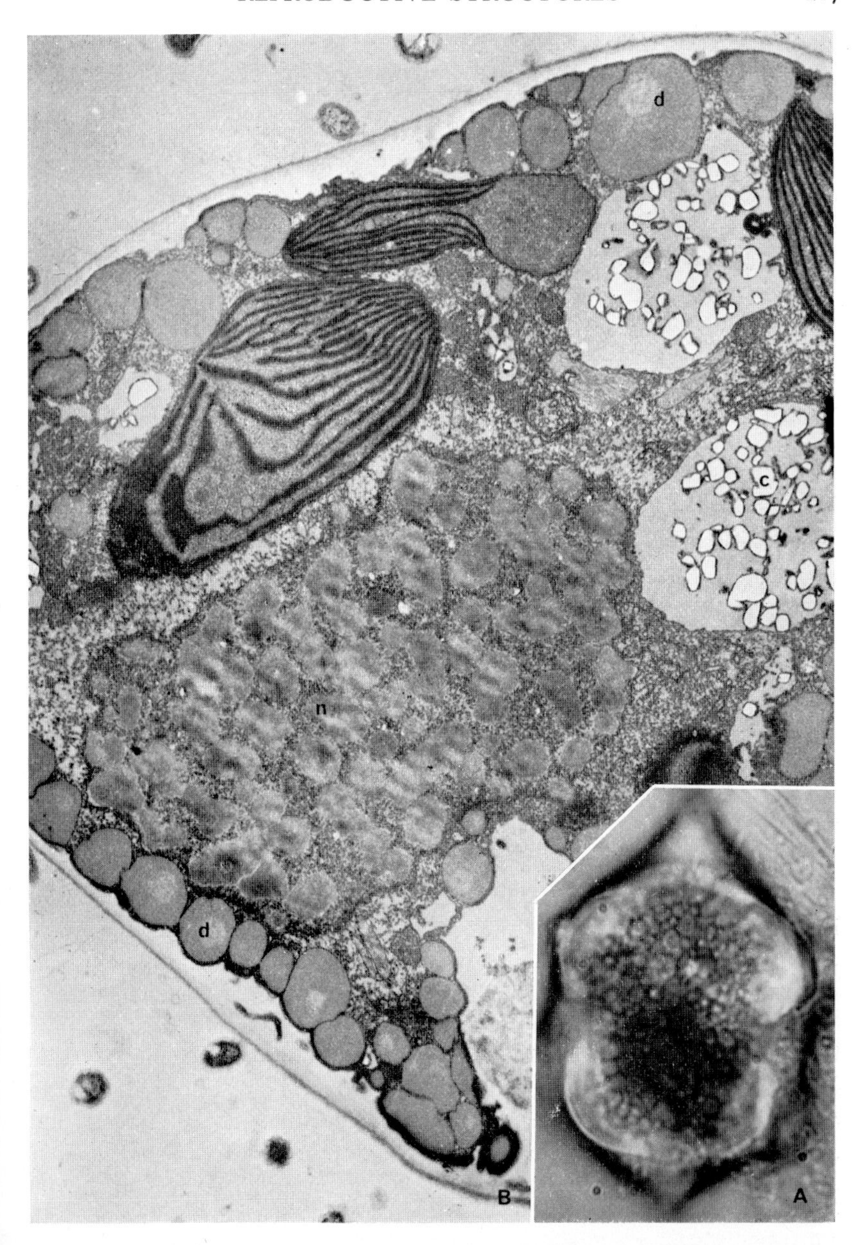

FIG. 11.7A–B. An almost mature cyst surrounded by a fairly thick wall and with an inner protective layer of lipid droplets (d) which can be seen both in the light micrograph A and the electron micrograph B. The other contents of the cyst are a nucleus (n), in which the chromosomes are no longer very distinct, a few chloroplasts, granular cytoplasm and large vacuoles containing crystals (c) which probably consist of guanine. (A, ×1500); (B, ×8400). (Figs 11.5–11.7 after Bibby and Dodge, 1972.)

was obtained about the fibres comprising the flagella and their roots. A later study of the zoospore of *Scytosiphon* (Manton, 1957), which did involve the use of sections, showed the zooid to be pear-shaped with two laterally inserted flagella, the anterior flagellum being garnished with two rows of thick hairs. The zoospore contained a large plastid with eyespot, a nucleus, mitochondrion, oil droplets and a structure which is probably a dictyosome. The fine structure of zoospores of *Cladophora* and *Chaetomorpha* has recently been studied with freeze-etching techniques (Barnett and Preston, 1970; Robinson and Preston, 1971a; Robinson, 1972; Robinson *et al.*, 1972). As these studies are mainly concerned with the structure of membranes and the development of cell walls they are reported elsewhere (Chapter 2).

In the red alga *Griffithsia* Peyrière (1969) has described some of the steps leading to the formation of tetraspores. The preliminary phase of nuclear enlargement in the tetracyst is accompanied by enrichment of the cytoplasm with ribosomes, increase in the number of plastids and mitochondria and extension of the Golgi complex. This is presumably followed by meiosis (not described) and the delimitation of the four spores. Tetrasporogenesis has now also been described in *Levringiella* (Kugrens and West, 1972b). The fine structure of the cystocarp has recently been described from another red alga, *Polysiphonia* (Tripodi, 1971).

IV. Encystment

This process has as yet been studied only in the freshwater dinoflagellate *Woloszynskia tylota* (Bibby and Dodge, 1972). Motile cells (Fig. 11.5) which look identical to vegetative cells but which may be planozygotes (von Stosch, pers. comm.) lose their flagella and cast off their theca as they develop a thin amorphous wall (Fig. 11.6). This wall thickens by the deposition of material on its inner face and the shape of the cells alters as various knobs develop. Whilst the wall is developing the number of chloroplasts and various other organelles gradually reduces but at the same time a large orange-red accumulation body develops and large vacuoles, containing what may be guanine crystals, appear in the cell (Fig. 11.6). The accumulation body is probably an autophagic vacuole or cytolosome. Towards the end of the process of organelle reduction the cytoplasm shrinks away from the wall and rounds up. A single layer of lipid droplets forms beneath the plasmalemma (Fig. 11.7). In this 'resistant' condition the cyst survives through the summer when the pond in which it lives may become completely dried up. It will germinate when conditions are again favourable. The combination of wall and lipid droplets would appear to provide a very good protection and the reduction of organelles presumably makes for a lowered metabolic activity.

12. *Symbiosis*

A number of algae, particularly relatives of *Chlorella* (Chlorophyceae) (Fig. 12.2) and the *Zooxanthellae* (Dinophyceae) (Fig. 12.1), are found living symbiotically within the cells of animals. Electron microscopical studies (Oschman, 1967; Karakashian *et al.*, 1968; Kevin *et al.*, 1969; Taylor, 1968b, 1969a) have shown that internally these algae have no abnormal structure related to their unusual situation although they may have a rather elaborate boundary layer made up of several membranes (Fig. 12.1). Rather unusually in one situation, the symbiosis between the dinoflagellate *Amphidinium klebsii* and the turbellarian *Amphiscolops* (Taylor, 1971b), the symbiont has a delicate covering (theca) which is exactly the same as that found when the alga is in the free-living state. It has recently been shown that in the Giant Clam older and senescent symbionts are culled from the population and are intracellularly digested by use of amoebocytic lysosomes (Fankboner, 1971). Similar removal of old cells probably takes place in all of these relationships (Karakashian *et al.*, 1968). However, the normal relationship, as studied autoradiographically by Taylor (1969b,c), is that carbon compounds pass from the autotrophic zooxanthella to the heterotrophic host and presumably nitrogenous material from host to symbiont.

Several studies have been made of the green alga which is found in the flatworm *Convoluta* (Oschman and Gray, 1965; Sarfatti and Bedini, 1965). The most detailed work is that of Parke and Manton (1967) which has shown that the symbiont here is *Platymonas convolutae* a member of the Prasinophyceae. This study has also included experimental resynthesis of the symbiotic relationship (Provasoli *et al.*, 1968), which gives added proof of the system.

Another association which has aroused interest is that between the ciliate *Mesodinium* and its red-brown symbiont. Electron microscopical studies (Taylor *et al.*, 1969, 1971) have shown that this phycobiont consists only of a chloroplast, a pyrenoid, and some mitochondria. The organism is bounded by a single membrane and the symbiont appears to lack a nucleus. The structure of the plastids and their pigment composition suggests a relationship with the Cryptophyceae, but the absence of a nucleus almost puts it on a par with the blue-green algal symbionts described below or the chloroplasts found in some opisthobranchs (see Chapter 4).

Many studies have been carried out on flagellates which contain blue-green algal symbionts, often termed cyanelles. In some cases this has been in an attempt to discover the true affinity of the host organism which has often been a matter of controversy. In *Cyanophora paradoxa* it has been shown (Hall and Claus, 1963;

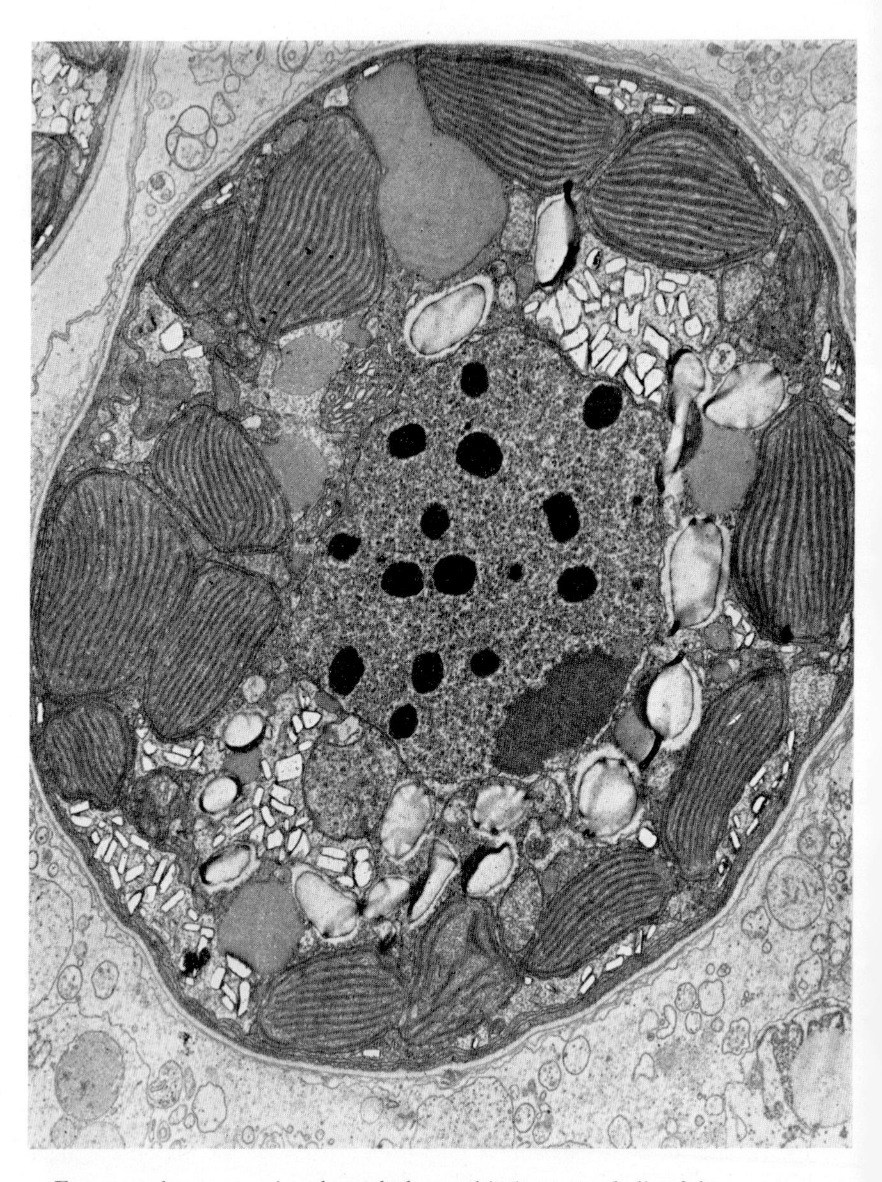

FIG. 12.1. A cross-section through the symbiotic zooxanthella of the sea anemone *Anemonia* which is a dinoflagellate variously named as *Symbiodinium* or *Gymnodinium*. Note the entire wall, outside of which are several layers of membrane belonging to the animal. The cell contains several chloroplasts, a nucleus, starch grains, fat droplets and many small crystals. (× 12,500)

Mignot *et al.*, 1969) that the symbionts are simple blue-green algae. The host was long thought to be a member of the Cryptophyceae but Mignot *et al.* (1969), on the basis of the insertion of the flagella and the structure of the cell covering, think that it may be a primitive dinoflagellate. Drum and Pankratz (1965a) have

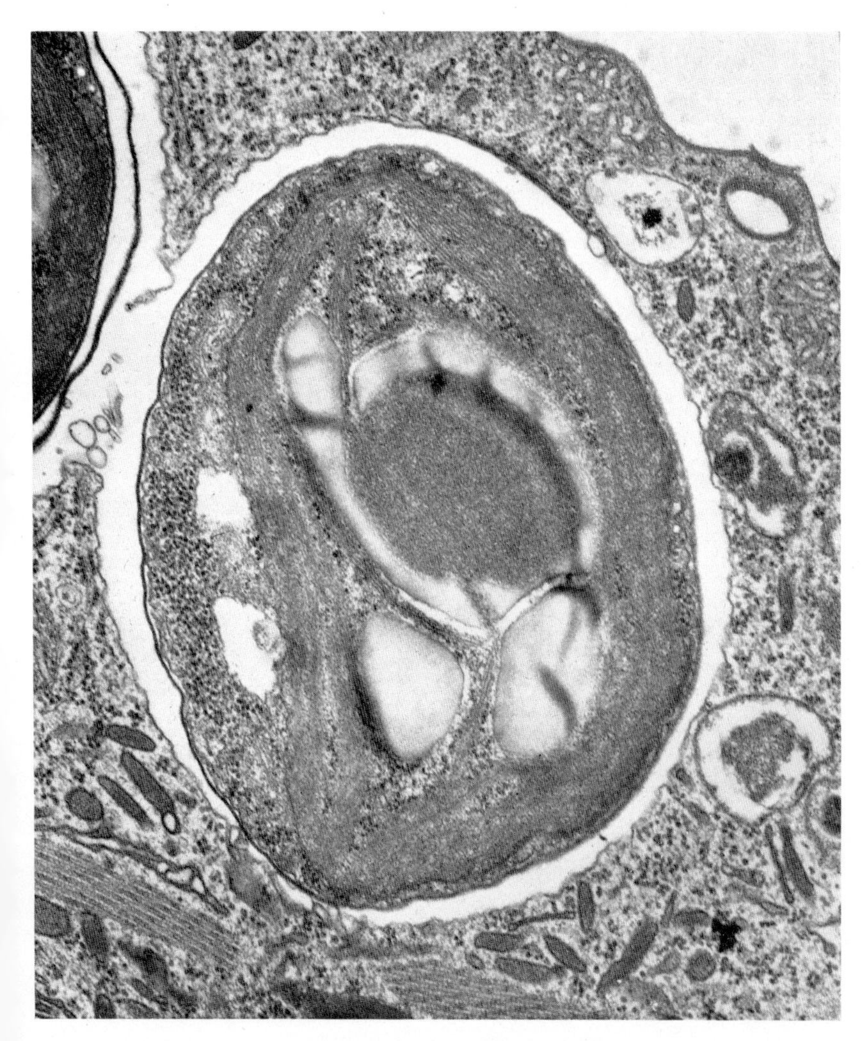

FIG. 12.2. A section through part of a 'green' *Paramecium* (Ciliata) cell showing one of the numerous symbiotic *Chlorella* cells present within the ciliate. The animal here encloses the alga by only a single membrane. (×36,300)

described 'an unusual cytoplasmic inclusion' in the diatom *Rhopalodia* but this clearly seems to be a simple blue-green alga. The colourless alga *Geosiphon* has been shown to have *Nostoc*-like symbionts (Schnepf, 1964).

Several studies have been concerned with *Glaucocystis nostochinearum* (Hall and Claus, 1967; Schnepf *et al.*, 1966; Echlin, 1967; Bourdu and Lefort, 1967; Robinson and Preston, 1971b,c) which, as its name implies, also has cyanophyte symbionts. Schnepf *et al.* (1966) decided that *Glaucocystis* could not be a green alga (which it was formerly thought to be) but they were unable to assign it to a definite systematic position. They did report the presence of flagellar bases during cell division, although this is a non-motile alga. Hall and Claus (1967) studied the detailed fine structure of the cyanelles and thought that they are bounded by

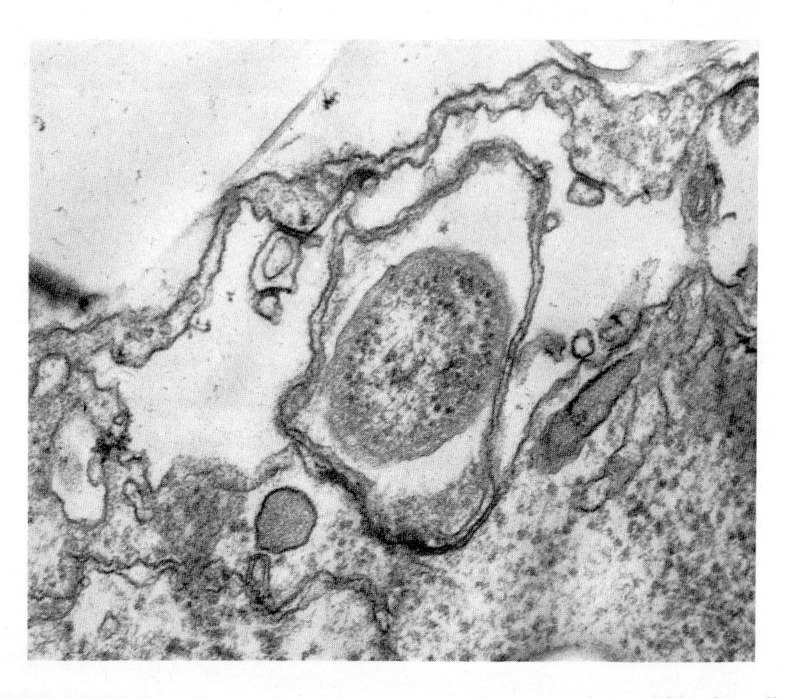

FIG. 12.3. A bacterium enclosed in a two-membrane envelope within the cell of *Katodinium glandulum* (Dinophyceae) and probably living symbiotically with the alga. (× 59,000)

only a single membrane but Echlin (1967) showed two membranes around them. The photosynthetic lamellae have been studied in great detail by Lefort (1965) and Bourdu and Lefort (1967) who showed that various types of particle were attached to the single thylakoids. Robinson and Preston (1971b,c) have studied the wall and plasmalemma of the host by freeze-etching and find that the wall is stratified and the complex plasmalemma consists of an outer membrane closely applied to the wall and an inner convoluted membrane. The wall was also studied by Schnepf (1965).

Another similar confused situation exists in *Glaucosphaera* (Richardson and Brown, 1970) where the presence of starch in the cytoplasm suggests that the host is not a green alga. An unusual feature here is the presence of eyespot-like structures in some of the cyanelles. All of these observations on organisms with blue-green algal symbionts are of considerable interest in view of the theories now being expounded (for review see Margulis, 1970; Taylor, 1970 and pp. 102–103) that chloroplasts have evolved from symbiotic cyanophytes.

Occasionally bacteria are seen in sections of algal cells and these appear to be living in symbiotic association with their host. The best documented case is that of the endonuclear bacteria in Euglenoid flagellates (Leedale, 1969). Here, numerous rod-shaped bacteria are found within the nucleus of several species belonging to the Euglenophyceae. Each bacterial cell is surrounded by a mucilaginous sheath outside its cell membrane. The bacteria appear to multiply to keep pace with the cell division of this alga. In the Dinophyceae bacteria have been observed in the cytoplasm of *Amphidinium herdmanii* and *Katodinium glandulum* (Fig. 12.3) and in *Bryopsis* (Chlorophyceae) they are common in the vegetative cells and gametes (Fig. 11.3A) (Burr and West, 1970).

The symbiotic relationship found in lichens has been the subject of a fair amount of study (Brown and Wilson, 1968; Chervin *et al.*, 1968; Walker, 1968; Ben-Shaul *et al.*, 1969; Jacobs and Ahmadjian, 1969, 1971; Galun *et al.*, 1970; and other references below) although, as yet, this has revealed little more than the individual ultrastructure of the respective phycobionts and mycobionts. In several studies the alga concerned has been *Trebouxia* (Chlorophyceae) (Peveling, 1969; Jacobs and Ahmadjian, 1971; Fisher and Lang, 1971a) and this alga isolated from lichens has also been studied in pure culture, mainly on account of its complex and interesting pyrenoids (Fisher and Lang, 1971a,b) (see Chapter 5). Galun *et al.* (1970) thought that in the lichen the *Trebouxia* prevented fungal penetration by developing a thick wall. However, fungal hyphae were seen to penetrate senescing cells which were probably in the older parts of the thallus. Much remains to be discovered about the structural aspects of lichen symbiosis when the difficult problems of fixation and embedding can be completely mastered.

13. Experimental Ultrastructure Studies

As yet comparatively little experimental ultrastructural work has been carried out with algae. No doubt much interesting material will be found amongst these organisms when molecular biologists need more complex organisms than bacteria and yeasts. Some of the experimental work relating to Chloroplast development is described in Chapter 4.

I. Effects of Chemicals

Studies involving the treatment of *Chlorella* with certain herbicides have shown that the effect on the structure of the alga is in some cases rather different to that caused in higher plants. With Atrazine, Ashton *et al.* (1966) found that there was inhibition of the normal increases in cell volume and chlorophyll content and no starch was laid down in the cells. However, there were no observable abnormalities in the structure of the cell. With the bipyridyl herbicide, diquat, Stokes *et al.* (1970) found a fairly rapid inhibition of photosynthesis and respiration. After 5 h the first signs of damage to the chloroplasts—a slight buckling of the lamellae—could be observed and by this time the nucleus and mitochondria had disappeared. After 10 h almost no starch remained in the cells and the chloroplast lamellae were completely broken down. *Chlorella* has also been treated with 2, 4-dichlorophenoxy-acetic acid (Bertagnolli and Nadakavukaren, 1970a). Here, after 6 h treatment the cells showed vesiculation of the plasma-membrane, swelling of the mitochondria and disruption of the chloroplast lamellae.

Several studies have been concerned with the structural changes which might be produced in the motile green alga *Dunaliella* by extracts from tobacco smoke and other more specific noxious chemicals. Using acrolein (a constituent of tobacco smoke) Le Baron-Marano and Izard (1968) found that cytoplasmic inclusions appeared in the nucleus and the chloroplast and Golgi were somewhat damaged. With a combination of acrolein and tobacco smoke it was found that there was a complete blockage of nuclear division at the stage where the dictyosomes and flagella had already replicated (Puiseux-Dao and Izard, 1968).

Ethidium bromide, an inhibitor of DNA synthesis has been applied to the cells of *Amphidinium* (Dinophyceae) (Matthys and Puiseux-Dao, 1968). It was found that after 24 h treatment the mitochondria had a very dense central fibrillar core but after prolonged treatment of up to 7 days this central area had become electron lucent and the mitochondrial cristae were also much reduced in number and size. The chloroplasts became smaller and had large areas devoid of lamellae. It appeared that the DNA in the mitochondria was first condensed and then broken down by this treatment.

Euglena has been treated with Chloramphenicol (Ben-Shaul and Markus, 1969), a drug which inhibits protein synthesis. It was found that multiplication of cells was not affected for 36 h after treatment, although the size of dividing cells decreased and their chlorophyll content was lowered. In cells grown in the dark and then brought into the light in the presence of chloramphenicol, chlorophyll synthesis was inhibited and the rate of plastid elongation and thylakoid formation was much decreased. Abnormal, much-branched mitochondria were produced. *Euglena* has also been treated with Streptomycin (Ben-Shaul and Ophir, 1970). This antibiotic, after a lag period of several generations, brought about a decrease in the size of the plastids and a reduction in their lamellar system. By the end of 11 generations in Streptomycin, in the light, the chloroplasts had become similar in structure to the pro-plastids of dark grown cells. Various concentric lamellar bodies were found in the cytoplasm.

The effect of Chloramphenicol on cell division and organelle structure has been studied during light-induced chloroplast development in the chrysophyte *Ochromonas* (Smith-Johannsen and Gibbs, 1972). To start with, the growth rate of treated cells is normal but chlorophyll synthesis and thylakoid production is slightly reduced. Later, prolamellar bodies and abnormal stacks of thylakoids appear in the chloroplasts and there is proliferation of the perinuclear reticulum. There is also a progressive loss of mitochondrial cristae but the mitochondrial ribosomes appear to be unaffected. Therefore, as with *Euglena*, the most noticeable structural effect of Chloramphenicol treatment is on the development of the chloroplast lamellar system. This is only to be expected as this drug has a specific inhibitory effect on protein synthesis by the 70S ribosomes of chloroplasts.

Several studies have been made into the effects of inhibitors of RNA and protein synthesis on normal and mutant *Chlamydomonas reinhardi* (Goodenough, 1971; Goodenough and Levine, 1971). With wild-type alga and *ac-20* mutant Rifampicin inhibited the production of chloroplast ribosomes. Streptomycin brought about the production of wide stacks of thylakoids beneath the chloroplast envelope but left most of the chloroplast with only scattered single thylakoids. Chloramphenicol also produced the wide bands but, in addition, the pyrenoid was disrupted and tubular vesicles were found in the centre of the chloroplast.

The mitotic inhibitor, colchicine, has been used to treat *Chlamydomonas* and *Chlorella*. In the former (Walne, 1967) a number of cytoplasmic changes took place. There was increased vesicle production by the Golgi; stratification of the cell walls; elongation of the mitochondria; increase in size of the nucleus and proliferation and convolution of the nuclear envelope. Many cell components were replicated. In *Chlorella* Wanka (1968) found that colchicine prevented cell division, and polyploid and polynucleate protoplasts were formed. In these the cell wall synthesis was upset and much surplus wall material formed, only to be sloughed off from the cell. The colchicine treated cells developed several pyrenoids which were often combined in clusters. If the cell had been treated whilst in division there was excessive development of endoplasmic reticulum and the number of Golgi bodies per cell became higher than in untreated material.

II. Effects of Environmental Conditions

A. NUTRITION

Studies on the effects of environmental factors such as nutrients and light have been almost entirely confined to *Chlorella* and most of these have involved some comparison of the structure of cells grown autotrophically in the light with those grown heterotrophically in the dark. The first work here was that of Lefort (1962) who noted the development of normal chloroplasts when cells were taken from heterotrophic (dark) conditions and allowed to grow autotrophically. Griffiths and Griffiths (1969) made a direct comparison of cells of *Chlorella vulgaris* (Emerson strain) grown under the two conditions. Heterotrophy produced 'giant' cells with large chloroplasts containing much starch and very few lamellae. Production of cell organelles did not keep pace with growth of the cells. Working with *C. pyrenoidosa* (which in fact owing to nomenclatural confusion, may be identical to *C. vulgaris*) Budd *et al.* (1969) found that in cells transferred from autotrophic conditions to darkness, in the presence of glucose, there was a gradual reduction in the number of thylakoids in the chloroplast until all that remained was a proplastid containing a structure like a prolamellar body. At the same time the cell wall increased in thickness. In another experiment cells were adapted to darkness plus glactose, and under these conditions, they developed much unusual storage material which, unlike normal starch, had osmiophilic properties. A variation of these experiments was provided by Rodríguez-López (1965) who grew *C. pyrenoidosa* in the light in the presence of glucose, fructose or mannose. 'Giant' cells were produced which each contained a single large chloroplast, distended by numerous starch grains. The percentages of carbohydrate and protein, 84% and 13% respectively, were found to differ considerably from those in cells grown autotrophically which were 23% carbohydrate and 48% protein.

Somewhat similar experiments have been carried out by Casselton, Crawford and Dodge (unpublished observations), but using *Chlorella protothecoides*. Here, growth in heterotrophic conditions in the dark resulted in reduction of the chloroplasts to small amyloplasts containing starch grains, fragments of the thylakoids, and abundant chloroplast ribosomes (Fig. 13.2). Growth in heterotrophic conditions in the light resulted in the production of cells with large chloroplasts which contained small profiles of 2-thylakoid lamellae in addition to numerous starch grains (Fig. 13.3). Control material grown autotrophically in the light had chloroplasts which were packed with lamellae and contained very few starch grains (Fig. 13.1).

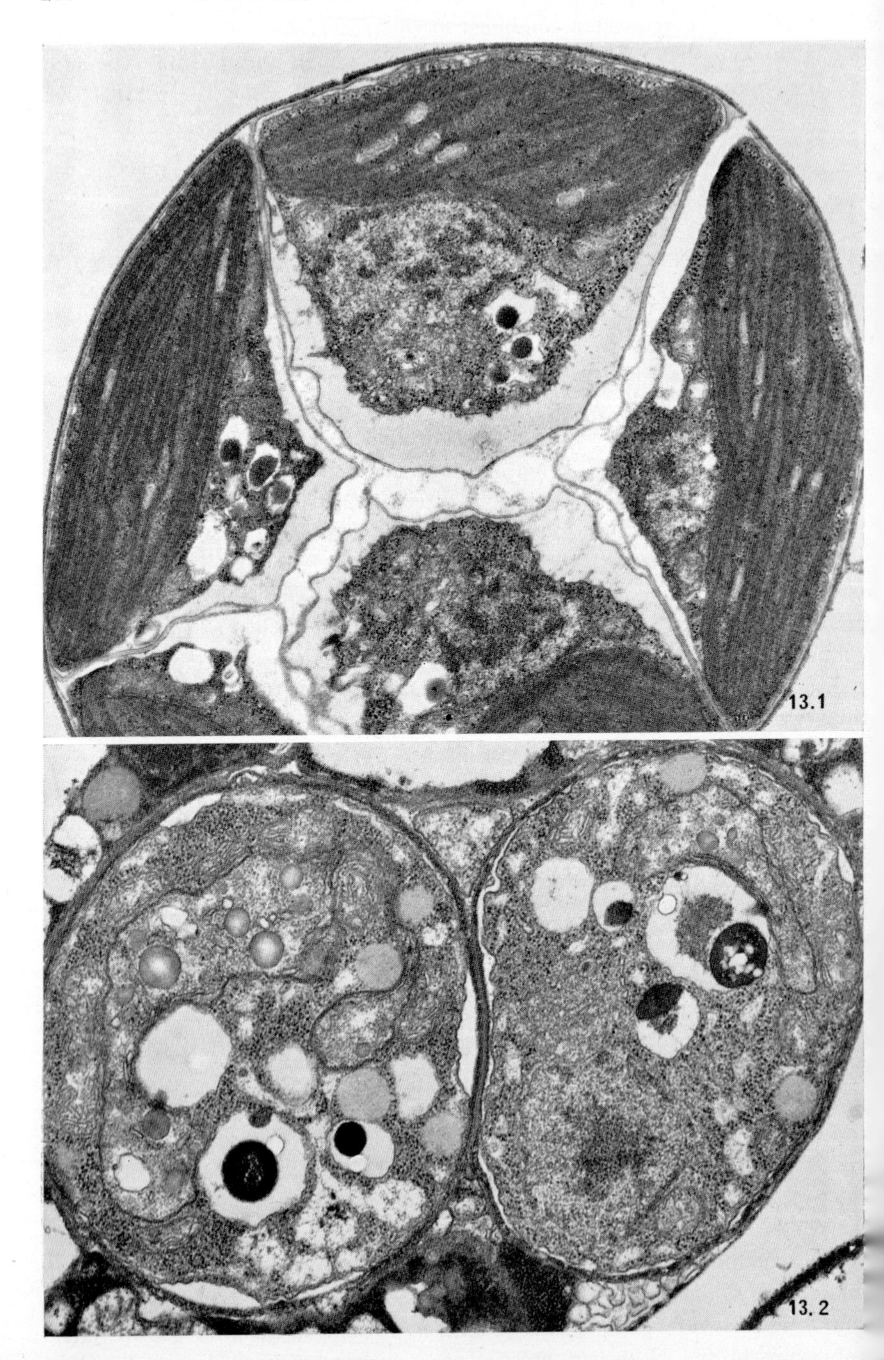

FIG. 13.1. *Chlorella protothecoides* grown autotrophically. Each autospore has a large chloroplast, packed with lamellae, and a nucleus, etc. (×19,000)

FIG. 13.2. Two autospores within a cell of *C. protothecoides* grown heterotrophically in the dark. The chloroplasts (c) are reduced to amyloplasts and the cells contain mainly ribosomes, mitochondria and various vesicular structures. (×21,000)

Thinh and Griffiths (1972a,b) have investigated the changes which took place when 'giant' cells of *C. vulgaris* were replaced in autotrophic conditions. There was a quick increase in the number of ribosomes in both chloroplast and cytoplasm and after 15 h many nuclear divisions took place to give an abnormally high number of autospores. Whilst this was happening the chloroplasts became repacked with thylakoids which seemed to grow out from the remnants of

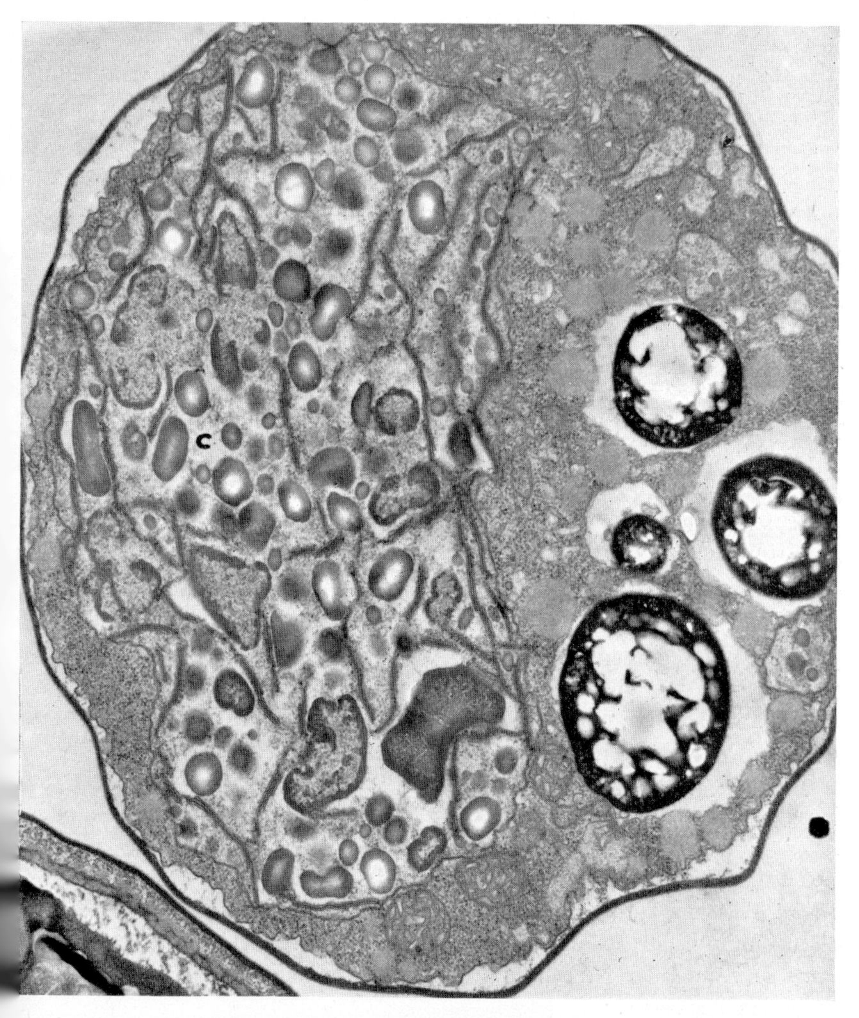

FIG. 13.3. *C. protothecoides* grown heterotrophically in the light. The cell has a arge chloroplast which contains much starch but a very reduced lamellar system. The rest of the cell contains many mitochondria, fat droplets and electron-dense bodies which may consist of polyphosphate (see Chapter 10). (×16,800) (Figures 3.1–13.3 were prepared by R. M. Crawford from material grown by P. J. Casselon.)

thylakoid present in the 'giant' cell. Chloramphenicol was not found to inhibit thylakoid regeneration although it did stop the thylakoids from associating together to form lamellae. Whilst the chloroplast was producing membranes the large store of starch accumulated by the 'giant' cells became used up.

Merrett (1969) and Wiessner and Amelunxen (1969a,b) have studied the structure of the chloroplast in the green alga *Chlamydobotrys* (= *Pyrobotrys*) under different nutritional conditions. This alga is an obligate phototroph which can nevertheless grow heterotrophically provided it is in the light. In photo-autotrophic conditions normal chloroplasts, with a fair amount of thylakoid stacking, were found. In photo-heterotrophic conditions (light+acetate) most thylakoids remained single and in these conditions no Hill reaction could be demonstrated. Weissner and Amelunxen were able to show that there was parallel development of the photosynthetic activity and the stacking together of developing thylakoids.

The effect of carbon dioxide supply on the chloroplast structure of *Chlorella* has been investigated by Gergis (1972), who found that CO_2 had a considerable effect on the structure of the lamellar system. When a supply of 0·03% CO_2 was provided, a distinct grana-intergrana system was found but when a much higher concentration of 3% CO_2 was supplied the chloroplast lamellae consisted of more or less regular stacks of 2–4 thylakoids. Other cell components did not seem to be affected by these treatments.

Malkoff and Buetow (1964) have investigated the effects of carbon starvation on *Euglena gracilis*. This treatment resulted in rapid disappearance of paramylon, massive vacuolization of the endoplasmic reticulum and segregation of areas of cytoplasm which were apparently sacrificed by the production of lysosomes. The nucleus remained unchanged as did the mitochondria, unless starvation was prolonged. Replenishment of the carbon source after 8 days resulted in complete reversal to normal morphology, except for the persistence of lysosome-like particles, and was characterized during the early stages by the presence of bizarre mitochondria. As a last stage of recovery paramylon grains were secreted directly by membranes of the E.R. cisternae.

In *Chylamydomonas reinhardi* Teichler-Zallen (1969) was able to produce cells incapable of photosystem II by growing them in a manganese deficient medium. After this treatment the chloroplasts showed much less stacking than normal.

Calcium deficiency has been studied using *Scenedesmus* (Walles and Kylin, 1972). This was found to cause a general degeneration of the cytoplasm and alteration in the structure of the cell wall. Whereas the normal cell has distinct inner and outer layers, cells grown in calcium deficient medium produced multiple outer layers.

B. LIGHT

Treharne *et al.* (1964) studied the effects of various light intensities on the plastid of *Chlorella*. They found that in low light (1076·4 lx) there was a marked increase in the number and density of the lamellae compared with when the alga was grown in light of high intensity (23,020 lx). Wanka and Mulders (1967) also working with *Chlorella* investigated the effect of light on DNA synthesis. They showed that the DNA normally duplicated four times between the 10th and 18th h after the start of the light period (in 16 h L: 8 h D-grown cells). Electron microscopy revealed one division of all the nuclei in the cell every 110 min. Premature transfer to dark, or to light of low intensity, brought forward the time of spore release.

Könitz (1965) grew *Euglena gracilis* on a fairly complete medium under a 12 h light: 12-h dark cycle. He compared the structure of cells fixed in the middle of the dark with those fixed in the middle of the light period and found considerable differences. In the dark-fixed cells the mitochondria had orderly longitudinally arranged cristae, there were no pyrenoids and very few Golgi bodies were present. In the light-fixed cells the mitochondria contained scattered, unordered, cristae, numerous pyrenoids were present and there seemed to be many Golgi bodies. This work suggests that cells grown in fluctuating conditions might be in a continual state of change in cellular structure besides the cyclical changes in their physiology.

C. OSMOTIC POTENTIAL

Dunaliella is an alga which may be found growing both in salt ponds with exceedingly high salinity and in pools with less than the normal salinity of the sea. It is therefore an interesting alga in which to investigate the structural changes which might be brought about by rapid change of the osmotic potential of its medium. Trezzi *et al.* (1965) found that when the OP of the medium was lowered the whole cell and most of its organelles became swollen. The mitochondria became distended and the chloroplast lamellae increased in size but the nucleus was very little affected. A high OP caused shrinkage of all parts of the cell with a consequent increase in their electron density. In most cases the alterations in structure were found to be reversible upon returning the alga to more normal conditions.

III. Senescence

Enforced senescence of cells in the dinoflagellate *Peridinium westii* has been found by Messer and Ben-Shaul (1970) to result in a considerable increase in lipid material in the cell and also what are described as collagen-like rods. These latter are almost certainly trichocysts which have discharged internally. One other change noted was that the theca had been replaced by a thick rigid envelope suggesting that the cells had encysted.

Observations have been made on the changes in structure, with ageing, in *Ochromonas* (Schuster *et al.*, 1968) and the green alga *Spongiochloris* (McLean, 1968). In the latter, unusual tubular structures were noted in the pyrenoid.

When cells of *Euglena granulata* were 'aged' they were found (Palisano and Walne, 1972) to lack flagella and to be spherical in shape. Internally they had accumulated numerous cytoplasmic vacuoles and lysosome-like structures containing heavily stained pigmented bodies and membrane fragments. There was considerable acid phosphatase activity at the maturing face of the dictyosomes and in the lysosome-like vacuoles. It would seem that the dense vacuoles are similar to the accumulation body which forms in encysting dinoflagellates (Bibby and Dodge, 1972) and is probably an autophagic vacuole responsible for the breakdown of surplus membranous structures and organelles into their basic components.

14. Postscript: Fine Structure and Phylogeny

At the present state of knowledge concerning algal fine structure, when in some classes very few organisms have been studied in any detail, it is perhaps a little too soon to try to build up hypotheses concerning the phylogenetic relationship of the groups. However, an attempt will be made to analyse the relationships between the classes on the basis of the information reviewed in this book. Even if the tentative conclusions presented here are eventually proved to be wrong they will perhaps have the effect of encouraging other workers to study some of the less well known groups of algae. It should be emphasized that whereas previous phylogenetic systems have been mainly based on pigment composition, with some consideration given to cell structure, the diagrams presented here are entirely based on the structure of cells. The schemes shown in Figs 14.1–14.3 were constructed by taking the lists of characters shown in Tables I, II, III and V and from these compiling charts showing the number of features held in

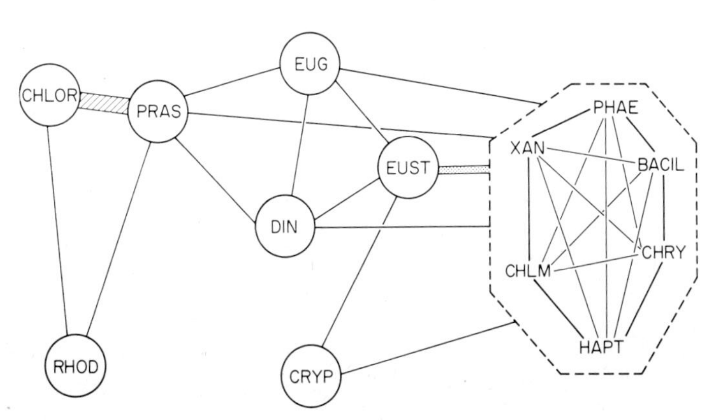

FIG. 14.1. A diagram to show the relationships of the classes on the basis of the structure of chloroplasts and eyespots. Similarity is indicated both by the distance apart on the diagram and the thickness of the connecting line. The six classes within the broken line have several features in common. (See text for details of methods used to construct the figure.)

In the figures in Chapter 14 the following abbreviations are used for the algal classes: BACIL, Bacillariophyceae; CHLOR, Chlorophyceae; CHLM, Chloromonadophyceae; CHRY, Chrysophyceae; CRYP, Cryptophyceae; DIN, Dinophyceae; EUG, Euglenophyceae; EUST, Eustigmatophyceae; HAPT, Haptophyceae; PHAE, Phaeophyceae; PRAS, Prasinophyceae; RHOD, Rhodophyceae; XAN, Xanthophyceae.

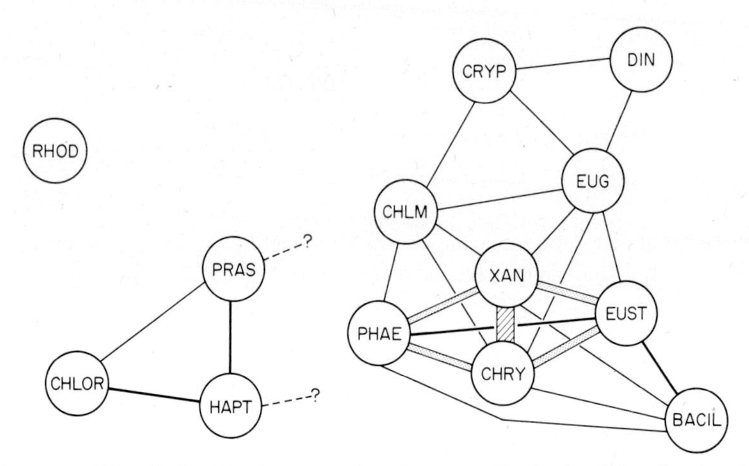

FIG. 14.2. The similarities between the classes on the basis of flagella structure. The three classes forming a triangle to the left have no features in common with the main group of classes, apart from the possession of flagella. The Rhodophyceae stands alone as the only class with no flagella of any type.

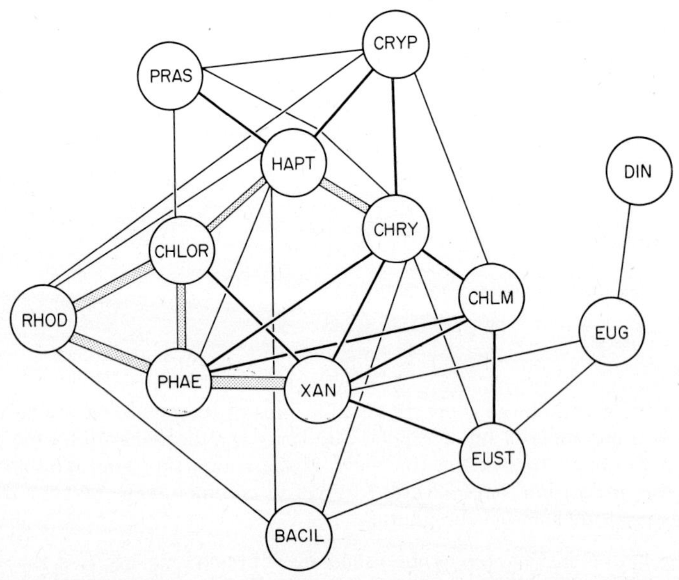

FIG. 14.3. A combined diagram showing similarities between the classes in the structure of the flagella, cell covering, nucleus, and ejectile organelles. To simplify the figure all single-similarities have been excluded. The thick dotted line indicates four similarities, the thick line three and the thin line two.

common by various pairs of classes. Using these data, figures were constructed with the classes having most similar features shown closest together and joined by a thick line and the other classes proportionately further away. Such figures should really be three-dimensional and in some cases it was difficult to fit certain classes into the two-dimensional diagrams.

In view of the current theories about the possibly symbiotic origin of chloroplasts (cf. Chapter 4) it is perhaps desirable to consider the characters of chloroplasts separately from other features. Eyespot structure was included here, for each class tends to have a distinctive type related to its chloroplast structure (Chapter 6). However, another structure always associated with chloroplasts, the pyrenoid, was not included for, as outlined in Chapter 5, the various types of pyrenoid rather surprisingly seem to cut across class distinctions.

Figure 14.1 shows that so far as chloroplast characters are concerned six classes are very closely related to each other with three or four features in common. Of these six, the Haptophyceae is perhaps less similar and more distinctive with its lack of a girdle lamella. This group is nearly identical with Christensen's (1962) Chromophyta. Five other classes have a more or less consistent similarity relationship, with all the members of the 'Chromophyta'. Most similar is the Eustigmatophyceae which has at least two features in common. The other four classes have less in common with the 'Chromophyta' but two, the Euglenophyceae and Dinophyceae also have some structures in common with the Prasinophyceae. This latter class is, not surprisingly, fairly similar to the Chlorophyceae. Quite at variance with schemes of phylogeny based on pigmentation, the Rhodophyceae appears most similar to the Chlorophyceae and Prasinophyceae and has nothing in common with the Cryptophyceae, the only other class to possess phycobilin pigments. This scheme appears not to support the monophyletic hypothesis of the origin of plastids from a colourless cryptophyte ancestor (Lee, 1972) but it does seem to indicate that a multiphyletic origin of the type proposed by Raven (1970) was possible.

A character much used in the classification of algae is the type and number of flagella. One of the earlier schemes divided algae into Isokontae (equal flagella), Heterokontae (dissimilar flagella) and Akontae (no flagella). The apparent relationships between the classes on the basis of flagella structure are shown in Fig. 14.2. Once again the strongest association is seen between some members of the 'Chromophyta'. Here, the Xanthophyceae, Chrysophyceae, Phaeophyceae and Chloromonadophyceae all have two flagella, one with stiff hairs, directed anteriorly, and the other, more or less smooth, is posteriorly orientated. Little work has been done on flagella in the Eustigmatophyceae and Bacillariophyceae but what facts we do have suggest that these classes are closely related and perhaps show a reduced condition. Once again the Cryptophyceae, Dinophyceae and Euglenophyceae—all with unique flagellation—are peripheral in position. Rather surprisingly, the Haptophyceae, with two smooth flagella, a stellate transition zone, and a haptonema, appears more similar to the Chlorophyceae and Prasinophyceae and to have little, if anything, in common with the 'Chromophyte' group. The Rhodophyceae stand on their own with no flagella at all.

The structural feature which might be expected to be most basic and therefore to give the most reliable phylogenetic information is the nucleus but, until more information is available about nuclear structure and division, we can make very few conclusions. Certainly, the nucleus of the Dinophyceae and, to a lesser extent, that of the Euglenophyceae, appears to be primitive.

A number of types of cell covering are distinctive but some, such as cell walls, are found in many classes. Several types of ejectile organelles are found and generally these are class-specific. As none of these characters is sufficient on its own to give a good idea of the relationships between the classes they have all been put together, with the flagella, to produce Fig. 14.3. This gives a much more complex figure than the previous ones in spite of the fact that all single-similarities between classes have been omitted. A striking feature of this scheme is the way in which the Chlorophyceae and Rhodophyceae, both fairly peripheral on flagella and chloroplast characters, now appear to have a close relationship to parts of the 'Chromophyta' group. As before, the Prasinophyceae, Cryptophyceae and Euglenophyceae are somewhat on the fringe and the most isolated class now appears to be the Dinophyceae. If the figure were drawn in three dimensions the Bacillariophyceae would be situated almost over the Chrysophyceae and the Prasinophyceae over the Haptophyceae, rather than in the positions shown in the diagram.

Tentative Conclusions

(1). No matter which characters are used the three classes Dinophyceae, Euglenophyceae and Cryptophyceae stand out as assemblages of unique organisms. The first two are perhaps more closely related to each other than they are to any other classes. The Euglenophyceae is not nearly as closely related to the Chlorophyceae as the presence, in both classes, of chlorophyll *b* has been taken to imply.

(2). The Chlorophyceae and the Prasinophyceae (which was split off from it in 1962), are closely related and very surprisingly the next nearest class appears to be the Rhodophyceae.

(3). The Haptophyceae occupies various positions according to which characters are being considered and is probably either an intermediate group or one which has not yet been investigated thoroughly enough.

(4). The core of Christensen's 'Chromophyta' (Phaeophyceae, Chrysophyceae and Xanthophyceae) is very closely interrelated and the Chloromonadophyceae and Eustigmatophyceae are fairly strongly attached to this core. The Bacillariophyceae, although as closely related on chloroplast structure and pigments, appears in a more distinct position when other factors are considered.

(5). On fine-structural characteristics no one class appears markedly more primitive than the others although the Dinophyceae seems to have nuclei nearest to those of the Procaryota.

Review Articles and Books of Special Interest

Brachet, J. and Bonotto, S. (Eds.) (1970). *Biology of* Acetabularia. Academic Press, New York and London.

Dodge, J. D. (1969a). A review of the fine structure of algal eyespots. *Br. phycol. J.* **4,** 199–210.

Dodge, J. D. (1971b). Fine structure of the Pyrrophyta. *Bot. Rev.* **37,** 481–508.

Dupraw, E. J. (1970). *DNA and Chromosomes.* Holt, Rinehart and Winston, New York.

Gibbs, S. P. (1970). The comparative ultrastructure of the algal chloroplast. *Ann. N.Y. Acad. Sci.* **175,** 454–473.

Griffiths, D. J. (1970). The pyrenoid. *Bot. Rev.* **36,** 29–58.

Joyon, L. and Mignot, J. P. (1969). Données récentes sur la structure de la cinétide chez les Protozoaires flagellés. *Ann. Biol.* **8,** 1–52.

Lang, N. J. (1969). The fine structure of blue-green algae. *Ann. Rev. Microbiol.* **22,** 15–46.

Leedale, G. F. (1967). *Euglenoid Flagellates.* Prentice Hall, New Jersey.

Leedale, G. F. (1970). Phylogenetic aspects of nuclear cytology in the algae. *Ann. N.Y. Acad. Sci.* **175,** 429–453.

Manton, I. (1965). Some phyletic implications of flagellar structure in plants. *Adv. Bot. Res.* **2,** 1–34.

Manton, I. (1966d). Some possibly significant structural relations between chloroplasts and other cell components. In: *The Biochemistry of Chloroplasts,* Vol. 1, pp. 23–47 (Ed. T. W. Goodwin). Academic Press, London and New York.

Margulis, L. (1970). *Origin of Eukaryotic Cells.* Yale University Press, New Haven.

Pickett-Heaps, J. D. (1969). The evolution of the mitotic apparatus: an attempt at comparative ultrastructural cytology in dividing plant cells. *Cytobios* **3,** 257–280.

Pickett-Heaps, J. D. (1972e). Variation in mitosis and cytokinesis in plant cells: its significance in the phylogeny and evolution of ultrastructural systems. *Cytobios* **5,** 59–77.

REFERENCES

Aaronson, S. (1971). *Limnol. Oceanogr.* **16,** 1–9.

Aaronson, S., Behrens, U., Orner, R. and Haines, T. H. (1971). *J. Ultrastruct. Res.* **35,** 418–430.

Afzelius, B. A. (1963). *J. Cell Biol.* **19,** 229–238.

Afzelius, B. A. (1965). *Expl Cell Res.* **37,** 516–539.

Anderson, E. (1962). *J. Protozool.* **9,** 380–395.

Arnott, H. J. and Brown, R. M. (1967). *J. Protozool.* **14,** 529–539.

Ashton, F. M., Bisalputra, T. and Risley, E. B. (1966). *Am. J. Bot.* **53,** 217–219.

Atkinson, A. W., Gunning, B. E. S., John, P. C. L. and McCullough, W. (1971). *Nature* (*New Biology*) **234,** 24–25.

Atkinson, A. W., Gunning, B. E. S. and John, P. C. L. (1972). *Planta* **107,** 1–32.

Bailey, A. and Bisalputra, T. (1969). *Phycologia* **8,** 57–63.

Bailey, A. and Bisalputra, T. (1970). *Phycologia* **9,** 83–101.

Barnett, J. R. and Preston, R. D. (1970). *Ann. Bot.* **34,** 1011–1017.

Barton, R., Davis, P. J. and Thomas, S. R. (1970). *Proc. 7th Int. Congr. E.M. Grenoble,* 433–434.

Belcher, J. H. (1968a). *Nova Hedwigia* **15,** 179–190.

Belcher, J. H. (1968b). *Nova Hedwigia* **16,** 131–139.

Belcher, J. H. (1968c). *Br. phycol. Bull.* **3,** 495–499.

Belcher, J. H. (1968d). *Arch. Mikrobiol.* **60,** 84–94.

Belcher, J. H. (1969a). *Br. phycol. J.* **4,** 105–117.

Belcher, J. H. (1969b). *Nova Hedwigia* **18,** 257–270.

Belcher, J. H. and Swale, E. M. F. (1967). *Br. phycol. Bull.* **3,** 257–267.

Belcher, J. H. and Swale, E. M. F. (1972a). *Br. phycol. J.* **7,** 53–59.

Belcher, J. H. and Swale, E. M. F. (1972b). *Br. phycol. J.* **7,** 335–346.

Bendix, S. (1960). *Bot. Rev.* **26,** 145–208.

Ben-Shaul, Y. and Markus, Y. (1969). *J. Cell Sci.* **4,** 627–644.

Ben-Shaul, Y. and Ophir, I. (1970). *Planta* **91,** 195–203.

Ben-Shaul, Y., Schiff, J. A. and Epstein, H. T. (1964). *Pl. Physiol., Lancaster* **39,** 231–240.

Ben-Shaul, Y., Epstein, H. T. and Schiff, J. A. (1965). *Can. J. Bot.* **43,** 129–136.

Ben-Shaul, Y., Paran, N. and Galun, M. (1969). *J. Microscopie* **8,** 415–422.

Bertagnolli, B. L. and Nadakavukaren, M. J. (1970a). *J. Phycol.* **7,** 98–100.

Bertagnolli, B. L. and Nadakavukaren, M. J. (1970b). *J. Cell Sci.* **7,** 623–630.

Bibby, B. T. and Dodge, J. D. (1972). *Br. phycol. J.* **7,** 85–100.

Bibby, B. T. and Dodge, J. D. (1973). *Planta* **112,** 7–16.

Billard, C. and Gayral, P. (1972). *Br. phycol. J.* **7,** 289–297.

Bisalputra, T. (1965). *Can. J. Bot.* **43,** 1549–1552.

Bisalputra, T. (1966). *Can. J. Bot.* **44,** 89–93.

Bisalputra, T. and Bisalputra, A. A. (1967a). *J. Ultrastruct. Res.* **17,** 14–22.

Bisalputra, T. and Bisalputra, A. A. (1967b). *J. Cell Biol.* **33,** 511–520.

Bisalputra, T. and Bisalputra, A. A. (1969). *J. Ultrastruct. Res.* **29,** 151–170.
Bisalputra, T. and Burton, H. (1969). *J. Ultrastruct. Res.* **29,** 224–235.
Bisalputra, T. and Weier, T. E. (1963). *Am. J. Bot.* **50,** 1011–1019.
Bisalputra, T. and Weier, T. E. (1964a). *Am. J. Bot.* **51,** 548–551.
Bisalputra, T. and Weier, T. E. (1964b). *Am. J. Bot.* **51,** 881–892.
Bisalputra, T., Ashton, F. M. and Weier, T. E. (1966). *Am. J. Bot.* **53,** 213–216.
Bisalputra, T., Rusanowski, P. C. and Walker, W. S. (1967). *J. Ultrastruct. Res.* **20,** 277–289.
Bisalputra, T., Shields, C. M. and Markham, J. W. (1971). *J. Microscopie* **10,** 83–98.
Black, M. (1965). *Endeavour* **24,** 131–137.
Black, M. and Barnes, B. (1961). *Jl R. microsc. Soc.* **80,** 137–147.
Bouck, G. B. (1962). *J. Cell Biol.* **12,** 553–569.
Bouck, G. B. (1965). *J. Cell Biol.* **26,** 523–537.
Bouck, G. B. (1969). *J. Cell Biol.* **40,** 446–460.
Bouck, G. B. (1971). *J. Cell Biol.* **50,** 362–384.
Bouck, G. B. and Sweeney, B. M. (1966). *Protoplasma* **61,** 205–223.
Bourdu, R. and Lefort, M. (1967). *C. r. hebd. Séanc. Acad. Sci., Paris* **265D,** 37–40.
Bourne, V. L. and Cole, K. (1968). *Can. J. Bot.* **46,** 1369–1375.
Bourne, V. L., Conway, E. and Cole, K. (1970). *Phycologia* **9,** 79–81.
Bourque, D. P., Boynton, J. E. and Gillham, N. W. (1971). *J. Cell Sci.* **8,** 153–183.
Bradley, D. E. (1964). *J. gen. Microbiol.* **37,** 321–333.
Bradley, D. E. (1965). *Q. Jl microsc. Sci.* **106,** 327–331.
Bradley, D. E. (1966). *Expl Cell Res.* **41,** 162–173.
Branton, D. and Park, R. B. (1967). *J. Ultrastruct. Res.* **19,** 283–303.
Bråten, T. (1971). *J. Cell Sci.* **9,** 621–635.
Bråten, T. and Nordby, Ø. (1973). *J. Cell Sci.* **13,** 69–81.
Brawerman, G. and Eisenstadt, J. M. (1964). *Biochim. Biophys. Acta* **91,** 477–485.
Brody, M. and Vatter, A. E. (1959). *J. Biophys. Biochem. Cytol.* **5,** 289–294.
Brown, H. P. (1945). *Ohio J. Sci.* **45,** 247–301.
Brown, R. M. (1969). *J. Cell Biol.* **41,** 109–123.
Brown, D. L. and Weier, T. E. (1970). *Phycologia* **9,** 217–235.
Brown, R. M. and Arnott, H. J. (1970). *J. Phycol.* **6,** 14–22.
Brown, R. M. and Franke, W. W. (1971). *Planta* **96,** 354–363.
Brown, R. M. and Wilson, R. (1968). *J. Phycol.* **4,** 230–240.
Brown, R. M., Johnson, C. and Bold, H. C. (1968). *J. Phycol.* **4,** 100–120.
Brown, R M., Franke, W. W., Kleinig, H., Falk, H. and Sitte, P. (1969). *Science, N.Y.* **166,** 894–896.
Brown, R. M., Franke, W. W., Kleinig, H., Falk, H. and Sitte, P. (1970). *J. Cell Biol.* **45,** 246–271.
Budd, T. W., Tjostem, J. L. and Duysen, M. E. (1969). *Am. J. Bot.* **56,** 540–545.
Buetow, D. E. (1968). *The Biology of* Euglena. Academic Press, New York and London.
Burczyk, J., Grzybek, H., Banaś, J. and Banaś, E. (1971). *Acta med. pol., Vars.* **12,** 143–146.
Burr, F. A. (1970). *Am. J. Bot.* **57,** 97–110.
Burr, F. A. and West, J. A. (1970). *Phycologia* **9,** 17–37.
Burr, F. A. and West, J. A. (1971a). *J. Ultrastruct. Res.* **35,** 476–498.
Burr, F. A. and West, J. A. (1971b). *J. Phycol.* **7,** 108–113.
Cachon, J., Cachon, M. and Greuet, C. (1970). *Protistologica* **6,** 467–476.
Caspar, S. J. (1972). *Arch. Protistenk.* **114,** 65–82.
Cavalier-Smith, T. (1970). *Nature, Lond.* **228,** 333–335.
Chapman, R. L., Chapman, M. R. and Lang, N. J. (1971). *J. Phycol.* (*Suppl.*) **7,** 3.

Chardard, R. (1971). *C. r. hebd. Séanc. Acad. Sci., Paris* **273D,** 1190–1192.
Chardard, R. (1972). *C. r. hebd. Séanc. Acad. Sci., Paris* **275D,** 1367–1370.
Cheignon, M. (1964). *C. r. hebd. Séanc. Acad. Sci., Paris* **258,** 676–678.
Chervin, R. E., Baker, G. E. and Hohl, H. R. (1968). *Can. J. Bot.* **46,** 241–245.
Chi, E. Y. (1971). *Protoplasma* **72,** 101–104.
Christensen, T. (1962). Alger. In *Botanik*, Vol. 2 (Eds. T. W. Böcher *et al.*). Munksgaard, Copenhagen.
Clarke, K. J. and Pennick, N. C. (1972). *Br. phycol. J.* **7,** 357–360.
Cohen-Bazire, G. and Lefort-Tran, M. (1970). *Arch. Mikrobiol.* **71,** 245–257.
Cole, K. (1969). *Phycologia* **8,** 101–108.
Cole, K. (1970). *Phycologia* **9,** 275–283.
Cole, K. and Lin, S-C. (1968). *Syesis* **1,** 103–119.
Cole, K. and Lin, S-C. (1970). *Can. J. Bot.* **48,** 265–268.
Coombs, J., Lauritis, J. A., Darley, W. M. and Volcani, B. E. (1968a). *Z. Pflanzenphysiol.* **59,** 124–152.
Coombs, J., Lauritis, J. A., Darley, W. M. and Volcani, B. E. (1968b). *Z. Pflanzenphysiol.* **59,** 274–284.
Crawford, R. M. (1971). *Br. phycol. J.* **6,** 175–186.
Crawford, R. M. (1973). *J. Phycol.* **9,** 50–61.
Crawford, R. M. and Dodge, J. D. (1973). *Nova Hedwigia* (In press).
Crawford R. M., Dodge, J. D. and Happey, C. M. (1971). *Nova Hedwigia* **19,** 825–840.
Crawley, J. C. W. (1963). *Expl Cell Res.* **32,** 368–378.
Crawley, J. C. W. (1964). *Expl Cell Res.* **35,** 497–506.
Crawley, J. C. W. (1966). *Planta* **69,** 365–376.
Cronshaw, J. and Preston, R. D. (1958). *Proc. R. Soc. B.* **148,** 137–148.
Cronshaw, J., Myers, A. and Preston, R. D. (1958). *Biochim. Biophys. Acta* **27,** 89–103.
Darley, W. M. and Volcani, B. E. (1969). *Expl Cell Res.* **58,** 334–342.
Dawes, C. J. (1965). *J. Phycol.* **1,** 121–127.
Dawes, C. J. (1966). *Ohio J. Sci.* **66,** 317–326.
Dawes, C. J. (1969). *Phycologia* **8,** 77–84.
Dawes, C. J. and Barilotti, D. C. (1969). *Am. J. Bot.* **56,** 8–15.
Dawes, C. J. and Rhamstine, E. L. (1967). *J. Phycol.* **3,** 117–126.
Dawes, C. J., Scott, F. M. and Bowler, E. (1961). *Am. J. Bot.* **48,** 925–934.
Deason, T. R. (1965). *J. Phycol.* **1,** 97–102.
Deason, T. R. (1971a). *J. Phycol.* **7,** 101–107.
Deason, T. R. (1971b). *Trans. Am. microsc. Soc.* **90,** 441–448.
Deason, T. R., Darden, W. H. and Ely, S. (1969). *J. Ultrastruct. Res.* **26,** 85–94.
Deflandre, G. (1934). *Annls Protist.* **4,** 31–54.
Descomps, S. (1963a). *C. r. hebd. Séanc. Acad. Sci., Paris* **256,** 1333–1335.
Descomps, S. (1963b). *C. r. hebd. Séanc. Acad. Sci., Paris* **257,** 727–729.
Descomps, S. (1972). *C. r. hebd. Séanc. Acad. Sci., Paris* **275D** 553–556.
Desikachary, T. V. (1956). *Jl R. microsc. Soc.* **76,** 9–36.
Dodge, J. D. (1965). *J. mar. biol. Ass. U.K.* **45,** 607–614.
Dodge, J. D. (1966). The Dinophyceae. In: *The Chromosomes of the Algae.* (Ed. M. B. E. Godward) pp. 96–115. Arnold, London.
Dodge, J. D. (1967). *Br. phycol. Bull.* **3,** 327–336.
Dodge, J. D. (1968). *J. Cell Sci.* **3,** 41–48.
Dodge, J. D. (1969b). *Arch. Microbiol.* **69,** 266–280.
Dodge, J. D. (1970). The Algae. In: *Electron Microscopy and Plant Ultrastructure.* (Ed. A. W. Robards). McGraw-Hill, New York and Maidenhead.
Dodge, J. D. (1971a). *Protoplasma* **73,** 145–157.

Dodge, J. D. (1972). *Protoplasma* **75**, 285–302.
Dodge, J. D. and Bibby, B. T. (1973). *Bot. J. Linn. Soc.* **67** (In press).
Dodge, J. D. and Crawford, R. M. (1968). *Protistologica* **4**, 231–242.
Dodge, J. D. and Crawford, R. M. (1969a). *New Phytol.* **68**, 613–618.
Dodge, J. D. and Crawford, R. M. (1969b). *J. Cell Sci.* **5**, 479–493.
Dodge, J. D. and Crawford, R. M. (1970a). *Bot. J. Linn. Soc.* **63**, 53–67.
Dodge, J. D. and Crawford, R. M. (1970b). *J. Phycol.* **6**, 137–149.
Dodge, J. D. and Crawford, R. M. (1971a). *Bot. J. Linn. Soc.* **64**, 105–115.
Dodge, J. D. and Crawford, R. M. (1971b). *Protistologica* **7**, 295–304.
Dodge, J. D. and Crawford, R. M. (1972). *Protistologica* **7**, 399–409.
Dolzmann, R. and Dolzmann, P. (1964). *Planta* **61**, 332–345.
Dragesco, J. (1951). *Bull. Microsc. Appl.* **1**, 172–175.
Drawert, H. and Mix, M. (1962). *Planta* **58**, 50–74.
Drum, R. W. (1963). *J. Cell Biol.* **18**, 429–440.
Drum, R. W. (1966). *J. Ultrastruct. Res.* **15**, 100–107.
Drum, R. W. (1969). *J. Phycol.* **5**, 21–26.
Drum, R. W. and Hopkins, J. T. (1966). *Protoplasma* **62**, 1–33.
Drum, R. W. and Pankratz, H. S. (1963). *Science, N.Y.* **142**, 61–63.
Drum, R. W. and Pankratz, H. S. (1964). *Am. J. Bot.* **51**, 405–418.
Drum, R. W. and Pankratz, H. S. (1965a). *Protoplasma* **60**, 141–149.
Drum, R. W. and Pankratz, H. S. (1965b). *Nova Hedwigia* **10**, 315–317.
Echlin, P. (1967). *Br. phycol. Bull.* **3**, 225–239.
Ehret, C. F. and de Haller, G. (1963). *J. Ultrastruct. Res.* **6** (Suppl.) 2–42.
Engelman, T. W. (1882). *Pflügers Arch. ges Physiol.* **29**, 387–400.
Esser, K. (1967). *Z. Naturf.* **22b**, 993.
Ettl, H. and Manton, I. (1964). *Nova Hedwigia* **8**, 421–451.
Evans, L. V. (1966). *J. Cell Sci.* **1**, 449–454.
Evans, L. V. (1968). *New Phytol.* **67**, 173–178.
Evans, L. V. (1970). *Br. phycol. J.* **5**, 1–13.
Evans, L. V. and Christie, A. O. (1970). *Ann. Bot.* **34**, 451–466.
Falk, H. (1967). *Arch. Mikrobiol.* **58**, 212–227.
Falk, H. and Kleinig, H. (1968). *Arch. Mikrobiol.* **61**, 347–362.
Fankboner, P. V. (1971). *Biol. Bull.* **141**, 222–234.
Fernandez-Moran, H., Oda, T., Blair, P. V. and Green D. E. (1964). *J. Cell Biol.* **22**, 63–100.
Fischer, A. (1894). *Jb. wiss. Bot.* **26**, 187–235.
Fisher, K. A. (1971). *Phycologia* **10**, 177–182.
Fisher, K. A. and Lang, N. J. (1971a). *J. Phycol.* **7**, 25–37.
Fisher, K. A. and Lang, N. J. (1971b). *J. Phycol.* **7**, 155–165.
Floyd, G. L., Stewart, K. D. and Mattox, K. R. (1971). *J. Phycol.* **7**, 306–309.
Floyd, G. L., Stewart, K. D. and Mattox, K. R. (1972a). *J. Phycol.* **8**, 68–81.
Floyd, G. L., Stewart, K. D. and Mattox, K. R. (1972b). *J. Phycol.* **8**, 176–184.
Fott, B. (1962). *Preslia* **34**, 69–84.
Fott, B. (1967). *Acta Univ. Carol. Biol.* **1967**, 223–240.
Fowke, L. C. and Pickett-Heaps, J. D. (1969a). *J. Phycol.* **5**, 240–259.
Fowke, L. C. and Pickett-Heaps, J. D. (1969b). *J. Phycol.* **5**, 273–281.
Fowke, L. C. and Pickett-Heaps, J. D. (1971). *J. Phycol.* **7**, 285–294.
Franke, W. W. and Brown, R. M. (1971). *Arch. Mikrobiol.* **77**, 12–19.
Fraser, T. W. and Gunning, B. E. S. (1969). *Planta* **88**, 244–254.
Frei, E. and Preston, R. D. (1961). *Nature, Lond.* **192**, 939–943.
Frei, E. and Preston, R. D. (1964a). *Proc. R. Soc. B.* **160**, 293–313.
Frei, E. and Preston, R. D. (1964b). *Proc. R. Soc. B.* **160**, 314–327.
Frey-Wyssling, A. and Mühlethaler, K., (1960). *Schweiz. Z. Hydrol.* **22**, 122–130.

Friedmann, I. and Manton, I. (1960). *Nova Hedwigia* **1**, 443–462.
Friedmann, I., Colwin, A. L. and Colwin, L. H. (1968). *J. Cell Sci.* **3**, 115–128.
Fritsch, F. E. (1935 and 1945). *The Structure and Reproduction of the Algae*, Vols. I and II. Cambridge University Press, London.
Gaarder, K. R. and Markali, J. (1956). *Nytt Mag. Bot.* **5**, 1–5.
Galun, M., Paran, N. and Ben-Shaul, Y. (1970). *Protoplasma* **69**, 85–96.
Gantt, E. (1969). *Pl. Physiol., Lancaster* **44**, 1629–1638.
Gantt, E. (1971). *J. Phycol.* **7**, 177–184.
Gantt, E. and Conti, S. F. (1965). *J. Cell Biol.* **26**, 365–381.
Gantt, E. and Conti, S. F. (1966). *J. Cell Biol.* **29**, 423–434.
Gantt, E. and Conti, S. F. (1967). *Brookhaven Symp. Biol.* **19**, 393–405.
Gantt, E., Edwards, M. R. and Conti, S. F. (1968). *J. Phycol.* **4**, 65–71.
Gantt, E., Edwards, M. R. and Provasoli, L. (1971). *J. Cell Biol.* **48**, 280–290.
Gaudsmith, J. T. and Dawes, C. J. (1972). *Phycologia* **11**, 123–132.
Gawlik, S. R. and Millington, W. F. (1969). *Am. J. Bot.* **56**, 1084–1093.
Gergis, M. S. (1969). *Arch. Mikrobiol.* **68**, 187–190.
Gergis, M. S. (1971). *Planta* **101**, 180–184.
Gergis, M. S. (1972). *Arch. Mikrobiol.* **83**, 321–327.
Gerhardt, B. and Berger, C. (1971). *Planta* **100**, 155–166.
Gerrath, J. F. (1969). *Phycologia* **8**, 109–118.
Gerrath, J. F. (1970). *Phycologia* **9**, 209–215.
Gibbs, S. P. (1960). *J. Ultrastruct. Res.* **4**, 127–148.
Gibbs, S. P. (1962a). *J. Ultrastruct. Res.* **7**, 247–261.
Gibbs, S. P. (1962b). *J. Ultrastruct. Res.* **7**, 262–272.
Gibbs, S. P. (1962c). *J. Ultrastruct. Res.* **7**, 418–435.
Gibbs, S. P. (1962d). *J. Cell Biol.* **14**, 433–444.
Gibbs, S. P. (1962e). *J. Cell Biol.* **15**, 343–361.
Gibbs, S. P. (1968). *J. Cell Sci.* **3**, 327–340.
Gibbs, S. P., Lewin, R. A. and Philpott, D. E. (1958). *Expl Cell Res.* **15**, 619–622.
Gibor, A. and Granick, S. (1962). *J. Protozool.* **9**, 327–334.
Giesbrecht, P. (1962). *Zentbl. Bakt. Parasitkde l. Orig.* **196**, 516–519.
Glazer, A. N., Cohen-Bazire, G. and Stanier, R. Y. (1971). *Arch. Mikrobiol.* **80**, 1–18.
Godin, J. (1970). *C. r. hebd. Séanc. Acad. Sci., Paris* **271**, 2290–2292.
Godward, M. B. E. and Jordan, E. G. (1965). *J. R. microsc. Soc.* **84**, 347–360.
Goldberg, I. and Ohad, I. (1970). *J. Cell Biol.* **44**, 572–591.
Gooday, G. W. (1971). *J. exp. Bot.* **22**, 959–971.
Goodenough, U. W. (1970). *J. Phycol.* **6**, 1–6.
Goodenough, U. W. (1971). *J. Cell Biol.* **50**, 35–49.
Goodenough, U. W. and Levine, R. P. (1969). *Pl. Physiol., Lancaster* **44**, 990–1000.
Goodenough, U. W. and Levine, R. P. (1970). *J. Cell Biol.* **44**, 547–562.
Goodenough, U. W. and Levine, R. P. (1971). *J. Cell Biol.* **50**, 50–62.
Goodenough, U. W. and Staehelin, L. A. (1971). *J. Cell Biol.* **48**, 594–619.
Goodenough, U. W., Armstrong, J. J. and Levine, R. P. (1969). *Pl. Physiol., Lancaster* **44**, 1001–1012.
Gouhier, M. (1970). *Protoplasma* **69**, 301–311.
Graves, L. B., Hanzeley, L. and Trelease, R. N. (1971). *Protoplasma* **72**, 141–152.
Green, J. C. and Leadbeater, B. S. C. (1972). *J. mar. biol. Ass. U.K.* **52**, 469–474.
Green, J. C. and Manton, I. (1970). *J. mar. biol. Ass. U.K.* **50**, 1113–1130.
Greenwood, A. D. (1959). *J. exp. Bot.* **10**, 55–68.
Greenwood, A. D. (1964). *Abstr. Xth Int. Congr. Bot.* pp. 212–213, Edinburgh.
Greenwood, A. D., Manton, I. and Clarke, B. (1957). *J. exp. Bot.* **8**, 71–86.

Grell, K. G. and Wohlfarth-Bottermann, K. E. (1957). *Z. Zellforsch. mikrosk. Anat.* **47,** 7–17.
Greuet, C. (1965). *C. r. hebd. Séanc. Acad. Sci., Paris* **261,** 1904–1907.
Greuet, C. (1968). *Protistologica* **4,** 209–230.
Greuet, C. (1970). *Proc. 7th Int. Cong. EM, Grenoble,* pp. 385–386.
Greuet, C. (1971). *Protistologica* **7,** 345–355.
Greuet, C. (1972). *C. r. hebd. Séanc. Acad. Sci. Paris* **275D,** 1239–1242.
Griffiths, D. A. and Griffiths, D. J. (1969). *Pl. Cell Physiol.* **10,** 11–19.
Guérin-Dumartrait, E. (1970). *C. r. hebd. Séanc. Acad. Sci., Paris* **270,** 1977–1979.
Hall, W. T. and Claus, G. (1963). *J. Cell Biol.* **19,** 551–563.
Hall, W. T. and Claus, G. (1967). *J. Phycol.* **3,** 37–51.
Halldal, P. (1964). In *Biochemistry and Physiology of Protozoa* (Ed. S. H. Hutner). Vol. 3, pp. 277–296. Academic Press, New York.
Halldal, P. and Markali, J. (1955). *Det. Norske Vid.—Akad. I. Mat. Nat. Klass 1955* No. 1 pp. 1–30.
Haller, G. de (1959). *Archs Sci., Genève* **12,** 309–340.
Haller, G. de and Rouiller, C. (1961). *J. Protozool.* **8,** 452–462.
Hanic, L. A. and Craigie, J. S. (1969). *J. Phycol.* **5,** 89–102.
Harris, K. (1966). *J. gen. Microbiol.* **42,** 175–184.
Harris, K. and Bradley, D. E. (1956). *Jl R. Microsc. Soc.* **76,** 37–48.
Harris, K. and Bradley, D. E. (1960). *J. gen. Microbiol.* **22,** 750–777.
Hasle, G. R. and Heimdal, B. R. (1970). *Beih. Nova Hedwigia* **31,** 559–581.
Haupt, W. (1959). In *Encyclopedia of Plant Physiology* (Ed. W. Ruhland). Vol. 17, pt. 1, pp. 318–370. Springer, Berlin.
Hawkins, A. F. and Leedale, G. F. (1971). *Ann. Bot.* **35,** 201–211.
Heath, I. B. and Darley, W. M. (1972). *J. Phycol.* **8,** 51–59.
Heath, I. B., Greenwood, A. D. and Griffiths, H. B. (1970). *J. Cell Sci.* **7,** 445–461.
Helmcke, J. E. and Krieger, W. (1953–66). *Diatomeenschalen im Elektronenmikroskopischen Bild,* Vols I–VI. Transmere-Photo, Berlin and J. Cramer, Lehre.
Herth, W., Franke, W. W., Stadler, J., Bittiger, H., Keilich, G. and Brown, R. M. (1972). *Planta* **105,** 79–92.
Heywood, P. (1968). Ph.D. thesis. University of London.
Heywood, P. (1972). *J. Ultrastruct. Res.* **39,** 608–623.
Heywood, P. and Godward, M. B. E. (1972). *Chromosoma* **39,** 333–339.
Hibberd, D. J. (1970). *Br. phycol. J.* **5,** 119–143.
Hibberd, D. J. (1971). *Br. phycol. J.* **6,** 207–223.
Hibberd, D. J. (1973). *Arch. Mikrobiol.* **89,** 291–304.
Hibberd, D. J. and Leedale, G. F. (1970). *Nature, Lond.* **225,** 758–760.
Hibberd, D. J. and Leedale, G. F. (1971). *Br. phycol. J.* **6,** 1–23.
Hibberd, D. J. and Leedale, G. F. (1972). *Ann Bot.* **36,** 49–71.
Hibberd, D. J., Greenwood, A. D. and Griffiths, H. B. (1971). *Br. phycol. J.* **6,** 61–72.
Hill, G. J. C. and Machlis, L. (1968). *J. Phycol.* **4,** 261–271.
Hobbs, M. J. (1971). *Br. phycol. J.* **6,** 81–103.
Hobbs, M. J. (1972). *Br. phycol. J.* **7,** 347–355.
Hoffman, L. R. (1967). *J. Phycol.* **4,** 212–221.
Hoffman, L. R. (1968a). *Trans. Am. Microsc. Soc.* **87,** 178–185.
Hoffman, L. R. (1968b). *J. Phycol.* **4,** 212–218.
Hoffman, L. R. (1970). *Can. J. Bot.* **48,** 189–196.
Hoffman, L. R. and Manton, I. (1962). *J. exp. Bot.* **13,** 443–449.
Hoffman, L. R. and Manton, I. (1963). *Am. J. Bot.* **50,** 455–463.
Holdsworth, R. H. (1968). *J. Cell Biol.* **37,** 831–837.
Holdsworth, R. H. (1971). *J. Cell Biol.* **51,** 499–513.

Holt, S. C. and Stern, A. I. (1970). *Pl. Physiol., Lancaster* **45,** 475–483.
Hopkins, J. M. (1970). *J. Cell Sci.* **7,** 823–839.
Hori, T. and Ueda, R. (1967). *Sci. Rep. Tokyo Kyoika Diagaku B* **12,** 225–244.
Horne, R. W., Davies, D. R., Norton, K. and Gurney-Smith M. (1971). *Nature, Lond.* **232,** 493–495.
Houwink, A. L. (1951). *Proc. K. ned. Akad. Wet. C.* **54,** 132–137.
Hovasse, R. (1965). *Protoplasmatologia* **1,** 81–88.
Hovasse, R., Mignot, J. P. and Joyon, L. (1967). *Protistologica* **3,** 241–255.
Ichida, A. A. and Fuller, M. S. (1968). *Mycologia* **60,** 141–155.
Jacobs, J. B. and Ahmadjian, V. (1969). *J. Phycol.* **5,** 227–240.
Jacobs, J. B. and Ahmadjian, V. (1971). *J. Phycol.* **7,** 71–81.
Johnson, U. G. and Porter, K. R. (1968). *J. Cell Biol.* **38,** 403–425.
Jordan, E. G. (1970). *Protoplasma* **69,** 405–416.
Jordan, E. G. and Godward, M. B. E. (1969). *J. Cell Sci.* **4,** 3–15.
Joyon, L. (1963a). *Ann. Fac. Sci. Univ. Clermont* **22,** 1–96.
Joyon, L. (1963b). *Archs Zool. exp. gén.* **102,** 199–200.
Joyon, L. and Fott, B. (1964). *J. Microscopie* **3,** 159–166.
Kalley, J. P. and Bisalputra, T. (1970). *J. Ultrastruct. Res.* **31,** 95–108.
Kalley, J. P. and Bisalputra, T. (1971). *J. Ultrastruct. Res.* **37,** 521–531.
Karakashian, S. J., Karakashian, M. W. and Rudzinska, M. A. (1968). *J. Protozool.* **15,** 113–128.
Karim, A. G. A. and Round, F. E. (1967). *New Phytol.* **66,** 409–412.
Kawaguti, S. and Yamasu, T. (1965). *Biol. J. Okayama Univ.* **11,** 57–65.
Kellenberger, E. (1964). *Genetics Today*, 309–321.
Kevin, M. J., Hall, W. T., McLaughlin, J. J. A. and Zahl, P. A. (1969). *J. Phycol.* **5,** 341–350.
Kiermayer, O. (1968). *Planta* **83,** 223–236.
Kiermayer, O. (1970). *Protoplasma* **69,** 97–132.
Kiermayer, O. and Staehelin, L. A. (1972). *Protoplasma* **74,** 227–237.
Kies, L. (1970a). *Protoplasma* **70,** 21–47.
Kies, L. (1970b). *Protoplasma* **71,** 139–146.
Kirk, J. T. O. and Tilney-Bassett, R. A. E. (1967). *The Plastids.* Freeman, London and San Francisco.
Kito, H. (1972). *Bull. Tohoku Reg. Fish Res. Lab.* **32,** 77–82.
Kivic, P. and Vesk, M. (1972). *Planta* **105,** 1–14.
Kivic, P. A. and Vesk, M. (1973). *J. exp. Bot.* (In press).
Klaveness, D. and Paasche, E. (1971). *Arch. Mikrobiol.* **75,** 382–385.
Klein, S. and Neuman, J. (1966). *Pl. Cell Physiol.* **7,** 115–123.
Klein, S., Schiff, J. A. and Holowinsky, A. W. (1972). *Dev. Biol.* **28,** 253–273.
Koch, W. and Schnepf, E. (1967). *Arch. Mikrobiol.* **57,** 196–198.
Kochert, G. and Olson, L. W. (1970a). *Arch. Mikrobiol.* **74,** 19–30.
Kochert, G. and Olson, L. W. (1970b). *Arch. Mikrobiol.* **74,** 31–40.
Komårek, J. (1970). *Ann. Rep. Lab. Algol (Trebõn)* **1969,** 29–34.
Könitz, W. (1965). *Planta* **66,** 345–373.
Kowallik, K. V. (1969). *J. Cell Sci.* **5,** 251–269.
Kowallik, K. V. (1971). *Arch. Mikrobiol.* **80,** 154–165.
Kowallik, K. V. and Haberkorn, G. (1971). *Arch. Mikrobiol.* **80,** 252–261.
Kramer, D. (1970). *Z. Naturforsch. B.* **256,** 1017–1020.
Kreger, D. R. and van der Veer, J. (1970). *Acta Bot. Neerl.* **19,** 401–402.
Kristiansen, J. (1969a). *Öst. bot. Z.* **116,** 70–84.
Kristiansen, J. (1969b). *Bot. Tidsskr.* **64,** 162–168.
Kristiansen, J. (1972a). *Br. phycol. J.* **7,** 1–12.
Kristiansen, J. (1972b). *Svensk bot. Tidskr.* **66,** 184–190.

Kubai, D. F. and Ris, H. (1969). *J. Cell Biol.* **40,** 508–528.
Kugrens, P. and West, J. A. (1972a). *J. Phycol.* **8,** 187–191.
Kugrens, P. and West, J. A. (1972b). *J. Phycol.* **8,** 370–383.
Lamport, D. T. A. (1970). *Ann. Rev. Pl. Physiol.* **21,** 235–270.
Lang, N. J. (1963a). *Am. J. Bot.* **50,** 280–300.
Lang, N. J. (1963b). *J. Cell Biol.* **19,** 631–634.
Lang, N. J. (1963c). *J. Protozool.* **10,** 333–339.
Lauritis, J. A., Coombs, J. and Volcani, B. E. (1968). *Arch. Mikrobiol.* **62,** 1–16.
Leadbeater, B. S. C. (1967). Ph.D. thesis, University of London.
Leadbeater, B. S. C. (1969). *Br. phycol. J.* **4,** 3–17.
Leadbeater, B. S. C. (1970). *Br. phycol. J.* **5,** 57–69.
Leadbeater, B. S. C. (1971a). *Ann. Bot.* **35,** 429–439.
Leadbeater, B. S. C. (1971b). *J. Cell Sci.* **9,** 443–451.
Leadbeater, B. S. C. (1972). *Sarsia* **49,** 65–80.
Leadbeater, B. S. C. and Dodge, J. D. (1966). *Br. phycol. Bull.* **3,** 1–17.
Leadbeater, B. S. C. and Dodge, J. D. (1967a). *J. gen. Microbiol.* **46,** 305–314.
Leadbeater, B. S. C. and Dodge, J. D. (1967b). *Arch. Mikrobiol.* **57,** 239–254.
Leadbeater, B. S. C. and Manton, I. (1969). *Arch. Mikrobiol.* **68,** 116–132.
Leadbeater, B. S. C. and Manton, I. (1971). *Arch. Mikrobiol.* **78,** 58–69.
Le Baron-Marano, F. and Izard, C. (1968). *C. r. hebd. Séanc. Acad. Sci., Paris* **267,** 137–139.
Lecal, J. (1965). *Protistologica* **1,** 63–70.
Lecal, J. (1966). *Protistologica* **2,** 57–70.
Lecal, J. (1972). *Bull. Soc. hist. nat. Toulouse* **108,** 302–324.
Ledbetter, M. C. and Porter, K. R. (1963). *J. Cell Biol.* **19,** 239–250.
Lee, R. E. (1971a). *Br. phycol. J.* **6,** 29–38.
Lee, R. E. (1971b). *J. Cell Sci.* **8,** 623–631.
Lee, R. E. (1972). *Nature, Lond.* **237,** 44–46.
Lee, R. E. and Fultz, S. A. (1970). *J. Phycol.* **6,** 22–28.
Leedale, G. F. (1964). *Br. phycol. Bull.* **2,** 291–306.
Leedale, G. F. (1966). *Advmt Sci., Lond.* for 1966, 22–37.
Leedale, G. F. (1968). In *The Biology of Euglena*, Vol. 1 (Ed. D. E. Buetow). Academic Press, New York and London.
Leedale, G. F. (1969). *Öst. Bot. Z.* **116,** 279–294.
Leedale, G. F., Meeuse, B. J. D. and Pringsheim, E. G. (1965). *Arch. Mikrobiol.* **50,** 68–102.
Leedale, G. F., Leadbeater, B. S. C. and Massalski, A. (1970). *J. Cell Sci.* **6,** 701–719.
Lefort, M. (1962). *C. r. hebd. Séanc. Acad. Sci. Paris* **254,** 2414–2416.
Lefort, M. (1964). *3rd Eur. Reg. Conf. on Elec. Micpy, Prague*, pp. 593–594.
Lefort, M. (1965). *C. r. hebd. Séanc. Acad. Sci. Paris* **261,** 233–236.
Lembi, C. A. and Herndon, W. R. (1966). *Can. J. Bot.* **44,** 710–712.
Lembi, C. A. and Lang, N. J. (1965). *Am. J. Bot.* **52,** 464–477.
Lembi, C. A. and Walne, P. L. (1971). *J. Cell Sci.* **9,** 569–579.
Lessie, P. E. and Lovatt, J. S. (1968). *Am. J. Bot.* **55,** 220–236.
Leyon, H. (1954). *Expl Cell Res.* **6,** 497–505.
Leyon, H. and Von Wettstein, D. (1954). *Z. Naturforsch.* **9,** 471–475.
Lewin, R. A. and Meinhart, J. O. (1953). *Can. J. Bot.* **31,** 711–717.
Lichtlé, C. and Giraud, G. (1970). *J. Phycol.* **6,** 281–288.
Liddle, L. B. and Neushul, M. (1969). *J. Phycol.* **5,** 4–12.
Loeblich, A. R. (1970). *Proc. N. Am. Paleont. Conv.* (1969), 867–929.
Loeffler, F. (1889). *Zentbl. Bakt.* **6,** 209–224.
Loiseaux, S. and West, J. A. (1970). *Trans. Am. microsc. Soc.* **89,** 524–532.

Løvlie, A. and Bråten, T. (1968). *Expl Cell Res.* **51,** 211–220.
Løvlie, A. and Bråten, T. (1970). *J. Cell Sci.* **6,** 109–129.
Lucas, I. A. N. (1970a). *J. Phycol.* **6,** 30–38.
Lucas, I. A. N. (1970b). *Br. phycol. J.* **5,** 29–37.
Lyon, T. L. (1969). *J. Phycol.* **5,** 380–382.
McBride, G. E. (1967). *Nature, Lond.* **216,** 939.
McBride, G. E. (1970). *Arch. Protistenk.* **112,** 365–375.
McBride, D. L. and Cole, K. (1969). *Phycologia* **8,** 177–186.
McBride, D. L. and Cole, K. (1971). *Phycologia* **10,** 49–61.
McBride, D. L. and Cole, K. (1972). *Phycologia* **11,** 181–192.
McCully, M. E. (1968). *J. Cell Sci.* **3,** 1–16.
McDonald, K. (1972). *J. Phycol.* **8,** 156–166.
McIntyre, A. and Bé, A. W. H. (1967). *Deep Sea Res.* **14,** 561–597.
Mackie, W. and Preston, R. D. (1968). *Planta* **79,** 249–253.
McLachlan, J. and Parke, M. (1967). *J. mar. biol. Ass. U.K.* **47,** 723–733.
McLean, R. J. (1968). *J. Phycol.* **4,** 277–283.
McLean, R. J. and Pessoney, G. F. (1970). *J. Cell Biol.* **45,** 522–531.
McVittie, A. (1972). *J. gen. Microbiol.* **71,** 525–540.
Maiwald, M. (1971). *Arch. Protistenk.* **113,** 334–344.
Malkoff, D. B. and Buetow, D. E. (1964). *Expl. Cell Res.* **35,** 58–63.
Manning, J. E., Wolstenholme, D. R., Ryan, R. S., Hunter, J. A. and Richards, O. C. (1971). *Proc. natn. Acad. Sci. U.S.A.* **68,** 1169–1173.
Manton, I. (1952). *Symp. Soc. exp. Biol.* **6,** 306–319.
Manton, I. (1954). *Proc. Int. Conf. E.M., London,* 594–599.
Manton, I. (1955). *Proc. Leeds Phil. Soc.* **6,** 306–316.
Manton, I. (1956). In *Cellular Mechanisms in Differentiation and Growth* (Ed. D. Rudnick). Princeton University Press, New Jersey.
Manton, I. (1957). *J. exp. Bot.* **8,** 294–303.
Manton, I. (1959a). *J. mar. biol. Ass. U.K.* **38,** 319–333.
Manton, I. (1959b). *J. exp. Bot.* **10,** 448–461.
Manton, I. (1964a). *New Phytol.* **63,** 244–254.
Manton, I. (1964b). *J. exp. Bot.* **15,** 399–411.
Manton, I. (1964c). *J. R. microsc. Soc.* **82,** 279–285.
Manton, I. (1964d). *Arch. Mikrobiol.* **49,** 315–330.
Manton, I. (1964e). *J. R. microsc. Soc.* **83,** 317–325.
Manton, I. (1966a). *J. Cell Sci.* **1,** 187–192.
Manton, I. (1966b). *J. Cell Sci.* **1,** 375–380.
Manton, I. (1966c). *J. Cell Sci.* **1,** 429–438.
Manton, I. (1967a). *Nova Hedwigia* **14,** 1–11.
Manton, I. (1967b). *J. Cell Sci.* **2,** 265–272.
Manton, I. (1967c). *J. Cell Sci.* **2,** 411–418.
Manton, I. (1968a). *Protoplasma* **66,** 35–53.
Manton, I. (1968b). *Proc. Linn. Soc. Lond.* **179,** 147–152.
Manton, I. (1969). *Öst. Bot. Z.* **116,** 378–392.
Manton, I. (1972a). *Br. phycol. J.* **7,** 21–35.
Manton, I. (1972b). *Br. phycol. J.* **7,** 235–248.
Manton, I. and Clarke, B. (1950). *Nature, Lond.* **166,** 973–974.
Manton, I. and Clarke, B. (1951a). *Ann. Bot.* **15,** 461–471.
Manton, I. and Clarke, B. (1951b). *J. exp. Bot.* **6,** 242–246.
Manton, I. and Clarke, B. (1956). *J. exp. Bot.* **7,** 416–432.
Manton, I. and Ettl, H. (1965). *J. Linn. Soc. (Bot.)* **59,** 175–184.
Manton, I. and Harris, K. (1966). *J. Linn. Soc. (Bot.)* **59,** 397–403.
Manton, I. and Leedale, G. F. (1961a). *J. mar. biol. Ass. U.K.* **41,** 145–155.

Manton, I. and Leedale, G. F. (1961b). *Phycologia* **1,** 37–57.
Manton, I. and Leedale, G. F. (1963a). *Arch. Mikrobiol.* **45,** 285–303.
Manton, I. and Leedale, G. F. (1963b). *Arch. Mikrobiol.* **47,** 115–136.
Manton, I. and Leedale, G. F. (1969). *J. mar. biol. Ass. U.K.* **49,** 1–16.
Manton, I. and Parke, M. (1960). *J. mar. biol. Ass. U.K.* **39,** 275–298.
Manton, I. and Parke, M. (1962). *J. mar. biol. Ass. U.K.* **42,** 565–578.
Manton, I. and Parke, M. (1965). *J. mar. biol. Ass. U.K.* **45,** 743–754.
Manton, I. and Peterfi, L. S. (1969). *Proc. R. Soc. B* **172,** 1–15.
Manton, I. and Stosch, H. A. von (1966). *J. R. microsc. Soc.* **85,** 119–134.
Manton, I., Clarke, B. and Greenwood, A. D. (1953). *J. exp. Bot.* **4,** 319–329.
Manton, I., Clarke, B. and Greenwood, A. D. (1955). *J. exp. Bot.* **6,** 126–128.
Manton, I., Oates, K. and Parke, M. (1963). *J. mar. biol. Ass. U.K.* **43,** 225–238.
Manton, I., Rayns, D. G., Ettl, H. and Parke, M. (1965). *J. mar. biol. Ass. U.K.* **45,** 241–255.
Manton, I., Kowallik, K. and Stosch, H. A. von (1969a). *J. Microscopy* **89,** 295–320.
Manton, I., Kowallik, K. and Stosch, H. A. von (1969b). *J. Cell Sci.* **5,** 271–298.
Manton, I., Kowallik, K. and Stosch, H. A. von (1970a). *J. Cell Sci.* **6,** 131–157.
Manton, I., Kowallik, K. and Stosch, H. A. von (1970b). *J. Cell Sci.* **7,** 407–443.
Marčenko, E. (1973). *Arch. Mikrobiol.* **88,** 153–161.
Marchant, H. J. (1972). *Br. phycol. J.* **7,** 81–84.
Marchant, H. J. and Pickett-Heaps, J. D. (1970). *Aust. J. biol. Sci.* **23,** 1173–1186.
Marchant, H. J. and Pickett-Heaps, J. D. (1971). *Aust. J. biol. Sci.* **24,** 471–486.
Marchant, H. J. and Pickett-Heaps, J. D. (1972a). *Aust. J. biol. Sci.* **25,** 265–278.
Marchant, H. J. and Pickett-Heaps, J. D. (1972b). *Aust. J. biol. Sci.* **25,** 279–291.
Marchant, R. and Robards, A. W. (1968). *Ann. Bot.* **32,** 457–471.
Massalski, A. and Leedale, G. F. (1969). *Br. phycol. J.* **4,** 159–180.
Matthys, E. and Puiseux-Dao, S. (1968). *C. r. hebd. Séanc. Acad. Sci., Paris* **267,** 2123–2125.
Mayer, F. (1969). *J. Ultrastruct. Res.* **28,** 102–111.
Menke, W. (1962). *Ann. Rev. Pl. Physiol.* **13,** 27–44.
Menke, W. and Fricke, B. (1962). *Port. Acta Biol.* **6A,** 243–252.
Mercer, F. W., Bogorad, L. and Mullens, R. (1962). *J. Cell Biol.* **13,** 393–403.
Merrett, M. (1969). *Arch. Mikrobiol.* **64,** 1–11.
Messer, G. and Ben-Shaul, Y. (1969). *J. Protozool.* **16,** 272–280.
Messer, G. and Ben-Shaul, Y. (1970). *Proc. 7th Int. Cong. E.M. Grenoble,* 411–412.
Messer, G. and Ben-Shaul, Y. (1971). *J. Ultrastruct. Res.* **37,** 94–104.
Micalef, H. (1972). *C. r. hebd. Séanc. Acad. Sci., Paris* **275,** 2481–2484.
Mignot, J. P. (1963). *C. r. hebd. Séanc Acad. Sci., Paris* **257,** 2530–2533.
Mignot, J. P. (1965). *J. Microscopie* **4,** 239–252.
Mignot, J. P. (1966). *Protistologica* **2,** 51–117.
Mignot, J. P. (1967). *Protistologica* **3,** 5–23.
Mignot, J. P. (1970). *Protistologica* **6,** 267–281.
Mignot, J. P., Joyon, L. and Pringsheim, E. G. (1968). *Protistologica* **4,** 493–506.
Mignot, J. P., Joyon, L. and Pringsheim, E. G. (1969). *J. Protozool.* **16,** 138–145.
Mignot, J. P., Hovasse, R. and Joyon, L. (1970). *J. Microscopie* **9,** 127–132.
Mignot, J. P., Brugerolle, G. and Metenier, G. (1972). *J. Microscopie* **14,** 327–342.
Millington, W. F. and Gawlik, S. R. (1967). *Nature, Lond.* **216,** 68.
Millington, W. F. and Gawlik, S. R. (1970). *Am. J. Bot.* **57,** 552–561.
Mix, M. (1966). *Arch. Mikrobiol.* **55,** 116–133.
Mix, M. (1972). *Arch. Mikrobiol.* **81,** 197–220
Moestrup, Ø. (1970a). *Planta* **93,** 295–308.
Moestrup, Ø. (1970b). *J. mar. biol. Ass. U.K.* **50,** 513–523.
Moestrup, Ø. (1972). *Br. phycol, J.* **7,** 169–183.

Mollenhauer, H. H., Evans, W. and Kogut, C. (1968). *J. Cell Biol.* **37,** 579–583.
Moner, J. G. and Chapman, G. B. (1960). *J. Ultrastruct. Res.* **4,** 26–42.
Moriber, L. G., Hershenov, B., Aaronson, S. and Bensky, B. (1963). *J. Protozool.* **10,** 80–86.
Moore, R. T. and McAlear, J. H. (1961). *Mycologia* **53,** 194–200.
Mornin, L. and Francis, D. (1967). *J. Microscopie* **6,** 759–772.
Murakami, S., Morimura, Y. and Takamiya, A. (1963). In *Microalgae and Photosynthetic Bacteria* (*Plant and Cell Physiol.*) pp. 65–83.
Myers, A., Preston, R. D. and Ripley, G. W. (1956). *Proc. R. Soc. B.* **144,** 450–459.
Myers, A. D., Preston, R. D. and Ripley, G. W. (1959). *Ann. Bot.* **23,** 257–260.
Neushul, M. (1970). *Am. J. Bot.* **57,** 1231–1239.
Neushul, M. (1971). *J. Ultrastruct. Res.* **37,** 532–543.
Neushul, M. and Dahl, A. L. (1972a). *Am. J. Bot.* **59,** 393–400.
Neushul, M. and Dahl, A. L. (1972b). *Am. J. Bot.* **59,** 401–410.
Neushul, M. and Liddle, L. (1968). *Am. J. Bot.* **55,** 1068–1073.
Nichols, H. W., Ridgeway, J. E. and Bold, H. C. (1966). *Ann. Missouri Bot. Gdn* **53,** 17–27.
Nicolai, E. (1957). *Nature, Lond.* **180,** 491–493.
Nicolai, E. and Preston, R. D. (1959). *Proc. R. Soc. B.* **151,** 244–255.
O'Donnell, E. H. J. (1965). *Cytologia* **30,** 118–154.
Oey, J. L. and Schnepf, E. (1970). *Arch. Mikrobiol.* **71,** 199–213.
Ohad, I., Siekevitz, P. and Palade, G. E. (1967a). *J. Cell Biol.* **35,** 521–552.
Ohad, I., Siekevitz, P. and Palade, G. E. (1967b). *J. Cell Biol.* **35,** 553–584.
Olson, L. W. and Kochert, G. (1970). *Arch. Mikrobiol.* **74,** 31–40.
Oschman, J. L. (1967). *J. Phycol.* **3,** 221–228.
Oschman, J. L. and Gray, P. (1965). *Trans. Am. microsc. Soc.* **84,** 368–374.
Ott, D. W. and Brown, R. M. (1972). *Br. phycol. J.* **7,** 361–374.
Outka, D. E. and Williams, D. C. (1971). *J. Protozool.* **18,** 285–297.
Palisano, J. R. and Walne, P. L. (1972). *J. Phycol.* **8,** 81–88.
Park, R. B. and Pfeifhofer, A. O. (1969). *J. Cell Sci.* **5,** 299–311.
Parke, M. (1971). *Proc. II Plank, Conf. Roma*, 929–937.
Parke, M. and Adams, I. (1960). *J. mar. biol. Ass. U.K.* **39,** 263–274.
Parke, M. and Dixon, P. S. (1968). *J. mar. biol. Ass. U.K.* **48,** 783–832.
Parke, M. and Manton, I. (1962). *J. mar. biol. Ass. U.K.* **42,** 391–404.
Parke, M. and Manton, I. (1965). *J. mar. biol. Ass. U.K.* **45,** 525–536.
Parke, M. and Manton, I. (1967) *J. mar. biol. Ass. U.K.* **47,** 445–464.
Parke, M. and Rayns, D. G. (1964). *J. mar. biol. Ass. U.K.* **44,** 209–217.
Parke, M., Manton, I. and Clarke, B. (1955). *J. mar. biol. Ass. U.K.* **34,** 579–609.
Parke, M., Manton, I. and Clarke, B. (1956). *J. mar. biol. Ass. U.K.* **35,** 387–414.
Parke, M., Manton, I. and Clarke, B. (1958). *J. mar. biol. Ass. U.K.* **37,** 209–228.
Parke, M., Lund, J. W. G. and Manton, I. (1962). *Arch. Mikrobiol.* **42,** 333–352.
Parke, M., Green, J. C. and Manton, I. (1971). *J. mar. biol. Ass. U.K.* **51,** 927–941.
Parker, B. C. and Huber, J. (1965). *J. Phycol.* **1,** 172–179.
Parker, B. C. and Leeper, G. F. (1969). *Planta* **87,** 86–94.
Parker, J. and Philpott, D. E. (1961). *Bull. Torrey bot. Club* **88,** 85–90.
Parker, B. C., Preston, R. D. and Fogg, G. E. (1963). *Proc. R. Soc. B* **158,** 435–445.
Pennick, N. C. and Clarke, K. J. (1972). *Br. phycol. J.* **7,** 45–48.
Peterfi, L. S. and Manton, I. (1968). *Br. phycol. Bull.* **3,** 423–440.
Petrocelis, B. de, Siekevitz, P. and Palade, G. E. (1970). *J. Cell Biol.* **44,** 618–634.
Peveling, E. (1969). *Planta* **87,** 69–85.
Peyrière, M. (1963). *C. r. hebd. Séanc. Acad. Sci., Paris* **257,** 730–732.
Peyrière, M. (1968). *C. r. hebd. Séanc. Acad. Sci., Paris* **266D,** 2253–2255.
Peyrière, M. (1969). *C. r. hebd. Séanc. Acad. Sci., Paris* **269D,** 2332–2334.

Pickett-Heaps, J. D. (1968a). *Planta* **81,** 193–200.
Pickett-Heaps, J. D. (1968b). *Aust. J. biol. Sci.* **21,** 655–690.
Pickett-Heaps, J. D. (1970a). *Planta* **90,** 174–190.
Pickett-Heaps, J. D. (1970b). *Protoplasma* **70,** 325–347.
Pickett-Heaps, J. D. (1970c). *Cytobios* **6,** 69–78.
Pickett-Heaps, J. D. (1971a). *Protoplasma* **72,** 275–314.
Pickett-Heaps, J. D. (1971b). *Planta* **100,** 357–359.
Pickett-Heaps, J. D. (1972a). *J. Phycol.* **8,** 44–47.
Pickett-Heaps, J. D. (1972b). *Protoplasma* **74,** 149–167.
Pickett-Heaps, J. D. (1972c). *Protoplasma* **74,** 169–193.
Pickett-Heaps, J. D. (1972d). *Protoplasma* **74,** 195–212.
Pickett-Heaps, J. D. (1972f). *Ann. Bot.* **36,** 693–701.
Pickett-Heaps, J. D. and Fowke, L. C. (1969). *Aust. J. biol. Sci.* **22,** 857–894.
Pickett-Heaps, J. D. and Fowke, L. C. (1970a). *Aust. J. biol. Sci.* **23,** 71–92.
Pickett-Heaps, J. D. and Fowke, L. C. (1970b). *Aust. J. biol. Sci.* **23,** 93–113.
Pickett-Heaps, J. D. and Fowke, L. C. (1970c). *J. Phycol.* **6,** 189–215.
Pickett-Heaps, J. D. and Fowke, L. C. (1971). *J. Phycol.* **7,** 37–50.
Pienaar, R. N. (1969). *J. Cell Sci.* **4,** 561–567.
Pienaar, R. N. (1971a). *Protoplasma* **73,** 217–224.
Pienaar, R. N. (1971b). *Proc. S. Africa E.M. Soc., Pretoria* 7–8.
Pitelka, D. R. and Child, F. M. (1964). In *Biochemistry and Physiology of Protozoa,* Vol. 3 (Ed. S. H. Hutner), pp. 131–198. Academic Press, New York.
Pitelka, D. R. and Schooley, C. N. (1955). *Univ. Calif. Publs. Zool.* **61,** 79–128.
Ploaie, P. E. (1971). *Rev. Roum. Biol.-Bot.* **16,** 179–183.
Preston, R. D. (1964). In *Formation of Wood in Forest Trees* (Ed. M. H. Zimmermann), pp. 169–188. Academic Press, New York and London.
Preston, R. D. and Goodman, R. N. (1968). *J. R. microsc. Soc.* **88,** 513–527.
Preston, R. D. and Kuyper, B. (1951). *J. exp. Bot.* **2,** 247–255.
Provasoli, L., Yamasu, T. and Manton, I. (1968). *J. mar. biol. Ass. U.K.* **48,** 465–479.
Puiseux-Dao, S. (1966). *Proc. 6th Int. Cong. E.M. Kyoto,* pp. 377–378.
Puiseux-Dao, S. (1970). *Acetabularia and Cell Biology.* Logos Press, London.
Puiseux-Dao, S. and Izard, C. (1968). *C. r. hebd. Séanc. Acad. Sci., Paris* **267D,** 74–75.
Puiseux-Dao, S., Gibello, D. and Hoursiagou-Neubrun, D. (1967). *C. r. hebd. Séanc. Acad. Sci., Paris* **265,** 406–408.
Quatrano, R. S. (1972). *Expl. Cell Res.* **70,** 1–12.
Rae, P. M. M. (1970). *J. Cell Biol.* **46,** 106–113.
Ramus, J. (1969a). *J. Phycol.* **5,** 57–63.
Ramus, J. (1969b). *J. Cell Biol.* **41,** 340–345.
Ramus, J. (1972). *J. Phycol.* **8,** 97–111.
Raven, P. H. (1970). *Science, N.Y.* **169,** 641–646.
Ray, D. S. and Hanawalt, P. C. (1964). *J. molec. Biol.* **9,** 812–824
Reimann, B. E. F. (1964). *Expl Cell Res.* **34,** 605–608.
Reimann, B. E. F., Lewin, J. C. and Volcani, B. E. (1965). *J. Cell Biol.* **24,** 39–55.
Retallack, B. and Butler, R. D. (1970a). *J. Cell Sci.* **6,** 229–241.
Retallack, B. and Butler, R. D. (1970b). *Arch. Mikrobiol.* **72,** 223–237.
Retallack, B. and Butler, R. D. (1972). *Arch. Mikrobiol.* **86,** 265–280.
Richardson, F. L. and Brown, T. E. (1970). *J. Phycol.* **6,** 165–171.
Riedmüller-Schölm. H. E. (1972). *Protoplasma* **74,** 33–39.
Riedmüller-Schölm, H. E. and Allen, M. B. (1972). *Botanica mar.* **15,** 52–56.
Ringo, D. L. (1967a). *J. Ultrastruct. Res.* **17,** 266–277.
Ringo, D. L. (1967b). *J. Cell Biol.* **33,** 543–571.

Ris, H. and Kubai, D. (1972). *J. Cell Biol.* **55,** 217a.
Ris, H. and Plaut, W. (1962). *J. Cell Biol.* **13,** 383–391.
Roberts, K. and Northcote, D. H. (1970). *Nature, Lond.* **228,** 385–386.
Roberts, K., Gurney-Smith, M. and Hills, G. J. (1972). *J. Ultrastruct. Res.* **40,** 599–613.
Robinson, D. G. (1972). *J. Cell Sci.* **10,** 307–314.
Robinson, D. G. and Preston, R. D. (1971a). *J. Cell Sci.* **9,** 581–601.
Robinson, D. G. and Preston, R. D. (1971b). *J. exp. Bot.* **22,** 635–643.
Robinson, D. G. and Preston, R. D. (1971c). *Br. phycol. J.* **6,** 113–128.
Robinson, D. G. and Preston, R. D. (1972). *Planta* **104,** 234–246.
Robinson, D. G. and White, R. K. (1972). *Br. phycol. J.* **7,** 109–118.
Robinson, D. G., White, R. K. and Preston, R. D. (1972). *Planta* **107,** 131–144.
Rodríguez-López, M. (1965). *Arch. Mikrobiol.* **52,** 319–324.
Roelofsen, P. A. (1959). *The Plant Cell Wall.* G. Borntraeger (Pub), Berlin.
Roelofsen, P. A. (1966). *Adv. Bot. Res.* **2,** 67–149.
Rogers, T. D., Scholes, V. E. and Schlichting, H. E. (1972). *J. Protozool.* **19,** 133–139.
Rosen, W. G. and Siegesmund, K. A. (1961). *J. Biophys. Biochem. Cytol.* **9,** 910–914.
Rosenbaum, J. L., Moulder, J. E. and Ringo, D. L. (1969). *J. Cell Biol.* **41,** 600–619.
Ross, R. and Simms, P. A. (1970). *Beih. Nova Hedwigia* **31,** 49–88.
Ross, R. and Simms, P. A. (1971). In *Scanning Electron Mi roscopy: Systematic and Evolutionary Applications.* (Ed. V. H. Heywood), pp. 155–177. Academic Press, London and New York.
Ross, R. and Simms, P. A. (1972). *Br. phycol. J.* **7,** 139–163.
Roth, L. E. (1959). *J. Protozool.* **6,** 107–116.
Rouiller, C. H. and Fauré-Fremiet, E. (1958). *Expl Cell Res.* **14,** 47–67.
Round, F. E. (1971a). *Br. phycol. J.* **6,** 135–143.
Round, F. E. (1971b). *Br. phycol. J.* **6,** 235–264.
Sagan, L., Ben-Shaul, Y., Schiff, J. and Epstein, H. T. (1965). *Pl. Physiol., Lancaster* **40,** 1257–1259.
Sager, R. and Hamilton, M. G. (1967). *Science, N.Y.* **157,** 709.
Sager, R. and Palade, G. E. (1954). *Expl Cell Res.* **7,** 584–588.
Sager, R. and Palade, G. E. (1957). *J. biophys. biochem. Cytol.* **3,** 463–488.
Sarfatti, G. and Bedini, C. (1965). *Caryologia* **18,** 207–223.
Sassen, A., Van Eyden-Emons, A., Lamers, A. and Wanka, F. (1970). *Cytobiologie* **1,** 373–382.
Schiff, J. A. and Epstein, H. T. (1965). In *Reproduction: Molecular, Subcellular and Cellular* (Ed. M. Locke), pp. 131–189. Academic Press, New York.
Schiff, J. A., Zeldin, M. H. and Rubman, J. (1967). *Pl. Physiol., Lancaster* **42,** 1716–1725.
Schmitter, R. E. (1971). *J. C ll Sci.* **9,** 147–173.
Schmitz, F. (1882). *Die Chromatophoren der Algen.* Max Cohen und Sohn, Bonn.
Schnepf, E. (1964). *Arch. Mikrobiol.* **49,** 112–131.
Schnepf, E. (1965). *Planta* **67,** 213–224.
Schnepf, E. (1969). *Öst. bot. Z.* **116,** 65–69.
Schnepf, E. and Deichgräber, G. (1969). *Protoplasma* **68,** 85–106.
Schnepf, E. and Deichgräber, G. (1972). *Protoplasma* **74,** 411–425.
Schnepf, E. and Koch, W. (1966). *Arch. Mikrobiol.* **54,** 229–236.
Schnepf, E., Koch, W. and Deichgräber, G. (1966). *Arch. Mikrobiol.* **55,** 149–174.
Schnepf, E., Deichgräber, G. and Koch, W. (1968). *Arch. Mikrobiol.* **63,** 15–25.
Schnepf, E., Hegewald, E. and Soeder, C. J. (1971a). *Arch. Mikrobiol.* **75,** 209–229.
Schnepf, E., Deichgräber, G., Hegewald, E. and Soeder, C. J. (1971b). *Arch. Mikrobiol.* **75,** 230–245.

Schötz, F. (1972). *Planta* **102,** 152–159.
Schötz, F., Bathelt, H., Arnold, C-G. and Schimmer, O. (1972). *Protoplasma* **75,** 229–254.
Schuster, F. L. (1968). *Expl Cell Res.* **49,** 277–284.
Schuster, F. L. (1970). *J. Protozool.* **17,** 521–526.
Schuster, F. L., Hershenov, B. and Aaronson, S. (1968). *J. Protozool.* **15,** 335–346.
Schwelitz, F. D., Evans, W. R., Mollenhauer, H. H. and Dilley, R. A. (1970). *Protoplasma* **69,** 341–349.
Schwelitz, F. D., Dilley, R. A. and Crane, F. L. (1972). *Pl. Physiol., Lancaster* **50,** 166–170.
Seckbach, J. (1972). *Microbios* **5,** 133–142.
Seckbach, J. and Ikan, R. (1972). *Pl. Physiol., Lancaster* **49,** 457–459.
Selman, G. G. and Jurand, A. (1970). *J. gen. Microbiol.* **60,** 365–372.
Simon-Bichard-Bréaud, J. (1971). *C. r. hebd. Séanc. Acad. Sci., Paris* **273D,** 1272–1275.
Simon-Bichard-Bréaud, J. (1972). *C. r. hebd. Séanc. Acad. Sci., Paris* **274D,** 1796–1799.
Slankis, T. and Gibbs, S. P. (1968). *J. Cell Biol.* **39,** 126a.
Slankis, T. and Gibbs, S. P. (1972). *J. Phycol.* **8,** 243–256.
Smith-Johannsen, H. and Gibbs, S. P. (1972). *J. Cell Biol.* **52,** 598–614.
Soeder, C. J. (1964). *Arch. Mikrobiol.* **47,** 311–324.
Soeder, C. J. (1965). *Arch. Mikrobiol.* **50,** 368–377.
Sommer, J. R. (1965). *J. Cell Biol.* **24,** 253–257.
Sommer, J. R. and Blum, J. J. (1964). *Expl Cell Res.* **35,** 423–425.
Sommer, J. R. and Blum, J. J. (1965a). *J. Cell Biol.* **24,** 235–251.
Sommer, J. R. and Blum, J. J. (1965b). *Expl Cell Res.* **39,** 504–527.
Sommerfeld, M. R. and Leeper, G. F. (1970). *Arch. Mikrobiol.* **73,** 55–60.
Soyer, M. O. (1969). *J. Microscopie* **8,** 569–580.
Speer, H. L., Dougherty, W. and Jones, R. F. (1964). *J. Ultrastruct. Res.* **11,** 84–89.
Sprey, B. (1970). *Protoplasma* **71,** 235–250.
Staehelin, L. A. (1966). *Z. Zellforsch. mikrosk. Anat.* **74,** 325–350.
Staehelin, L. A. (1968). *Proc. R. Soc. B.* **171,** 249–259.
Staehelin, L. A. and Kiermayer, O. (1970). *J. Cell Sci.* **7,** 787–792.
Stein, J. R. and Bisalputra, T. (1969). *Can. J. Bot.* **47,** 233–236.
Steward, F. C. and Mühlethaler, K. (1953). *Ann. Bot.* **17,** 295–316.
Stewart, K. D., Floyd, G. L., Mattox, K. R. and Davis, M. E. (1972). *J. Cell Biol.* **54,** 431–434.
Stoermer, E. F. and Pankratz, H. S. (1964). *Am. J. Bot.* **51,** 986–990.
Stoermer, E. F., H. S. Pankratz, H. S. and Drum, R. W. (1964). *Protoplasma* **59,** 1–13.
Stoermer, E. F., Pankratz, H. S. and Bowen, C. C. (1965). *Am. J. Bot.* **52,** 1067–1078.
Stokes, D. M., Turner, J. S. and Markus, K. (1970). *Aust. J. biol. Sci.* **23,** 265–274.
Strugger, S. and Peveling, E. (1961). *Protoplasma* **54,** 254–262.
Swale, E. M. F. (1969). *Br. phycol. J.* **4,** 65–86.
Swale, E. M. F. and Belcher, J. H. (1968). *Proc. Linn. Soc. Lond.* **179,** 77–81.
Swale, E. M. F. and Belcher, J. H. (1971). *Br. Phycol. J.* **6,** 41–50.
Swift, E. and Remsen, C. C. (1970). *J. Phycol.* **6,** 79–86.
Taylor, D. L. (1968a). *J. mar. biol. Ass. U.K.* **48,** 1–15.
Taylor, D. L. (1968b). *J. mar. biol. Ass. U.K.* **48,** 349–366.
Taylor, D. L. (1969a). *J. Phycol.* **5,** 336–340.
Taylor, D. L. (1969b). *J. mar. biol. Ass. U.K.* **49,** 1057–1065.

Taylor, D. L. (1969c). *J. Cell Sci.* **4,** 751–762.
Taylor, D. L. (1970). *Int. Rev. Cytol.* **27,** 29–64.
Taylor, D. L. (1971a). *J. mar. biol. Ass. U.K.* **51,** 227–234.
Taylor, D. L. (1971b). *J. mar. biol. Ass. U.K.* **51,** 301–313.
Taylor, D. L. (1972). *Arch. Mikrobiol.* **81,** 136–145.
Taylor, D. L. and Lee, C. C. (1971). *Arch. Mikrobiol.* **75,** 269–280.
Taylor, F. J. R. (1971). *J. Phycol.* **7,** 249–258.
Taylor, F. J. R., Blackbourn, D. J. and Blackbourn, J. (1969). *Nature, Lond.* **224,** 819–821.
Taylor, F. J. R., Blackbourn, D. J. and Blackbourn, J. (1971). *J. Fish. Res. Bd Can.* **28,** 391–407.
Teichler-Zallen, D. (1969). *Pl. Physiol., Lancaster* **44,** 701–710.
Thinh, L. V. and Griffiths, D. J. (1972a). *Arch. Mikrobiol.* **87,** 47–60.
Thinh, L. V. and Griffiths, D. J. (1972b). *Arch. Mikrobiol.* **87,** 61–75.
Thompson, E. W. and Preston, R. D. (1968). *J. exp. Bot.* **19,** 690–697.
Throndsen, J. (1971). *Norw. J. Bot.* **18,** 47–64.
Tikhonenko, A. S. and Zavarzina, N. B. (1966). *Mikrobiologiya* **35,** 850–852.
Toth, R. and Wilce, R. T. (1972). *J. Phycol.* **8,** 126–130.
Trainor, F. R. and Massalski, A. (1971). *Can. J. Bot.* **49,** 1273–1276.
Treharne, R. W., Melton, C. W. and Roppel, R. M. (1964). *J. molec. Biol.* **10,** 57–62.
Trench, R. K., Greene, R. W. and Bystrom, B. G. (1969). *J. Cell Biol.* **42,** 404–417.
Trezzi, F., Galli, M. G. and Bellini, E. (1964). *G. Bot. Ital.* **71,** 127–136.
Trezzi, F., Galli, M. G. and Bellini, E. (1965). *G. Bot. Ital.* **72,** 255–263.
Tripodi, G. (1971). *J. submicros. Cytol.* **3,** 71–79.
Tripodi, G., Pizzolongo, P. and Giannattasio, M. (1972). *J. Cell Biol.* **55,** 530–532.
Tsekos, I. and Schnepf, E. (1972). *Naturwissenschaften* **59,** 272–273.
Turner, F. R. (1968). *J. Cell Biol.* **37,** 370–393.
Ueda, K. (1960). *Cytologia* **25,** 8–16.
Ueda, K. (1961). *Cytologia* **26,** 344–358.
Urban, P. (1969). *J. Cell Biol.* **42,** 606–611.
Valkenburg, S. D. van (1971a). *J. Phycol.* **7,** 113–118.
Valkenburg, S. D. van (1971b). *J. Phycol.* **7,** 118–132.
Veer, J. van der (1970). *Acta Bot. Neerl.* **19,** 616–636.
Veer, J. van der (1972). *Nova Hedwigia* **23,** 131–159.
Walker, A. T. (1968). *Am. J. Bot.* **55,** 641–648.
Walles, B. and Kylin, A. (1972). *Z. Pflanzenphysiol.* **66,** 197–205.
Walne, P. L. (1967). *Am. J. Bot.* **54,** 564–577.
Walne, P. L. (1971). In *Contributions in Phycology* (Eds. B. C. Parker and R. M. Brown), pp. 107–120. Allen Press, Lawrence, Kansas.
Walne, P. L. and Arnott, H. J. (1967). *Planta* **77,** 325–353.
Wanka, F. (1968). *Protoplasma* **66,** 105–130.
Wanka, F. and Mulders, P. F. M. (1967). *Arch. Mikrobiol.* **58,** 257–269.
Warr, J. R., McVittie, A., Randall, J. and Hopkins, J. M. (1966). *Genet. Res.* **7,** 335–351.
Webster, D. A., Hackett, D. P. and Park, R. B. (1968). *J. Ultrastruct. Res.* **21,** 514–523.
Wecke, J. and Giesbrecht, P. (1970). *7th Int. Congr. EM, Grenoble,* 233–234.
Wehrmeyer, W. (1970a). *Protoplasma* **70,** 295–315.
Wehrmeyer, W. (1970b). *Arch. Mikrobiol.* **71,** 367–383.
Wehrmeyer, W. (1971). *Arch. Mikrobiol.* **75,** 121–139.
Weier, T. E., Englebrecht, A. H. P., Harrison, A. and Risley, E. B. (1965). *J. Ultrastruct. Res.* **13,** 92–111.

Weier, T. E., Bisalputra, T. and Harrison, A. (1966). *J. Ultrastruct. Res.* **15**, 38–56.
Werz, G. (1964). *Planta* **62**, 255–271.
Werz, G. and Kellner, G. (1970). *Protoplasma* **69**, 351–364.
Wiessner, W. and Amelunxen, F. (1969a). *Arch. Mikrobiol.* **66**, 14–24.
Wiessner, W. and Amelunxen, F. (1969b). *Arch. Mikrobiol.* **67**, 357–369
Wilbur, K., Colinvaux, L. and Watabe, N. (1969). *Phycologia* **8**, 27–35.
Wilson, K. (1951). *Ann. Bot.* **15**, 279–288.
Wilson, H. J., Wanka, F. and Linskens, H. F. (1973). *Planta* **109**, 259–267.
Witman, G. B., Carlson, K., Berlinger, J. and Rosenbaum, J. L. (1972a). *J. Cell Biol.* **54**, 507–539.
Witman, G. B., Carlson, K. and Rosenbaum, J. L. (1972b). *J. Cell Biol.* **54**, 540–555.
Wolken, J. J. (1956). *J. Protozool.* **3**, 211–221.
Wolken, J. J. and Palade, G. E. (1952). *Nature, Lond.* **170**, 114–115.
Wolken, J. J. and Palade, G. E. (1953). *Ann. N.Y. Acad. Sci.* **56**, 873–881.
Woodcock, C. L. F. (1971). *J. Cell Sci.* **8**, 611–621.
Woodcock, C. L. F. and Bogorad, L. (1970). *J. Cell Biol.* **44**, 361–375.
Wujek, D. E. (1968). *Ohio J. Sci.* **68**, 187–191.
Wujek, D. E. (1969). *Cytologia* **34**, 71–79.
Wujek, D. E. (1971). *Michigan Academician* **3**, 59–62.
Wujek, D. E. and Hamilton, R. (1972). *Mich. Botanist* **11**, 51–59.
Wygash, J. (1963). *Z. Naturf.* **18**, 827–830.
Yokomura, E. (1967a). *Cytologia* **32**, 361–377.
Yokomura, E. (1967b). *Cytologia* **32**, 378–389.
Ziegler, H. and Ruck, I. (1967). *Planta* **73**, 62–73.
Zingmark, R. G. (1970). *Am. J. Bot.* **57**, 586–592.

AUTHOR INDEX

Numbers in *italics* refer to pages on which references are listed at the end of this book.

SUBJECT INDEX

Numbers in heavy type refer to pages carrying illustrations.

R

S